Astrophysical Lasers

Astrophysical Lasers

Vladilen LETOKHOV
Institute of Spectroscopy
Russian Academy of Sciences

Sveneric JOHANSSON
Institute of Astronomy
Lund University

OXFORD
UNIVERSITY PRESS

Great Clarendon Street, Oxford OX2 6DP

Oxford University Press is a department of the University of Oxford.
It furthers the University's objective of excellence in research, scholarship,
and education by publishing worldwide in

Oxford New York

Auckland Cape Town Dar es Salaam Hong Kong Karachi
Kuala Lumpur Madrid Melbourne Mexico City Nairobi
New Delhi Shanghai Taipei Toronto

With offices in

Argentina Austria Brazil Chile Czech Republic France Greece
Guatemala Hungary Italy Japan Poland Portugal Singapore
South Korea Switzerland Thailand Turkey Ukraine Vietnam

Oxford is a registered trade mark of Oxford University Press
in the UK and in certain other countries

Published in the United States
by Oxford University Press Inc., New York

© Vladilen Letokhov and Sveneric Johansson 2009

The moral rights of the authors have been asserted
Database right Oxford University Press (maker)

First Published 2009

All rights reserved. No part of this publication may be reproduced,
stored in a retrieval system, or transmitted, in any form or by any means,
without the prior permission in writing of Oxford University Press,
or as expressly permitted by law, or under terms agreed with the appropriate
reprographics rights organization. Enquiries concerning reproduction
outside the scope of the above should be sent to the Rights Department,
Oxford University Press, at the address above

You must not circulate this book in any other binding or cover
and you must impose the same condition on any acquirer

British Library Cataloguing in Publication Data

Data available

Library of Congress Cataloging in Publication Data

Data available

Typeset by Newgen Imaging Systems (P) Ltd., Chennai, India
Printed in the UK
on acid-free paper by
the MPG Books Group

ISBN 978–0–19–954827–9 (Hbk)

10 9 8 7 6 5 4 3 2 1

Bengt EDLÉN

We dedicate this volume to the late Professor Bengt Edlén (1906–1993) to the memory of his 100th birthday. To the astronomical community Edlén is best known for this explanation of the strong emission lines in the solar corona spectrum.

Edlén started his scientific career in Siegbahn's spectroscopy laboratory in Uppsala. Siegbahn had received the Nobel Prize for physics in 1924 for his studies of X-ray spectra, and Edlén got the task of bridging the wavelength gap between X-ray and optical spectroscopy. His first important results came as early as 1929, when he published experimental results for He-like ionic systems, i.e. Li^+, Be^{2+}, B^{3+} etc., which at that time had become a target for the application of quantum mechanics.

Edlén's doctoral thesis from 1934 contains extraordinarily extensive and careful analyses of the spectra and energy levels of various ions of the eight lightest elements (H to O). The famous astrophysicist H. N. Russel evaluated it as follows: "It is superb work and so complete that it should be the definitive treatise on the subject. I need not emphasize how great its astrophysical value has been and will continue to be". Still today this thesis is the only source for experimental wavelengths for many fundamental resonance lines in astrophysically important ions. In fact, Edlén pointed out a small difference between measured and calculated energies in hydrogen-like ions, and this was the first detection of the Lamb shift, later predicted by quantum electro dynamics.

Edlén had a great interest in astronomy and astrophysics. His pioneering work on spectra of ionized carbon, nitrogen and oxygen led to extensive line identifications in spectra of Wolf–Rayet stars, and the early (1933) division of these stars into two parallel sequences: WC(C-rich) and WN (N-rich). The optical lines from these ionized spectra originated from higher quantum states, as the resonance lines appear in the far-UV region. From the low energy levels of Fe VII, which had been determined in an experimental analysis, Edlén derived wavelengths for forbidden transitions, applying the same technique as Bowen had used for O^{2+} about 10 years earlier. Edlén and Bowen published in 1939 a joint paper on the identification of forbidden Fe^{6+} lines in Nova Pictoris. This was at that time the highest ionization state found in any astrophysical object.

Thus, the spectral lines from highly ionized atoms that could be detected in the optical region were either allowed transitions between high quantum states or forbidden transitions between low quantum states. This fact came to be the solution behind Edlén's identification of the strong emission lines, observed during a solar eclipse in 1969. The well-known astronomer Harlow Shapley stated in 1933 that "the coronal lines are our most conspicuous riddle in astronomical spectroscopy". From extrapolations of his own laboratory measurements Edlén concluded in 1942 that the lines originated from forbidden transitions in very high ionization stages of Ar, Ca, Fe and Ni. The strongest line, the green coronal line at 5303 Å, is emitted by Fe^{13+}. The impressive fact about the identifications of the coronal lines is that forbidden transitions do not appear in laboratory sources, and that the conclusion was based on extrapolations from measurements in lighter elements. The intriguing fact was that the temperature in the solar corona must be about 2 million degrees, which seemed unlikely in comparison to the 6000 degrees on the solar surface.

Edlén was a person of high integrity, free from tactical and political considerations about scientific matters. He was very accurate and had high demands on quality. Edlén was a member of many academies and served on the Nobel Committee for Physics during 1961–1976. He received the Gold Medal of the Royal Astronomical Society in London, which at that time was regarded as the highest sign of recognition in astronomy.

Edlén's contribution to astrophysics is to a great extent associated with identifications of prominent lines in spectra of various astrophysical objects. Spectral lines are the outstanding source of detailed information about these objects. They also offer a great challenge. One of the authors (S. J.) had the privilege of being a PhD student in his group, and much material in this volume comes from a major research project on iron, suggested by Edlén.

Preface

This book is based on the early research by V. L. in 1972, which was continued after almost 30 years together with S. J., when V. L. was the Tage Erlander Professor at Lund University in 2000 and later through support by grants from the Royal Academy of Sciences, the Nobel Foundation and the Wenner-Gren Foundations. V. L. is grateful to the Lund Observatory for hospitality, and to the Russian Foundation for Basic Research for support. The research project on astrophysical lasers was also supported by a grant (S. J.) from the Swedish National Space Board. The Royal Physiographic Society, Lund, kindly supported the final preparation of this book.

The Oxford University Press and particularly Dr Sönke Adlung were so kind to support the idea of publishing this book. We would like to thank Professor Vladimir Minogin for reading a few Chapters. Finally we would like to thank Mrs Ol'ga Tatyanchenko for her highly qualified and fast processing of our manuscript, Mrs Alla Makarova and Mrs Elena Nikolaeva, librarians at the Institute of Spectroscopy, for their valuable help and Mr Alexandr Korniushkin for graphic work.

Lund, Sweden, 2008
Troitsk, Russia, 2008

Contents

1 **Introduction**   1
  1.1   Historical remark   2
  1.2   From astrophysical masers to astrophysical lasers   4
  1.3   Amplification conditions for an atomic ensemble   6
  1.4   Structure of the book   10

2 **Elements of radiative quantum transitions**   12
  2.1   Spontaneous and stimulated emission   12
      2.1.1   Thermodynamical equilibrium – Einstein coefficients A and B   12
      2.1.2   Semiclassical approach   16
      2.1.3   Quantum theory   18
  2.2   Broadening of spectral lines   22
      2.2.1   Radiative (natural) broadening   22
      2.2.2   Collisional broadening   23
      2.2.3   Doppler (nonhomogeneous) broadening   25
  2.3   Resonance scattering of radiation   27
      2.3.1   Coherence of scattering in the atomic frame   27
      2.3.2   Doppler redistribution of frequency   29
      2.3.3   Number of resonance scattering events – escaping of photons   30

3 **Elements of atomic spectroscopy**   36
  3.1   Basic concepts   36
      3.1.1   Structure and interactions   36
      3.1.2   LS coupling   37
      3.1.3   Terms, levels, and transitions in LS coupling – equivalent electrons   37
      3.1.4   Complex spectra   38
      3.1.5   Metastable states, pseudo-metastable states – forbidden lines   39
  3.2   One-electron systems   39
      3.2.1   One-electron atoms and ions   39
      3.2.2   Alkali and alkali-like spectra   40
  3.3   Two-electron systems   42
      3.3.1   Alkaline earth elements and iso-electronic spectra   42
      3.3.2   Elements with two electrons or two holes in the p-shell   43
  3.4   Complex spectra – with emphasis on Fe II   45

		3.4.1 Definition and properties of complex spectra	45
		3.4.2 Astrophysical importance of Fe II	45
		3.4.3 The atomic structure of Fe II	46
4	**Elementary excitation processes in rarefied plasmas**		**51**
	4.1	Photoionization of atoms	51
	4.2	Recombination	53
	4.3	Electron excitation and ionization	55
	4.4	Total and local thermodynamic equilibrium (TE and LTE) in plasma	55
	4.5	Non-LTE astrophysical media	58
	4.6	Non-LTE: photoselective excitation	59
5	**Astrophysical rarefied gas/plasma**		**62**
	5.1	Low-density gas nearby a hot star – Strömgren sphere	62
	5.2	Planetary nebulae	66
		5.2.1 Forbidden transitions	69
		5.2.2 Accidental coincidences of spectral emission and absorption lines	71
	5.3	Gas condensations (blobs) in the vicinity of a hot star	73
	5.4	Surroundings of symbiotic stars	76
6	**Basic elements of laser physics**		**78**
	6.1	Amplification coefficient	78
	6.2	Saturation of amplification	79
		6.2.1 Homogeneous broadening	80
		6.2.2 Doppler broadening	81
	6.3	Laser with resonant optical feedback (cavity)	84
		6.3.1 Simple consideration	85
		6.3.2 Quantum considerations	86
	6.4	Coherent properties of laser light	91
		6.4.1 Spatial coherence	91
		6.4.2 Temporal coherence and spectral bandwidth	92
		6.4.3 Coherent state of light field	95
	6.5	Laser as an amplifier	97
		6.5.1 Amplification of coherent light	98
		6.5.2 Amplification of spontaneous emission	100
7	**Introduction to astrophysical lasers**		**102**
	7.1	Amplification under non-LTE conditions	102
	7.2	Astrophysical predecessors of the laser	103
	7.3	How is laser action manifested under astrophysical conditions?	107
		7.3.1 Integral intensity of a spectral line	108
		7.3.2 Width of a spectral line	112
		7.3.3 Divergence of radiation	112

8	**Basics of collisionally pumped astrophysical lasers**		115
	8.1	Proposals of population inversion by collisional pumping	115
		8.1.1 Fine-structure levels of ions	116
		8.1.2 He II and He I in stellar envelopes	118
		8.1.3 OH radical and H_2O molecule in star-forming regions	118
	8.2	Hydrogen recombination far-IR laser in MWC 349	119
	8.3	IR CO_2 laser in the atmospheres of Mars and Venus	121
9	**Basics of optically pumped astrophysical lasers**		124
	9.1	Bowen accidental resonances and fluorescence lines	124
	9.2	Inversion of population by accidental resonance pumping	127
10	**Anomalous spectral effects in the Weigelt blobs of Eta Carinae**		132
	10.1	Structure of H II/H I regions in the Weigelt blobs	134
	10.2	High effective H Lyα temperature in the Weigelt blobs	139
11	**Astrophysical lasers on the Fe II lines**		144
	11.1	Radiative excitation and relaxation of Fe II levels	145
	11.2	Spectral peculiarities in Fe II	148
	11.3	Population inversion and amplification in Fe II transitions	150
	11.4	Radiative cycling with stimulated emission	152
		11.4.1 Strong radiative cycle	153
		11.4.2 Weak radiative cycle	156
		11.4.3 Combination of strong and weak cycles	157
	11.5	Effective temperature of Lyα from radiative cycle	159
	11.6	Sources of Lyα radiation pumping of Fe II in the Weigelt blobs	162
	11.7	Andromeda spectral puzzle	164
12	**Astrophysical laser in the O I 8446 Å line**		168
	12.1	Spectral anomalies in O I in stars	168
	12.2	Excitation mechanism of O I in the Weigelt blobs	170
	12.3	Inverted population and amplification coefficient of the 8446 Å line	172
	12.4	Spatial features of the O I laser in the Weigelt blobs	176
	12.5	Possible O I lasers in stellar atmospheres and Orion Nebulae	179
13	**Narrowing of spectral lines in astrophysical lasers**		181
	13.1	Role of geometry in the saturation regime on spectral narrowing	181

13.2 How to measure the narrow spectral line width
of astrophysical lasers   185

14 Possibility of scattering feedback in astrophysical masers/lasers   190
14.1 Noncoherent scattering feedback   191
14.2 Resonant scattering feedback   197
14.3 Space masers with scattering feedback   199
14.4 Space lasers with resonance scattering feedback   203

15 Nonlinear optical effects in astrophysical conditions   207
15.1 Photoionization processes in radiation-rich
astrophysical plasma   207
15.2 Rate of resonance-enhanced two-photon ionization in
bichromatic radiation   210
15.3 RETPI of Si II in the Weigelt blobs of Eta Carinae   215
15.4 Successive RETPI schemes for some light elements   220

16 Laser and interstellar communications   227

References   231

Appendix: Useful units   249

Index   251

1
Introduction

Astrophysical lasers (natural lasers in space) present a challenging scientific problem lying at the intersection of several fields of physics: astrophysics, atomic physics, and laser physics. They have just recently become the subject of not only theoretical but also experimental studies owing the rapid progress in observational astronomy. The decisive role in this progress can be ascribed to one of the most powerful instruments in the history of astronomy – The Hubble Space Telescope (HST), successfully operating since 1990. The Space Telescope Imaging Spectrograph (STIS), which was later installed on HST, has been utilized with great success for seven years. Among many fruitful observations of a diversity of astrophysical objects, HST/STIS spectra have been obtained in the UV, visible, and near-IR ranges of one of the brightest objects in our Galaxy, Eta Carinae and its surrounding nebula. The high spatial and spectral resolution of HST/STIS made it possible to study the spectra of the star and a number of ejected gas condensations separately from each other. A great number of anomalous spectral lines were registered in these spectra, and the interpretation of them led us to conclusions about the action of astrophysical lasers and the existence of nonlinear optical effects in gas condensations. We will describe these phenomena in the present book, which explains the reason why a photo of Eta Carinae and its surrounding nebula (Homunculus) is presented on the front cover of the book.

The discovery of astrophysical lasers described in this book is a natural step in the progress starting with the invention of laboratory masers and lasers to the discovery of natural masers in Space. Thus, lasers and masers are "programmed" by Nature, and their discovery seems to be quite logical. In reality, however, the way to create or detect them is not programmed at all but is rather non-direct, individual and dramatic. To understand the history of quantum electronics, the great discovery of the 20th century, the reader is referred to the memoirs of the laser pioneers (Townes, 1994, 1999; Maiman, 2000).

The history of masers and lasers shows that some scientific phenomena are discovered in the laboratory before they are observed in Nature, but the order can also be reversed. For example, thermonuclear fusion of light nuclei was proposed to operate inside stars before anyone managed to initiate such nuclear reactions on Earth in the form of the thermonuclear explosion of a hydrogen bomb. But, the nuclear fission chain reaction was utilized first in nuclear reactors and bombs and later discovered in Nature (the natural nuclear reactor in Oklo, Africa). An analogous development has occurred with masers and lasers. After the maser had been invented in the laboratory the maser effect was discovered in space in the form of stimulated emission of microwave radiation

2 *Astrophysical lasers*

in molecules (OH, H_2O etc.). As a consequence, the search for maser microwave radiation from space became the main method to search for molecules in space by using radio astronomy. This was simply based on the fact that stimulated emission is many orders of magnitude more intense than spontaneous emission in the microwave region. Recently we have discovered the laser effect at visible and near infrared wavelengths in an astrophysical plasma. This means that both the maser and laser effects, first created under laboratory conditions, have later been discovered to operate in certain environments in Space.

1.1 Historical remark

The principles behind the operation of masers and lasers were discovered during the decade between 1950 and 1960 (Gordon et al., 1954; Basov and Prokhorov, 1954; Schawlow and Townes, 1958). These fundamental pioneering studies received the highest international recognition. The experimental demonstration of lasers and the subsequent exponential progress in both the development of lasers themselves and their applications confirmed the significance of this discovery.

It is fairly interesting to consider the origin of quantum electronics, which appeared because of the development and synthesis of two avenues of research in physics: classical electromagnetic physics (we can call it "radio") and quantum physics (we can call it "optics"). Both these fields are combined by Maxwell's equations. The creation of quantum physics was based on atomic physics and optical spectroscopy and it later became the one of foundations of physics as a whole.

Fig. 1.1 shows a selected sequence of key discoveries, ideas, and experiments, which resulted in the creation of quantum electronics. The main sequence of events, which led to the discovery of the laser is shown by solid arrows. The possibility of discovering a laser by amplification of light in a medium having an inverted level population took never place, and that is indicated by the question mark in Fig. 1.1. Although lasers could be discovered along the optical avenue, they were in reality discovered after the construction of masers, i.e. because of the progress along the radio avenue. At first, the concepts of quantum physics and the invention of microwave sources of electromagnetic radiation for radar led to the development of radio spectroscopy, and then masers (molecular generators) were created. Thereafter the extension into the optical region of the concept of the resonant cavity feedback, where the idea of the inverted population of energy levels involved in optical transitions was proposed long ago, resulted in the creation of a laser. This confirms once more the extremely important role of the interrelation and synthesis of various fields of science and technology in scientific and technical progress. Although lasers could be discovered independently of radio, this did not happen. It confirms the inevitability of the discovery itself despite the fact that its specific scenario is unpredictable and always has its own personal and subjective aspect.

In this connection, it is interesting to note that natural masers operating under astrophysical conditions on the OH radical at $\lambda = 18.5$ cm (Weaver et al., 1965) and on the H_2O molecule at $\lambda = 1.35$ cm (Cheung et al., 1969) were discovered in radio astronomy in the mid 1960s, ten years after the invention of the first laboratory masers.

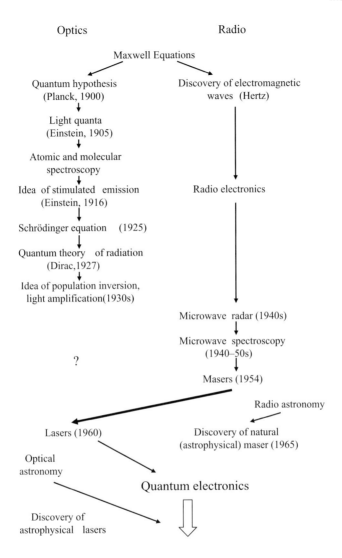

Fig. 1.1 Sequence of discoveries in optics and radio resulting in the birth of quantum electronics.

Since the brightness temperature of the microwave radiation of astrophysical masers is as high as 10^{12}–10^{15} K, it can only be explained as amplification caused by stimulated emission in a medium with an inverted population of low-lying molecular levels. This is another scenario of a discovery that could have been made earlier. Astrophysical lasers emitting in the optical range, which are discussed in this book, were discovered quite recently, forty years after the creation of the laboratory laser. However, their roots can be traced back as early as in the 1930s–1940s, i.e. long before the creation of lasers and masers. In those years, astrophysicists explained the appearance of anomalous spectral emission lines of an element as a result of fluorescence due to a wavelength coincidence with an absorption line of another element – the Bowen mechanism (Bowen, 1934, 1935). They also used the relations between the pump and emission rates, which in modern terms is equivalent to the conditions for inverted population (Thackeray, 1935). This is another unrealized scenario of the discovery of laser action based on astrophysical observations.

1.2 From astrophysical masers to astrophysical lasers

The first interstellar maser transition included the ground electronic state of the OH hydroxyl radical at a wavelength of 18.5 cm but it was not initially recognized as a maser (Weaver *et al.*, 1965; Weinreb *et al.*, 1965). However, the laboratory masers had already been invented (Gordon *et al.*, 1954; Basov and Prokhorov, 1954), which allowed the understanding of the physical mechanism behind this phenomenon to be applied to the space maser (Davies *et al.*, 1967). Townes (1993) remarked: *"...the first suggestion I can remember that it (OH mysterium) might be maser came from Shklovskii. This was just informal conversation with me. We talked about it, and the possibility certainly seemed reasonable ... One of the things I've always admired about Skhlovskii is that he would challenge anything ... You should all read his book."* (Shklovskii, 1991).

Many observations characterized the emission of OH masers to be time-variable, polarized (linearly or circularly), and having narrow line widths. These characteristics are typical for most astrophysical masers (Elitzur, 1992; Gray, 1999). Masers in space have been found in transitions of OH, water (Cheung *et al.*, 1969), SiO, ammonia, methanol and many other molecules (Townes, 1997) and also in recombination lines of hydrogen in the submillimeter range (Strelnitskii *et al.*, 1996). Astrophysical masers occur in several places in the Universe: in the vicinity of newly born stars and in regions of ionized hydrogen (H II regions), in circumstellar shells around stars at the end of their life (red giants and supergiants), in shock-regions where supernova remants are expanding into an adjacent molecular cloud, and in the nuclei and jets of active galaxies with various mechanisms of pumping. Detailed studies of masers has proceeded along with the development of very long baseline interferometry (VLBI), which provides angular resolutions better than 0.1 mas (milliarcsec).

C. H. Townes (1993) also made an interesting remark on the early years of research on astronomical masers:

"Masers and lasers have of course been of economic and technical importance, as well as of scientific interest. I sometimes point out to politicians and other administrative types that if radio astronomy had been sponsored more strongly in the United States we probably would have had masers and lasers sooner. Masers could have been detected long ago in the sky – probably as early as the 1930s, and certainly immediately after or during World War II they could easily have been detected, but nobody was looking. The United States as a whole did not do very well by radio astronomy shortly after the war, as compared to the British and the Australians. They generally did a wonderful job, but still not quite enough.

Suppose someone had indeed looked, and seen this peculiar maser radiation. What would they have thought if they detected such radiation before masers were thought of in the lab? Would they have attributed it to Little Green Men, just as Little Green Men was initially used as the origin of pulsars because nobody understood them. The maser radiation would have been seen to be modulated, of all things. A very successful SETI operation? No such phenomena could happen naturally according to thinking of that time. However, I'm sure that someone would have figured it out before long and concluded yes, that is what's happening – the molecular states are inverted, and we get amplification. That would probably have suggested trying out such an idea in the laboratory. So masers and lasers – highly applied technology, in a sense – could well have come out of astronomy".

The first proposals suggesting laser action in stellar atmospheres in non-local thermodynamical equilibrium (non-LTE) were presented at the end of the 1960s (Menzel, 1969; Smith, 1969; Letokhov, 1972a). Menzel (1969) considered in a general form the radiation transfer equation and suggested, from some assumptions on the ratio of occupation numbers (populations) for the upper and lower levels, the possibility of negative absorption (amplification) of light. Smith (1969) examined theoretically the possibility of obtaining population inversion in the simple atomic p-shell configurations ($2p^q$ and $3p^q$, $q = 2,3,4$) under astrophysical plasma conditions assuming excitation by electron collisions. Letokhov (1972) presented explanations of numerous line intensity anomalies observed in optical stellar spectra suggesting the possibility of stimulated emission due to inverse population of the transitions, particularly the 8446 Å line of O I. The excitation mechanism considered for the oxygen atoms was optical pumping caused by an accidental wavelength coincidence between an O I absorption line and H Lyβ at 1025.72 Å (Bowen, 1934). A detailed study of the possibility of stimulated emission in stellar atmospheres, including line narrowing and scattering feedback, and the O I case in particular, was presented by Lavrinovich and Letokhov (1974).

The first experimental observation of non-thermal emission (laser action) in the infrared (IR) range was done for the 10 µm CO_2 lines in the atmosphere of Mars and Venus (Johnson et al., 1976). The direct line absorption of mid-IR solar flux by CO_2 and N_2 with collisional transfer of energy to CO_2 were proposed as plausible excitation mechanisms. Subsequent experiments confirmed laser action in the CO_2 molecule in Martian and Venusian atmospheres (Mumma et al., 1981).

The discoveries of astrophysical lasers in the optical range for Fe^+ ions (Johansson and Letokhov, 2002; 2003a; 2004c) and O atoms (Johansson and Letokhov, 2005c) are quite non-trivial features, since the spontaneous radiation dominates the visible range in space and masks possible stimulated radiation. The occurrence of inverted population of quantum states, which is necessary for the laser effect, is much less probable for atomic energy levels in the optical region than for low-lying molecular states in the microwave range. A similar situation is true for non linear optical effects, which are produced quite easily in the laboratory, but occur only under very special conditions in astrophysical plasmas. We found evidence of the existence of resonance-enhanced two-photon ionization (RETPI) in Space in the vicinity of a hot blue star (Johansson and Letokhov, 2001a; Johansson et al., 2006).

The sites for our initial discoveries were special gas condensations in the close vicinity of the very massive LBV (luminous blue variable) star η Carinae. This did not occur accidentally since η Carinae is one of the most luminous stars of the Galaxy.

Table 1.1 Maser/laser action in space

– microwaves	molecules	OH, 18.5 cm	Weaver et al., 1965
		H_2O, 1.35 cm	Cheung et al., 1969
	> 100 molecular species		Townes, 1997
– submillimeters	atom	H**	Strelnitskii et al., 1996
– IR	molecule (Mars, Venus)	CO_2, 10 µm	Johnson et al., 1976
			Mumma et al., 1981
– optical waves	ion, atom	$Fe^+ \simeq 1$ µm	Johansson, Letokhov, 2002, 2003a
		O I, 8446 Å	
	Eta Carinae, gas condensations		Johansson, Letokhov, 2005c

The environment of this extended object can be observed in very high angular (spatial) and spectral resolution simultaneously using the Hubble Space Telescope (HST) with its Space Telescope Imaging Spectrograph (STIS) on board (Kimble et al., 1998). This unique instrument has provided a huge volume of information about Eta Carinae and its environment in the form of high-resolution spectra between 1150 and 10400 Å (Gull et al., 2001). Parts of this spectral information were difficult to interpret, which attracted our attention. It finally led to the discovery of such unusual features as laser effect and RETPI occurring in astrophysical environments, and they are described in this book. In the future, when observations at very high spectral and angular resolution will be performed with large ground-based telescopes, such phenomena can presumably be studied in gas clouds associated with other objects (LBV, hot stars, symbiotic stars, stellar chromospheres etc.).

In Table 1.1 we present the milestones of progress in the discoveries of astrophysical masers and lasers from the microwave to the optical wavelength regions. However, before we proceed with a detailed description of these "new" effects we give some basic elements of atomic physics, astrophysics and quantum electronics. This may be useful and valuable to the reader, since astrophysics and quantum electronics have long-standing but poorly known links.

1.3 Amplification conditions for an atomic ensemble

Electromagnetic radiation can be amplified by an ensemble of quantum particles only when the thermodynamic equilibrium is strongly violated, i.e., when at least one of the following equilibrium conditions is violated.

(i) The equilibrium Boltzmann distribution of the population of the quantum states of particles in the ensemble.
(ii) The equilibrium state of the electromagnetic radiation, in which an ensemble of particles with a given equilibrium temperature is located, i.e. the absence of external radiation.

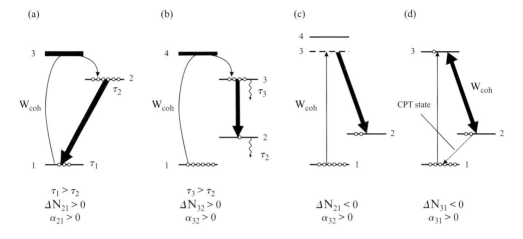

Fig. 1.2 Various methods for obtaining amplification, based on producing population inversion in three-level (a) and four-level (b) schemes, by means of Raman scattering without the population inversion (c), and using the coherent superposition of states 1 and 2 without the population inversion for levels 2 and 3 (d). To obtain amplification in schemes (a) and (b), incoherent pumping at a moderate rate – especially in scheme (b) – is required. To obtain amplification in scheme (c), pumping with a very high spectral intensity is required (not necessarily coherent). To obtain amplification in scheme (d), coherent pumping in the $3 \rightarrow 1$ transition is required and CPT states 1 and 2 should be present.

(iii) The equilibrium (random) distribution of the phases of the wave functions of quantum states, i.e., the absence of the non-zero (non-fluctuation) average polarisability or coherence in the ensemble.

The violation of the first condition, resulting in the population inversion for some pair of levels, can be achieved by many methods, for example, by selective incoherent pumping of an excited level in a three- or four-level atom scheme (Figs. 1.2a, b, respectively). It is used in virtually all lasers (Siegman, 1986).

A strong violation of the second equilibrium condition can cause amplification even without an inverted population. It occurs in a Raman laser (Woodbury and Ng, 1962) where a transparent medium in a strong laser field becomes amplifying at the Stokes frequency of Raman scattering. The equilibrium population of the levels is determined only by the presence of intense monochromatic radiation in the ensemble (Fig. 1.2c)[1]. Note that, in this case, the laser pumping radiation only has to be intense and can have a broad spectrum (of the order of the Raman line width), whereas a high coherence is not required.

When amplification is produced by violating the first and second equilibrium conditions, the active medium is in an incoherent state, which is natural because the

---

[1] We do not consider the possibility of inversion-free amplification due to the recoil effect, when emission and absorption lines have different frequencies. Free relativistic electron lasers are a spectacular example of such a situation.

phase relaxation time in an ensemble is usually much shorter than the population relaxation time. Coherence is produced only when the active medium is placed into an optical cavity with the selection of a finite number of spatial modes, in which a high rate of stimulated emission is maintained. In other words, the coherence of radiation is produced spontaneously owing to the "phase transition" in the "active medium + intracavity radiation" system (Lamb and Scully, 1967).

A substantially different situation takes place in the case of the third equilibrium condition, which can be violated only when special measures are taken. The equilibrium (random) distribution of phases (more exactly, phase differences) of the wave functions of a pair of quantum states can be violated under the action of an external coherent field, which matches the phase difference of the wave functions of two states interacting with this field (one- or two-frequency). Owing to such a matching, a quantum system in the ensemble can be prepared in a superposition antisymmetric state of two stationary quantum states (1 and 2 in Fig. 1.2d). Such a state is called a coherently population trapped (CPT) state (Alzetta et al., 1976; Arimondo, 1996). A particle in the CPT state absorbs radiation neither at the $1 \rightarrow 2$ transition nor at the $2 \rightarrow 3$ transition (the $1 \rightarrow 2$ transition is assumed forbidden) because of the quantum destructive interference of two opposite quantum transition routes $1 \rightarrow 3 \rightarrow 2$ and $2 \rightarrow 3 \rightarrow 1$. This effect is related to the electromagnetically induced transparency (Harris, 1997), when a coherent electromagnetic field produces the CPT state from which radiation is not absorbed. This effect was used to demonstrate the inversion-free laser operating due to the violation of the third equilibrium condition in the Λ-scheme shown in Fig. 1.2d (Zibrov et al., 1995) and in the V-scheme (Padmabandu et al., 1996).

The first equilibrium condition can be easily violated in astrophysical conditions. In this case, the inverted population appears because of the spatial inhomogeneity of density, temperature and other parameters of the astrophysical medium (gas clouds near star, etc.), when the medium is in a stationary but not an equilibrium state. By discussing this possibility for astrophysical media, we should consider separately the pulsed and continuous or quasi-stationary regimes for producing the population inversion. In the pulsed regime, the pumping rate of a level should exceed its relaxation rate, which for optical transitions is usually greater than 10^8 s^{-1} (while for forbidden transitions the relaxation rate sometimes reaches 1 s^{-1}). Under astrophysical conditions, this situation is unlikely because the characteristic time of pulsed astrophysical phenomena, which can be used for pumping, is much longer than the population relaxation time for quantum particles.

In the continuous wave (CW) laser regime, the effect of a faster relaxation of the lower level of the lasing transition compared to that of the upper level is used (Javan et al., 1961) (Fig. 1.2b). In this case, population inversion always exists in such a transition, and the pumping rate should provide a certain density of the inverted level population, which is required for obtaining the threshold amplification. This regime is used in all CW lasers, and it seems to be the most probable for astrophysical lasers. It is realized most clearly when the radiative lifetimes of the two corresponding levels are substantially different (in an appropriate way). In this case, the inverted population appears in an isolated quantum system without any collisions, which is typical for rarefied gaseous circumstellar or interstellar matter. Astrophysical lasers[2] operating according to such a scheme in medium-excited (5–10 eVpgevels of Fe II at a wavelength of ~1 μm have recently been discovered in gas condensations close to one of the brightest and most massive stars, η Car, in the Galaxy (Johansson and Letokhov, 2002, 2003a).

However, although collisions of particles at higher densities in stellar atmospheres return the ensemble of particles back to the equilibrium state, collisions can also play a positive role by significantly depleting the lower level. It may seem strange, but the role of collisions in the enhancement of the intensity of emission lines in stellar atmospheres according to the typical laser scheme (Gould, 1965) was discussed in astrophysical papers back in the 1930s (Thackeray, 1935). Since the questions related to stimulated emission were discussed in these papers, they are considered separately in Chapter 7.

To violate the second equilibrium condition, an intense monochromatic excitation is required, which can usually only be generated by a laser. A Raman line is always amplifying, and the gain is proportional to the intensity of the external radiation within the Raman line width (Woodbury and Ng, 1962). One specific observation of Raman scattering under astrophysical conditions is of special interest, viz. the scattering of the O VI resonance lines at $\lambda = 1032$ and 1038 Å by the 1s – 3p transition in neutral hydrogen H I (close to resonance with H Lyβ at 1026 Å) reported by Schmid (1989). Owing to such scattering, the hydrogen atom was found in the 2s state, and the Raman line was observed in the visible region redshifted from H Ba α (at 6825 and 7082 Å). This effect was observed under special astrophysical conditions in symbiotic stars, where strong VUV emission lines of highly ionized atoms irradiate the high-density regions of neutral hydrogen. The gain in Raman lines is rather low, and only spontaneous scattering can occur.

The spontaneous (without any coherent field) violation of the third equilibrium condition, i.e. the condition of a random distribution of the phase differences of the wave functions of quantum states of independent particles in the ensemble, seems unlikely. (The only exception is an ensemble of an ultracold gas of boson atoms of a sufficient density, where the Bose–Einstein condensation is observed). Nevertheless, Sorokin and Glownia (2002) put forward a bold hypothesis about the possibility of inversion-free amplification of the 3p – 2s transition in hydrogen (H Ba α) under astrophysical conditions. The relaxation of the phases of quantum states in rarefied gases (of density lower than 10^8 cm^{-3}) can indeed occur very slowly (for minutes and longer), but in the excited 2s state of H I it only occurs for 0.12s (because of spontaneous two-photon decay). However, to produce coherence of the 1s and 2s states (or, more exactly, CPT states) of hydrogen, the mechanism of spontaneous appearance of the CPT state should exist. Sorokin and Glownia (2002) assume that such a process can be induced by two intense incoherent emission lines, H Lyα at 1215 Å and H Baα at 6563 Å, by spontaneous decay of excited states of an atom in two completely different channels, although the probability is extremely low. It would be very interesting to

---

[2]We call an astrophysical laser amplifier the astrophysical laser (APL). Conditions under which the APL is transformed to a generator owing to an incoherent feedback produced by scattering are discussed in Chapter 14.

detect this process, because it is an example of an unusual phase transition of an ensemble of quantum particles from an incoherent to a coherent state. At present, only one example of such a phase transition is known, viz. when an amplifying incoherent laser medium is placed in an open resonator, which drastically restricts the number of high-quality spatial modes of stimulated emission, i.e. laser action with positive resonance feedback.

1.4 Structure of the book

The subject of this book is evidently interdisciplinary by its character as it lies at the intersection of astrophysics, atomic and molecular spectroscopy, and laser physics. Historically, all these fields have been connected by common roots, whereas astrophysical lasers just confirm these ancient ties. Within the framework of this book it is impossible and pointless to consider all these fields in detail, but we will try to give the necessary elements in simple language. This is done in a concise form in Chapters 2–6. Then, in Chapter 7, we consider the general questions of astrophysical lasers including the astrophysical predecessors of lasers dating back to the 1930s.

The pumping mechanisms of astrophysical lasers are based on the use of radiation energy from the nearest star. The radiation energy of a star can be absorbed in space clouds, which causes a spatially inhomogeneous distribution of kinetic and internal energy of particles (atoms, ions, radicals, molecules) and, finally, an inverse population in certain quantum transitions in the process of particle collisions. Astrophysical lasers of this type can be called collisionally-pumped lasers, some of which were predicted theoretically. However, some collisionally-pumped lasers have been discovered experimentally: IR CO_2 lasers in the atmospheres of Mars and Venus, and hydrogen far-IR lasers in the Orion Nebula. Collisionally-pumped astrophysical lasers are briefly discussed in Chapter 8. We expect that the progress of getting higher angular and spectral resolution in IR astronomy will make it possible to discover other collisionally-pumped lasers in Space.

Chapters 9–12 are concerned with astrophysical lasers based on a Bowen-type of resonant pumping due to accidental wavelength coincidences of emission and absorption lines. It is just such lasers, radiating in the red and near-IR wavelength regions in Fe II and O I transitions, that were discovered in gas condensations close to η Carinae. With the development of optical telescopes having high angular resolution and high-resolution spectrometers these lasers may be discovered indirectly in the vicinity of hot stars in spite of the obvious difficulties of overcoming the spontaneous radiation in the optical range.

The studies of astrophysical lasers have just begun since the spectral resolution of the existing optical telescopes is still insufficient to prove directly the laser effect by measuring the true sub-Doppler spectral width. This challenging problem, discussed in Chapter 13, is common for all astrophysical lasers. Unlike the microwave range, the resonance scattering in the optical range is of great importance. Resonance scattering may prove to be essential in converting an astrophysical laser, acting as an amplifier, into an astrophysical laser with scattering feedback, which is due to the return of radiation energy into the amplifying region (Letokhov, 1967a, 1996). This interesting problem of laser physics is under consideration in Chapter 14, the material of which may prove to be useful for future investigations of astrophysical lasers.

Chapters 15–16 deal with two problems, which are not connected directly with astrophysical lasers but are closely related to them: Firstly, the possible existence of nonlinear optical effects in Space, particularly the resonance-enhanced two-photon ionization under astrophysical conditions (Johansson and Letokhov, 2001a), and secondly the connection of lasers for interstellar communications (Schwartz and Townes, 1960; Townes, 1983).

2
Elements of Radiative Quantum Transitions

This Chapter presents only some basic elements of the theory of radiative quantum transitions in a two-level system which are needed to describe the properties of astrophysical lasers as well as their distinctions from astrophysical masers and laboratory lasers. A more detailed description can be found in many classical textbooks referred to below.

2.1 Spontaneous and stimulated emission

The concept of two types of radiative processes – spontaneous and stimulated emission – was introduced by Einstein (1916) as he analyzed the interaction between radiation quanta and atoms (molecules). This interaction is followed by an energy exchange which is necessary to keep thermodynamical equilibrium. In his next work, Einstein (1917) introduced the term momentum exchange in an elementary emission event or in radiation absorption. Both these processes – energy and momentum exchange – play a fundamental role in astrophysical conditions in general and in the formation of radiation in stellar atmospheres, in particular (Mihalas, 1978).

The next step was a semiclassical consideration of emission processes, where the electromagnetic field was classical and the atom quantum-mechanical. This consideration is not complete but it verifies Einstein's thermodynamic approach. Finally, the complete quantum theory of emission (spontaneous and stimulated) and absorption processes, when both the atom and the field are quantized, was elaborated by Dirac (1930, 1957). The quantization of an electromagnetic field results in a more successive description of the probabilities of spontaneous and stimulated emission both in thermodynamical equilibrium and under non-equilibrium conditions. The latter is particularly important in describing masers and lasers when the particle ensemble (atoms, molecules) as well as the electromagnetic field are in thermodynamically nonequilibrium conditions.

Let us now briefly consider all these approaches and introduce concepts and notations which are necessary for further discussions.

2.1.1 Thermodynamical equilibrium – Einstein coefficients A and B

Einstein (1916) introduced the concepts of spontaneous and stimulated light emission in studying the thermodynamical equilibrium between the ensemble of atoms (molecules) and thermal radiation. The spectrum of this black body radiation is described by the distribution $U(\nu)$ recognized by Planck 16 years before:

$$U(\nu) = \left(\frac{8\pi\nu^2}{c^3}\right) \frac{h\nu}{\exp\left(\frac{h\nu}{kT}\right) - 1} \left[\frac{\text{erg}}{\text{cm}^3\text{Hz}}\right], \quad (2.1)$$

where the spectral radiation energy density $U(\nu)$ varies with the temperature T and the frequency $\nu = \omega/2\pi$. Einstein obtained the Planck distribution using the following arguments which are given literally (except for some designations changed):

'Let us consider the gas from similar atomic particles being in statistical equilibrium ... with thermal radiation. Assume that each particle may be ... only in discrete states 1, 2, ... m, n..., etc. with energies E_1, E_2 ..., etc. By analogy with statistical mechanics or directly from the Boltzmann principle or, finally, from thermodynamic considerations it follows that the probability W_n of the state n and, respectively, the relative number of particles in the state n is

$$W_n = g_n e^{-\frac{E_n}{kT}}, \quad (2.2)$$

where k is the Boltzmann constant, and g_n is the statistical "weight" of the state n, i.e. a constant characteristic of a quantum state and independent of gas temperature T.

Assume now that a particle can pass from state m to state n absorbing radiation with a certain frequency $v = v_{nm}$ and from state n to state m emitting radiation of the same frequency. The change in energy due to radiation in this case comes to $E_n - E_m$ ... (Fig. 2.1).

At thermal equilibrium the number of particles passing from m to n per unit time in radiation absorption is equal to that of particles passing from n to m in radiation emission. Simple hypotheses can be established for these transitions which may be of two types:

(a) Spontaneous emission. This transition from n to m is with radiation emission with the energy $E_n - E_m$. This transition occurs without outer actions. One

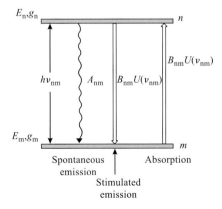

Fig. 2.1 Three main types of radiative quantum transitions between two quantum levels.

can hardly imagine that it is similar to something else but radioactive decay. Let the number of transitions per unit time be

$$A_{nm} N_n,$$

where A_{nm} is a constant belonging to a combination of m and n, N_n is the number of particles in state n.

(b) Stimulated emission. Stimulated emission is conditioned by radiation acting on a particle; it is proportional to the radiation density U with the corresponding frequency... In our case the radiation can induce both the transition $n \to m$ and the transition $m \to n$. The number of $n \to m$ transitions per unit time is expressed in this case as

$$B_{nm} N_n U,$$

and the number of $m \to n$ transitions is expressed as

$$B_{mn} N_m U,$$

where the constants B_{nm} and B_{mn} relate to a combination of the states n and m.

Thus, the equation

$$(A_{nm} + B_{nm} U) N_n = B_{mn} N_m U \qquad (2.3)$$

can be used as a condition of statistical equilibrium of the transitions $n \to m$ and $m \to n$. On the other hand, (2.2) gives

$$\frac{N_n}{N_m} = \frac{g_n e^{-\frac{E_n - E_m}{kT}}}{g_m}. \qquad (2.4)$$

From Eq. (2.3) and (2.4) we have

$$A_{nm} g_n = U \left(B_{nm} g_n e^{-\frac{E_n - E_m}{kT}} - B_{mn} g_m \right). \qquad (2.5)$$

U is the density of radiation with frequency ν which is emitted or absorbed at transitions $n \to m$ and $m \to n$. For this frequency our equation gives a relationship between T and U. Assuming that U grows without limit as T increases, we must have

$$B_{nm} g_n = B_{mn} g_m \qquad (2.6)$$

As a result, we get

$$U = \left(\frac{A_{nm}}{B_{nm}}\right) \frac{1}{\exp\left(\frac{E_n - E_m}{kT}\right) - 1}. \qquad (2.7)$$

This is the Planck relation between the radiation energy density and the temperature T, the constants being still undetermined. The constants A_{nm} and B_{nm} could be calculated directly if we have electrodynamics and mechanics modified for the quantum hypothesis.

The fact that A_{nm}/B_{nm} and $(E_n - E_m)$ can depend not on a special property of a particle but only on the corresponding frequency ν follows from the condition that the radiation density U must be a universal function of T and ν. Furthermore, from Wien's displacement law

$$U(\nu) = \nu^3 f\left(\frac{\nu}{T}\right) \tag{2.8}$$

it follows that the relation A_{nm}/B_{nm} must be proportional to ν^3 and $(E_n - E_m)$ is in proportion to ν. Accordingly we have

$$E_n - E_m = h\nu_{nm},$$

where h denotes the Planck constant.

Of course, I willingly admit that the three hypotheses dealing with spontaneous and stimulated emission do not become truthful results at all due to the fact that they culminate in the Planck radiation formula. But their simplicity, common character, easy consideration as well as a natural transition to a limiting case of Planck linear oscillator (in terms of classical electrodynamics and mechanics) enable me to think it rather probable this consideration will form the basis for future theoretical representation.'

Shortly before Planck, Rayleigh derived the following expression for equilibrium radiation energy fluency in the classical limit $kT \gg h\nu$:

$$U = 8\pi \frac{\nu^2}{c^3} kT. \tag{2.9}$$

The Planck quantum distribution (2.7) must coincide in this limit with (2.9). Hence we can express the ratio of the Einstein coefficients A_{nm}/B_{nm} as:

$$\frac{A_{nm}}{B_{nm}} = h\nu \left(\frac{8\pi\nu^2}{c^3}\right), \tag{2.10a}$$

where the factor on the right of (2.10a) is the Rayleigh–Jeans expression for field oscillator density known in classical physics:

$$\rho(\nu) = \frac{8\pi\nu^2}{c^3} [\text{Hz}^{-1}\ \text{cm}^{-3}]. \tag{2.11a}$$

The radial frequency $\omega = 2\pi\nu$ (rad/sec) is used also. In these terms Eq. (2.10a) and (2.11a) have the form

$$\frac{A_{nm}}{B_{nm}} = \hbar\omega \left(\frac{\omega^2}{\pi^2 c^3}\right) \tag{2.10b}$$

$$\rho(\omega) = \frac{\omega^2}{\pi^2 c^3}, \tag{2.11b}$$

respectively.

The values of the Einstein coefficient A_{nm} for a given transition $n \to m$ of an atomic particle depend on the properties of the particle and must not depend on the type (configuration) of the radiation field (unless the particle can be considered a free particle).

2.1.2 Semiclassical approach

In a semiclassical approach the atom is described quantum-mechanically with the Schrödinger equation, and the electromagnetic field is considered classically, and the interaction of a quantized atom with a classical field is described with the interaction Hamiltonian H_{int} in the Schrödinger equation. The field–atom interaction energy is considered low and, consequently, the external field slightly disturbs the stationary quantum states of the atom. In a dipole approximation, when the field amplitude $\mathbf{E}_0$ varies a little on the atomic dimensions (a_0 is the Bohr radius $\ll \lambda$, the electromagnetic field wavelength $\lambda = c/\nu$), the Hamiltonian H_{int} can be presented in simple form

$$H_{int} = -\mathbf{d}\mathbf{E}_0 \cos \omega t, \quad \omega = 2\pi\nu, \tag{2.12}$$

where $\mathbf{d}$ is the full electric dipole atomic moment of an atom

$$\mathbf{d} = -e \sum_i \mathbf{r}_i \tag{2.13}$$

and $\mathbf{r}_i$ is the i-th electron's coordinate.

The semiclassical approach allows us to calculate the Einstein coefficient B for a stimulated quantum transition. These calculations can be found in many classical books (Fermi, 1932; Bethe, 1964; Rybicki and Lightman, 1979; Loudon, 1983) to which we refer readers interested in the details. The quantum-mechanical calculation in the present approximations for the Einstein coefficient B gives the expression

$$B_{nm} = \frac{\pi}{3} \frac{|\mathbf{d}_{nm}|^2}{\hbar^2}, \tag{2.14}$$

where $\mathbf{d}_{nm}$ is the dipole moment matrix element determined by the standard expression

$$\mathbf{d}_{nm} = \int \Psi_n^* \mathbf{d} \Psi_m d\nu, \tag{2.15}$$

and Ψ_n and Ψ_m denote the wave functions of the stationary atomic states n and m. The integration is performed over coordinates of all the electrons of the atom. Averaging for all possible orientations of the dipole momentum $\mathbf{d}$ results in the factor $1/3$ in (2.14).

Thus, if we know the wave functions of the stationary atomic states n and m we can find the Einstein coefficients B_{nm} and B_{mn}. If in a dipole approximation $\mathbf{d}_{nm} = 0$ (for example, for quantum atomic transitions between levels with the same symmetry), we can perform a quantum-mechanical calculation in the next order multipole approximations (quadrupole and magneto-dipole transitions) (Chapter 3).

The semiclassical approach makes it impossible to calculate the Einstein coefficient A_{nm} for spontaneous emission because it requires field quantization discussed in subsection 2.1.3. But the expression for A_{nm} can be obtained from (2.10) and (2.14). This combination results in (Sobel'man, 1979)

$$A_{nm} = \frac{4\omega_{nm}^3 \left|\tilde{\mathbf{d}}_{nm}\right|^2}{3\hbar c^3} \frac{g_m}{g_n}. \tag{2.16}$$

So, the mean lifetime of the excited state n with respect to the spontaneous transition to m (but not to other possible states!) is expressed as

$$\tau_{nm} = \frac{1}{A_{nm}}. \tag{2.17}$$

In a dipole approximation, the matrix element of the transition $n \to m$ in (2.14) is determined from (2.15). The selection rules are dictated by the conditions under which $\mathbf{d}_{mn}$ becomes equal to zero. For example, in the case of the one-electron hydrogen atom, $\mathbf{d}_{mn}$ is zero for transitions between spherically symmetrical S-states. This transition is forbidden to emit one photon. Transitions between S- and P-states are allowed as $\mathbf{d}_{mn}$ for them is nonzero. S- and P-states are states with the orbital angular momentum of electron $L = 0$ and $L = 1$, respectively, (Chapter 3) and, hence, the selection rule states that $\Delta L = \pm 1$.

For a many-electron atom, the matrix elements Eq. (2.15) for all the electrons are summed. In this case the selection rules are dictated by the change in the total momentum J of all the electrons. The matrix element is nonzero when the final state J_m takes one of three values of $J_n \pm 1$ or J_n, i.e. the selection rule assumes the form $\Delta J = 0, \pm 1$.

The parity of wave functions is very convenient for understanding selection rules. This property describes the variation of the wave function at a mirror reflection of all the coordinates of the electrons. When

$$\Psi(-\mathbf{r}_1, -\mathbf{r}_2, \ldots) = \Psi(\mathbf{r}_1, \mathbf{r}_2, \ldots), \tag{2.18a}$$

the wave function is even, and when

$$\Psi(-\mathbf{r}_1, -\mathbf{r}_2, \ldots) = -\Psi(\mathbf{r}_1, \mathbf{r}_2, \ldots), \tag{2.18b}$$

it is odd. As the dipole momentum $\mathbf{d}$ is an odd function, the matrix element $\mathbf{d}_{nm}$ is zero when the parity is the same for the initial n and the final m states. Hence in allowed transitions the parity must change.

Forbidden transitions in many-electron atoms have a non-zero probability as the parity in them is determined by the sum of the electronic orbital momenta. Such transitions are much less probable than allowed transitions but quite observable if there are not other collisional (nonradiative) deactivation mechanisms of the excited atoms. This situation is, for example, quite typical for rarefied gases, especially under astrophysical conditions (Chapter 5).

The selection rules in the higher (after dipole) approximations (electrical quadrupole, magnetic-dipole) for many-electron atoms are briefly discussed in Chapter 3 and in detail, for instance, in monographs (Condon and Shortley, 1963; Shore and Menzel, 1968; Sobel'man, 1979).

2.1.3 Quantum theory

A comprehensive description of spontaneous and stimulated emission is given by the quantum radiation theory (quantum electrodynamics) by Dirac (1930). In its simplest form it was formulated by Fermi in 1932. According to this formulation, the atom and the radiation are considered in a big (as compared to the atom) cube. In classical physics such a cube (a cavity) is regarded as a system having natural oscillations called harmonic oscillators (modes). The mode density is described by the Rayleigh–Jeans distribution (2.11). A manifold of such harmonic oscillators (modes) interacts with the matter (atoms).

In quantum theory, harmonic oscillators are thought to be quantised rather than classical. The energy of each oscillator in this case is

$$E_n = \left(n + \tfrac{1}{2}\right)\hbar\omega, \tag{2.19}$$

where $n = 0, 1, 2, \ldots$, so the ground state energy of the oscillator equals $(1/2)\hbar\omega$. The interaction of an atom with oscillators causes the number of photons to change by ± 1 for emission and absorption respectively. The relation with Maxwell's equations for a classical field is realized by normalizing the classical field amplitude. To do this, the radiation energy density U is taken equal to the product of $\hbar\omega$ and the probability of presence of one photon in a unit volume. The photons obey Bose–Einstein statistics that admit an unlimited number of photons in a quantum state, and the statistical weight n of identical photons is equal to $n!$ rather than to 1 as it must be in a classical approximation.

As an example, let us point out some details of the quantum theoretical treatment of atom–radiation interaction, which will be necessary below in considering spontaneous and stimulated emission not only for 4π steradians but also into a limited solid angle in a certain frequency range. Some details of this complete consideration can be found in many books (e.g., Loudon, 1983; Sobel'man, 1979).

The electric field strength $\mathbf{E}$ can be expressed in the form of expansion in terms of plane waves $e^{i(\mathbf{kr}-\omega t)}$:

$$\mathbf{E} = \sum_{\mathbf{k}}\sum_{\rho=1,2} ik\mathbf{e}_{\mathbf{k}\rho}\left(a_{\mathbf{k}\rho}e^{i\mathbf{kr}} + a^*_{\mathbf{k}\rho}e^{-i\mathbf{kr}}\right), \tag{2.20}$$

where the wave amplitudes $a_{\mathbf{k}\rho}$ are determined from the condition of radiation density normalization. The Hamiltonian of the interaction of an atom with each of the plane waves in (2.20) has the form

$$H_{\text{int}} = -\frac{e}{m_e c}\mathbf{p}\mathbf{e}_{\mathbf{k}\rho}\left(a_{\mathbf{k}\rho}e^{i\mathbf{kr}} + a^*_{\mathbf{k}\rho}e^{-i\mathbf{kr}}\right), \tag{2.21}$$

where $\mathbf{p}$ is the atomic electron momentum. In case of several electrons $\mathbf{p} = \sum_i \mathbf{p}_i$.

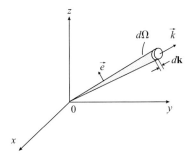

Fig. 2.2 Illustration of $dkd\Omega$ element of $\vec{k}$-space which contains N oscillators (modes) of electromagnetic field.

The wave vector $\mathbf{k}$ can be oriented in space within 4π steradians. The number of field oscillators (modes), with $\mathbf{k}$ confined within the solid angle element $d\Omega$ and the absolute value of the wave vector confined in the interval $(\mathbf{k}, \mathbf{k}+d\mathbf{k})$ (Fig. 2.2), is expressed as

$$dN = \frac{V}{(2\pi)^3} d\mathbf{k} = \frac{V}{(2\pi)^3} k^2 dk d\Omega, \qquad (2.22)$$

i.e. the number of oscillators per unit volume $N_0 = dN/V$ is

$$N_0 = \frac{k^2}{(2\pi)^3} dk d\Omega = \frac{\omega^2}{8\pi^3 c^3} d\omega d\Omega. \qquad (2.23)$$

This results in the above-derived expression (2.11b), derived above for mode density $\rho(\omega) = dN/d\omega$ in the case $d\Omega = 4\pi$ and for one photon polarization in a mode. The expression (2.23) is more convenient considering spontaneous and stimulated emission into a limited number of modes and within given spectral $d\omega$ and angular $d\Omega$ intervals.

The matrix element H_{int} with respect to the wave functions of the "atom + field" system is nonzero only for quantum transitions where the number of photons n changes only by unity. Our calculations give the following key expression for the probability of emission of a photon polarized by $\mathbf{e}$ to an element of the solid angle $d\Omega$ at atomic quantum transition between the states n and m (Fig. 2.2) (Sobel'man, 1979):

$$dW_{\text{nm}} = \frac{e^2 \omega}{2\pi \hbar c^3 m_e^2} |\mathbf{e}\langle n| \mathbf{p} e^{i\mathbf{k}\mathbf{r}} |m\rangle|^2 (\langle n \rangle + 1) d\Omega, \qquad (2.24)$$

where $\langle n \rangle$ is the average number of photons of a given polarization within the interval $\mathbf{k}, \mathbf{k}+d\mathbf{k}$, m_e is the electron mass, and the matrix element for $n \to m$ is written in a more compact form than (2.15). By analogy, the probability of absorption at the same quantum transition is expressed as

$$dW_{\text{mn}} = \frac{e^2 \omega}{2\pi \hbar c^3 m_e^2} |\mathbf{e}\langle n| \mathbf{p} k^{i\mathbf{k}\mathbf{r}} |m\rangle|^2 \langle n \rangle d\Omega. \qquad (2.25)$$

Expressions (2.24) and (2.25) confirm strictly Einstein's hypothesis on the elementary processes of emission and absorption. In the absence of external photons in field oscillators ($\langle n \rangle = 0$) only spontaneous emission is possible. With $\langle n \rangle \neq 0$ stimulated emission may occur and it is related to spontaneous emission by the fundamental Dirac law

$$w_{\text{em}} = w_{\text{sp}}(1 + \langle n \rangle). \tag{2.26}$$

It should be underlined that w_{sp} means spontaneous emission only to a certain range of frequencies and angles, i.e. to a field oscillator (mode).

The relation between the average number of photons $\langle n \rangle$ in the interval $(\omega, \omega + d\omega)$ and $d\Omega$ and the spectral intensity $J(\omega)$ (in erg/cm^2) can be found in a simple way. The value of $J(\omega)d\omega d\Omega$ must produce an energy flux, within the interval $d\omega d\Omega$, which, in turn, is related to the number of field oscillators (modes) in this interval N_0, the average number of photons in a mode $\langle n \rangle$ and photon energy

$$J(\omega)d\omega d\Omega = N_0 \langle n \rangle c\hbar\omega \tag{2.27}$$

where N_0 is determined by (2.23). This results in a simple relation between $J(\omega)$ and $\bar{n}$:

$$\langle n \rangle = \frac{8\pi^3 c^2}{\hbar \omega^3} J(\omega). \tag{2.28}$$

In the case of isotropic equilibrium radiation with the temperature T the average number of photons in a field oscillator (mode) is given by the expression

$$\langle n \rangle = \frac{1}{e^{\hbar\omega/kT} - 1}. \tag{2.29}$$

In the general case of nonequilibrium, for example, directed monochromatic laser radiation $\langle n \rangle$ could be very large but only for a limited number of modes (only for one mode in an ideal case). The radiation of astrophysical lasers can be assigned to an intermediate case when the directivity is far from the diffraction limit (i.e. one mode) but, nevertheless, the radiation solid angle Ω may be much smaller than 4π (Chapter 6).

Now let us turn back to the general expressions (2.24) and (2.25) for the probabilities of photon emission and absorption in the transition $n \to m$. The atomic size a is much smaller than the radiation wavelength λ. Therefore, in a matrix element $\mathbf{kr} \ll 1$ and, hence, in a first (so-called dipole) approximation we can consider $e^{i\mathbf{kr}} \simeq 1$. In this case the matrix element $\langle n | \mathbf{p} e^{i\mathbf{kr}} | m \rangle = m_e \langle n | v_e | m \rangle = -i\omega m_e \langle n | \mathbf{r} | m \rangle$, and the value $e \langle n | \mathbf{r} | m \rangle = \mathbf{d}_{\text{nm}}$ is the matrix element of the electric dipole moment of the atomic transition $n \to m$, v_e is the atomic electron velocity. In this (dipole) approximation the expressions (2.24) and (2.25) become simpler

$$dW_{\text{nm}} = \frac{\omega^3}{2\pi c^3 \hbar} |\mathbf{ed}_{\text{nm}}|^2 (\langle n \rangle + 1) d\Omega \tag{2.30}$$

and

$$dW_{\text{mn}} = \frac{\omega^3}{2\pi c^3 \hbar} |\mathbf{ed}_{\text{mn}}|^2 \langle n \rangle d\Omega \tag{2.31}$$

The expression for spontaneous emission probability is usually derived for isotropic ($d\Omega = 4\pi$) naturally polarized (the sum of two polarization states) emission of randomly oriented dipole moments. This is given by the factor $4\pi \times 2 \times 1/3$ in (2.30) for $\langle n \rangle = 0$:

$$W^{\rm sp}_{\rm nm} = \frac{4\omega^3}{3\hbar c^3} |\mathbf{d}_{\rm nm}|^2 = A_{\rm nm}, \tag{2.32}$$

It naturally coincides with the expression (2.16) obtained above in a semiclassical approximation.

The expression for stimulated (induced) emission probability takes the form

$$W^{\rm ind}_{\rm nm} = W^{\rm sp}_{\rm nm} \langle n \rangle = W^{\rm sp}_{\rm nm} \frac{4\pi^3 c^2}{\hbar \omega^3} J(\omega) \tag{2.33}$$

where $\langle n \rangle$ is the mean number of photons determined from (2.28), and account is taken of the haling $\langle n \rangle$ for one polarization of naturally polarized radiation. Accordingly the absorption probability is determined by the relations

$$W^{\rm abs}_{\rm mn} = W^{\rm ind}_{\rm mn} = W^{\rm sp}_{\rm nm} \frac{4\pi^3 c^2}{\hbar \omega^3} J(\omega). \tag{2.34}$$

In calculations it is often more convenient to use a more simple expression for induced emission and absorption rates ($\rm s^{-1}$), where the cross-sections ($\rm cm^2$) of the corresponding elementary processes are used. The absorption cross-section $\sigma_{\rm mn}$, for example, is defined as the ratio between absorbed energy $\hbar \omega W^{\rm abs}_{\rm mn}$ and incident energy flux $J(\omega) d\omega$, i.e.

$$\sigma^{\rm abs}_{\rm mn}(\omega) = \frac{\hbar \omega W^{\rm abs}_{\rm mn}}{J(\omega) d\omega} = \pi \frac{\lambda^2}{4} \frac{A_{\rm nm}}{d\omega}, \tag{2.35}$$

where $d\Omega = 4\pi$ since the photon flux directivity is not essential for the total absorption probability (Fig. 2.4 – page 25). The expression (2.35) contains a spectral range of incident radiation $d\omega$. An atom can absorb (emit) over a certain frequency range depending on the absorption (emission) spectral line shape. The integral (total) probability of spontaneous emission appearing in (2.35) can be written as

$$A_{\rm nm} = \int a_{\rm nm}(\omega) d\omega, \tag{2.36}$$

where $a_{\rm nm}(\omega) d\omega$ is the probability of spontaneous emission in the frequency range $(\omega, \omega + d\omega)$, depending on the spectral line shape. The line shape is determined by several mechanisms discussed in Section 2.2. Here we present expressions for cross-sections in terms of statistical weights (degenerations), $g_{\rm n}$ and $g_{\rm m}$ for levels n and m, which for simplicity have been omitted so far:

$$\sigma^{\rm abs}_{\rm mn}(\omega) = \frac{\lambda^2}{4} \frac{g_{\rm n}}{g_{\rm m}} a_{\rm nm}(\omega), \tag{2.37}$$

$$g_{\rm m} \sigma^{\rm abs}_{\rm mn}(\omega) = g_{\rm n} \sigma^{\rm ind}_{\rm nm}(\omega). \tag{2.38}$$

In the next Section we will consider the nature of broadening of atomic spectral lines, i.e. the spectral shape $a_{\rm nm}(\omega)$.

2.2 Broadening of spectral lines

In astrophysical conditions the broadening of spectral lines is governed by radiative decay, collisions of atoms and electrons and Doppler effect due to thermal motion of the atoms. Below, these mechanisms are discussed to the minimal extent needed for the forthcoming Chapters.

2.2.1 Radiative (natural) broadening

In an approximation of stationary quantum atomic states the spectral lines must be infinitely narrow as the energy of these states is determined precisely. Owing to spontaneous emission, however, the excited states are decayed with a rate determined by the probability of spontaneous emission (2.16). The decay is described by the exponential law $e^{-\gamma_{nm} t}$, where $\gamma_{nm} = A_{nm}$. For an optical transition, for example, $1/\gamma_{nm}$ is of the order of 10^{-8} sec, i.e. much longer than the period of electron motion in an atom $\omega_{nm} \simeq 10^{-15}$ sec. That is the reason why the approximation of stationary quantum states is rather good.

At first glance, it is the transition rate γ_{nm} that must determine the spectral line width of the transition $n \to m$ according to the Fourier transform of the exponential decay law. In other words, weak (forbidden) spectral lines should be very narrow. But this is only true for the case where the final state m is the ground one, i.e. absolutely stable. In the general case, it is not true for transitions between excited states.

The problem of radiative broadening of spectral lines was solved by Weisskopf and Wigner (1930). They proved that the spontaneous emission spectrum depends on the rates of radiative decay of both the upper and the lowest states. It is quite natural because in the quantum theory of radiation the matrix elements of a quantum transition between n and m depend on the wave functions of both states which decay with their individual rates γ_n and γ_m. So the quantum calculation of the spectral distribution $a_{nm}(\omega)$ given above by the expression (2.36) results in

$$a_{nm}(\omega) = A_{nm} \frac{\gamma}{2\pi} \frac{1}{(\omega - \omega_0)^2 + (\gamma/2)^2}, \qquad (2.39)$$

which has a Lorentz contour with a total width of γ where

$$\gamma = \gamma_n + \gamma_m \qquad (2.40)$$

at the central frequency $\omega_0 = \omega_{nm}$. The formula (2.39) shows that the spectral line must be wide when the lifetime of the initial and/or the final state is short, no matter whether the intensity of the line itself is determined by γ_{nm}. Fig. 2.3 illustrates this situation. Naturally, γ_n and γ_m include all the decays to the low-lying states, with γ_{nm} included into γ_n and, as a rule, $\gamma_{nm} \ll \gamma_n$.

Radiative broadening is small as compared with Doppler broadening (§2.2.3) but it is fundamentally important because of its prevailing value in the spectral line wings, i.e. for $|\omega - \omega_{nm}| \gg (\gamma_n + \gamma_m)$. Hence the spontaneous decay of excited atomic levels results in a Lorentz shape of the spectral line (2.39) with its total width γ (2.40).

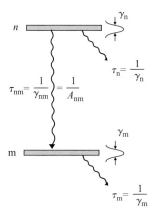

Fig. 2.3 Radiative broadening of spectral line due to radiative decay of upper (n) and lower (m) levels.

Stimulated transitions make a contribution to the radiative broadening of spectral lines, too. According to the quantum theory of radiation, radiative decay is explained by the finite lifetime of a stationary state (quantum level). Therefore, every level has its finite width according to the uncertainty principle. The decay rate Γ_n of the level n, however, depends not only on the spontaneous decay rate γ_n but also on the sum of rates for all the types of radiative transitions for levels n to m

$$\Gamma_n = \gamma_n + W_{nm}^{ind} + W_{mn}^{abs}. \qquad (2.41)$$

For the laser transition $n \to m$ the rate of induced emission due to radiation even to several modes may exceed the rate of spontaneous decay A_{nm} and even γ_n. So, strictly speaking, the radiative broadening in laser operation, with due regard for stimulated transitions changing the level population, that is, under saturation condition, may be essential.

2.2.2 Collisional broadening

An atomic particle moving in a gas (plasma) collides from time to time with other atoms, ions and electrons. All the collisions can be classified as inelastic, when the internal energies of the colliding particles change, and elastic, with the internal energy remaining constant. The frequency of nonelastic collisions is usually much smaller than that of elastic ones. In elastic collisions the particle trajectory changes, as well as the phase of the wave function of the particle state changes, in a random way. In these cases the atom emits a sinusoidal wave where a phase shift by a random value in a random time interval occurs.

The Fourier transform $f_\tau(\omega)$ of the sequence of such sinusoidal waves with a random duration τ has the form

$$|f_\tau(\omega)|^2 = \frac{\sin^2\left[(\omega - \omega_0)\tau/2\right]}{2\pi\left[(\omega - \omega_0)/2\right]^2}. \qquad (2.42)$$

For independent random collisions the Poisson distribution $P(\tau)$ is valid ($\tau = \tau_{\text{coll}}$):

$$P(\tau) = \gamma_{\text{coll}} e^{-\gamma_{\text{coll}}\tau} \tag{2.43}$$

Using (2.41)–(2.43) we can obtain a spectral contour of radiation broadened owing to collisions

$$\Gamma(\omega) = \int_0^\infty P(\tau)\gamma_{\text{coll}} |f_\tau(\omega)|^2 d\tau, \tag{2.44}$$

which has the form of a simple Lorentz contour

$$\Gamma(\omega) = \frac{\gamma_{\text{coll}}}{\pi} \frac{1}{(\omega - \omega_0)^2 + \gamma_{\text{coll}}^2} \tag{2.45}$$

The corresponding width is given by

$$\Delta\omega_{\text{coll}} = 2\gamma_{\text{coll}}, \tag{2.46}$$

where γ_{coll} is the collision frequency.

In the simplest case the collision frequency γ_{coll} in a gas (plasma) having the density N_0 of colliding particles with the velocity v and the effective cross-section of collisions $\sigma_{\text{coll}}(v)$ can be expressed as

$$\gamma_{\text{coll}} = N_0 \int w(v)\sigma_{\text{coll}}(v) dv = N_0 \langle v\sigma_{\text{coll}} \rangle, \tag{2.47}$$

where $w(v)$ is the velocity distribution of colliding particles, and $\langle v\sigma_{\text{coll}}\rangle$ is the short notation of the averaging on velocity distribution. Collisions with different types of particles (neutral atoms and ions, electrons) have different effective cross-sections. The idea about them can be inferred from other books (Traving, 1968). It is convenient to use the effective collisional cross-section for neutral atoms $\sigma_{\text{coll}} = \pi a_0^2$, where $a_0 = 0.5 \cdot 10^{-8}$ cm denotes the Bohr orbit of the hydrogen atom. In this case we can use the average value of the free path length ℓ in

$$\ell = \left(N_0 \pi a_0^2\right)^{-1} \simeq 10^{16} N_0^{-1} \text{ (cm)}, \tag{2.48}$$

for hydrogen atoms with their concentration N_0.

The collisions are responsible for Lorentz contour broadening (2.45). Radiative and collisional broadening results in a total Lorentz contour with a homogenous half-width of the spectral line

$$\Delta\omega_{\text{hom}} = \gamma + \Delta\omega_{\text{coll}} = \gamma + 2\gamma_{\text{coll}}. \tag{2.49}$$

In an astrophysical rarefied plasma characteristic for hosting the astrophysical lasers discussed in this book the collisional broadening is much less than radiative broadening.

2.2.3 Doppler (nonhomogeneous) broadening

When an atom moves with velocity v at an angle φ relative to the direction of observation (Fig. 2.4a), the frequency of the observed radiation, ω is shifted owing to the Doppler effect

$$\omega = \omega_0 \frac{1 - (v/c)\cos\varphi}{\left[1 - (v/c)^2\right]^{1/2}} \simeq \omega_0 \left(1 - \frac{v}{c}\cos\varphi\right), \tag{2.50}$$

where ω_0 is the radiation frequency of an atom at rest. With $v \ll c$ the value of the Doppler shift $\Delta\omega = \omega - \omega_0$ is

$$\frac{\Delta\omega}{\omega_0} = \frac{v}{c}\cos\varphi, \tag{2.51}$$

and the spectral line width is determined by the so-called homogeneous width (2.49) as shown in Fig. 2.4b. If the atoms are at thermal equilibrium, the projection of their velocity in the chosen direction on an observation, for example the x-axis $v = v_x = v\cos\varphi$, obeys the Maxwellian distribution (Fig. 2.4c)

$$w(v_x) = \frac{1}{u\sqrt{\pi}} \exp\left[-\left(\frac{v_x}{u}\right)^2\right], \tag{2.52}$$

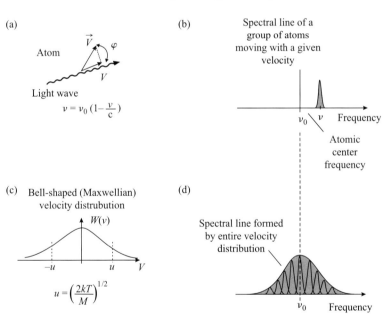

Fig. 2.4 Influence of the Doppler effect on the shape of a spectral line: (b) Doppler shift due to an atom, moving with velocity **v**. The Doppler effect shifts the emission frequency from ν_0 to ν; (c) thermal distribution of atom velocities and (d) corresponding Doppler inhomogeneously broadened spectral line. The upper curve (b) is the response of particles moving with velocity **v** as in (a). The lower curve (d) is the response due to atoms over the entire thermal distributions.

where u is the mean square velocity of particles with their mass M at the temperature T

$$u = \left(\frac{2kT}{M}\right)^{1/2} = 1.28 \cdot 10^4 \left(\frac{T}{A}\right)^{1/2} \quad [\text{cm/s}], \tag{2.53}$$

where A is the atomic weight of a particle in atomic units. Thus, random thermal motion causes a Doppler broadening with a Gaussian profile of the spectral line

$$\begin{aligned} g(\omega) &= \frac{1}{\sqrt{\pi}\delta\omega_D} \exp\left[-\left(\frac{\omega-\omega_0}{\delta\omega_D}\right)^2\right] \\ &= \left(2\sqrt{\frac{\ell n 2}{\pi}}\right)\frac{1}{\Delta\omega_D} \exp\left[-4\ell n 2\left(\frac{\omega-\omega_0}{\Delta\omega_D}\right)^2\right] \end{aligned} \tag{2.54}$$

where $\delta\omega_D = \omega_0(\frac{u}{c})$ is the Doppler half-width on the level $1/e$. The full Doppler width at half-maximum of the spectral line

$$\Delta\omega_D = 2\sqrt{\ell n 2}\delta\omega_D = 2\omega_0\left(2\ell n 2 \frac{kT}{Mc^2}\right)^{1/2} = 7.16 \cdot 10^{-7}\omega_0 (T/A)^{1/2}. \tag{2.55}$$

The total contour is a sum of many (convolution) Lorentz contours (Fig. 2.4d) according to the velocity distribution over a fixed direction of observation and the corresponding Doppler frequency shift $\omega_0 v/c$:

$$I(\omega) \simeq \frac{\gamma}{2\pi} \int_{-\infty}^{\infty} \frac{w(v)\, dv}{(\omega-\omega_0-\omega_0 v/c)^2 + (\gamma/2)^2}. \tag{2.56}$$

In case of Maxwellian velocity distribution $w(v)$ this total contour is known as the Voigt contour.

In astrophysics we use the so-called "damping factor", i.e. the ratio between the Lorentz homogeneous width $\Delta\omega_{\text{hom}}$ (2.49) and the Doppler width (2.55)

$$a = \frac{\Delta\omega_{\text{hom}}}{2\delta\omega_D} = \frac{\gamma + 2\gamma_{\text{coll}}}{2\delta\omega_D} = \sqrt{\ell n 2}\frac{\gamma}{\Delta\omega_D} \quad (\text{for } \gamma \gg \gamma_{\text{coll}}). \tag{2.57}$$

This factor is important in the theory of spectral line formation in stellar atmospheres and nebulae (Mihalas, 1978).

Doppler broadening is often called inhomogeneous as it masks the true homogeneous broadening governed by radiative decay and collisions (Fig. 2.4d). In linear interaction of light with a spectral transition the information on the true width is masked by the Doppler effect. But astrophysicists gain information on the homogeneous width by observing the far wings of the Lorentz contour which decreases as the frequency detuning $(\omega - \omega_0)$ is much slower than the Gaussian profile wings. In nonlinear interaction (under quantum transition saturation), particularly with anisotropic radiation, the difference between homogeneous and inhomogeneous broadening is of fundamental importance. This difference is essential for astrophysical lasers (Chapter 6) as it provides the basis for nonlinear laser spectroscopy free of Doppler broadening (Letokhov and Chebotayev, 1977).

2.3 Resonance scattering of radiation

Resonance scattering of radiation by atoms is one of the basic radiative processes responsible for the formation of spectral lines in gaseous astrophysical media (stellar atmospheres, nebulae, etc.). This rather complex process (which includes radiative decay, collisions, and thermal motion of atoms) has been studied in astrophysical literature for decades. The results of these studies are summed up in an excellent monograph by Mihalas (1978). Below we shall restrict ourselves just to a minimum of these results needed for further consideration.

2.3.1 Coherence of scattering in the atomic frame

Coherence of scattering has a wide meaning in astrophysics as the presence of correlation between the frequencies of incident and scattered radiation (*coherent scattering*) or its absence (*incoherent scattering*). For a motionless isolated atom (atom in its own frame of reference) only coherent scattering is possible which is dictated by the laws of conservation of energy and momentum – accurate enough to exclude the negligible but fundamentally important frequency shift due to the recoil effect (Minogin and Letokhov, 1987; Letokhov, 2007). In the atomic frame the presence of collisions makes both coherent and incoherent scattering possible, depending on the collision frequency. And, finally, the above statements remain true for a moving atom in its own frame. For isotropic scattering in the observer's coordinate system the frequency always changes, i.e. the scattering is incoherent. But, this statement is also strictly valid, too, for the central part of the Doppler contour when the mean frequency shift due to scattering is comparable with the Doppler width. When the radiation from the atoms with velocity $v \gg u$ is scattered, i.e. at the far wing of the Doppler contour, a certain frequency correlation is retained since the frequency shift in this case is comparatively small, i.e. of the order of $\delta\omega_D$. The scattering here is partially coherent.

For simplicity, let us consider resonance scattering in the three cases illustrated in Fig. 2.5. In case (a) the lowest state of an isolated (free) atom is the ground state, i.e. it has a zero radiative width. In this case a scattered photon with frequency ω, because of the absence of perturbations in the upper state, has the same frequency as an incident photon. The scattering cross-section σ_{sc} averaged over atomic orientations is determined by the expression (Loudon, 1983)

$$\sigma_{sc}(\omega) = \frac{\lambda^2}{4} a(\omega) = \frac{\lambda^2}{2\pi} \frac{(\gamma/2)^2}{(\omega - \omega_0)^2 + (\gamma/2)^2}, \tag{2.58}$$

where $\gamma = \gamma_n$, i.e. (2.58) coincides with (2.37) for the absorption cross-section at transition $0 \to n$. At the central frequency ω_0 the resonance scattering cross-section is determined by the simple expression

$$\sigma_{sc}(\omega_0) = \frac{\lambda^2}{2\pi} \tag{2.59}$$

In the case of an isolated atom we should distinguish the absorption probability of a photon and the excitation probability of an atom. At first sight it seems that a photon

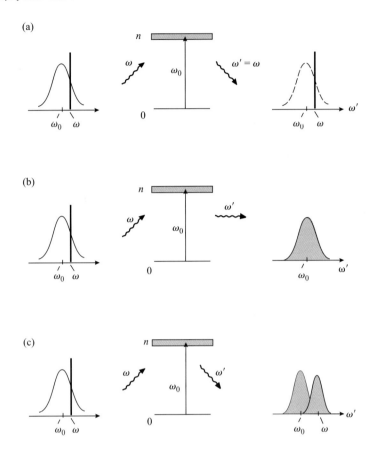

Fig. 2.5 Various cases of resonance coherent and noncoherent scattering of light by an atom with: (a) radiative decay of upper level n; (b) collisional decay of upper level; (c) radiative decays of both (upper and lower) levels. Spectra of incident light on frequency ω (left side) and spectra of scattered light with frequency ω' (right side) are shown also.

with this frequency $\omega \neq \omega_0$ in the Lorentz wing will excite the atom too but, according to (2.37) and (2.39), with a probability that is $((\omega - \omega_0)/\gamma)^2$ times less than at the centre of the line. But, that is not true since it contradicts the energy conservation law of the "atom + photon" system. An isolated atom cannot transfer the excess (or acquire the shortage) of its energy $\hbar(\omega - \omega_0)$. An atom can enter an excited state only in a two-photon absorption process (absorption of two photons simultaneously) with reemission of one photon with the frequency $\omega' = 2\omega - \omega_0$ (Makarov, 1983). When a photon is absorbed with the frequency ω such that $|\omega - \omega_0| \gg \gamma$, the atom scatters the photon but remains excited virtually for a very short time $1/|\omega - \omega_0|$. It is evident that, in the case of a non-isolated atom interacting with other particles (reservoir), the excitation of the atom becomes possible.

In the case (b) (Fig. 2.3) the lower level remains narrow but the excited level is broadened not only by radiative decay with the probability γ_n but also by collisions with a frequency much larger than γ_n. The spectral line shape remains, as before, Lorentz-like (§ 2.2.2) with its full width (2.49). In this ultimate case the frequency of the reemitted photon ω' cannot be correlated at all with that of the incident photon. The incoherent scattering cross-section is determined in this case by an expression like (2.58) where the full width γ must mean the full width due to radiative decay γ_n and collisions $2\gamma_{\text{coll}}$, i.e. (2.49). This case does not exist under the conditions of astrophysical lasers because they always have low density of atoms, ions and electrons. It is included here just to make the illustration of coherent and incoherent scattering complete.

In case (c) (Fig. 2.5) the atomic levels m and n are broadened by radiative decay but after absorbing a photon with the frequency ω the atom returns to the initial level. The expression for photon absorption probability $\hbar\omega$ with photon reemission $\hbar\omega'$ was derived by Weisskopf (1933) and later on by Woolley (1938). The expression for resonance scattering cross-section in this process is rather cumbersome. It is given in the monograph by Mihalas (1978). Here we consider only a specific feature of the reemitted photon spectrum. There are two peaks: at the incident photon frequency ω and at the frequency of the centre of spectral line ω_0 with characteristic radiative widths $\gamma = \gamma_n + \gamma_m$ and γ_n, respectively. The details of this spectrum are not essential here because, under astrophysical conditions, the Doppler effect causes a much larger broadening of the scattered radiation spectrum.

These are the properties of radiation scattered by a fixed or a moving atom in its own coordinate system. The change to the coordinate system of the observer who gets the light scattered by randomly moving atoms makes the picture much more complicated and interesting.

2.3.2 Doppler redistribution of frequency

In astrophysical gas media thermal motion results in an essential redistribution of the frequency of radiation by resonance scattering from moving atoms. Without going into details of the complicated calculations given, for example, by Mihalas (1978), we shall restrict ourselves to qualitative results.

Even if the scattering is coherent in the atomic coordinate system (the case (a) in Fig. 2.5), in the observer's coordinate system the frequency is redistributed owing to a random Doppler shift of frequency at scattering. In the Doppler contour nucleus ($|\omega - \omega_0| \lesssim 3\delta\omega_D$), for example, full frequency redistribution of scattered radiation occurs, i.e. the resonance scattering is fully noncoherent.

Outside the Doppler nucleus of the spectral line, particularly in far wings ($|\omega - \omega_0| \gg 3\delta\omega_D$), the resonance scattering is incoherent by approximately two-thirds and coherent by one-third. The conservation of a certain part of coherence is thus explained. With large frequency shifts of order $(\omega - \omega_0)$ in the Maxwellian distribution there are no fast atoms, which could compensate for this detuning by the Doppler broadened line. Therefore, the scattering in the far-removed wing of the Doppler line is more coherent than in the line core, where the resonance scattering is fully incoherent owing to the Doppler redistribution of frequency.

The spectral profile of the resonance scattering cross-section is essentially deformed. Instead of the Lorentz contour (2.39), a Gaussian profile is formed (described by $\delta\omega_D \gg \gamma$ in the expression (2.54)). The cross-section of resonance scattering at the centre of the line, ω_0 $\sigma_{sc}^D(\omega_0)$, for a Doppler-broadened transition is related to the scattering cross-section in the absence of Doppler broadening (i.e. at a radiation-broadened transition) as

$$\sigma_{sc}^D(\omega_0) = \left(\sqrt{ln2}\,\frac{\gamma}{\Delta\omega_D}\right)\sigma_{sc}^{rad}(\omega_0) = a\sigma_{sc}^{rad}(\omega_0), \qquad (2.60)$$

where $a = (\gamma/\Delta\omega_D)\sqrt{ln2}$ is the damping constant (2.57). For example, the value of this constant in such an important case as hydrogen resonance line Lyα (1215 Å) is $a = 4.7 \cdot 10^{-4}$ for $T = 10^4$ K.

In the far wings, however, the fraction of atoms moving with $v \gg u$ is negligible and they do not contribute to the cross-section of resonance scattering in contrast to atoms with low velocities $v \lesssim u$. The cross-section of resonance scattering from these slow atoms, with $|\omega - \omega_0| \gg \delta\omega_D$, is determined by radiative decay because the Lorentz profile in a wing is much more intense than the Doppler profile. So, the cross-section of resonance scattering on a spectral line wing is determined not by (2.60) but by the expression

$$\sigma_{sc}^D(|\omega-\omega_0| \gg \Delta\omega_D) = \sigma_{sc}^{(\omega_0)}\frac{(\gamma/2)^2}{(\omega-\omega_0)^2+(\gamma/2)^2}, \qquad (2.61)$$

where $\sigma_{sc}^{(\omega_0)}$ is described by (2.58). The same considerations, as illustrated in Fig. 2.6, are valid for the resonance absorption cross-section $\sigma_{abs}(\omega)$ as well. We shall make use of them below when discussing optically dense media.

The frequency redistribution at resonance scattering is essential when explaining the profile of spectral lines in stellar atmospheres (Mihalas, 1978). For the physics of astrophysical lasers acting at the centre of the Doppler contour this point is less important. The Doppler diffusion of frequency, however, is essential for evaluating the parameters of the hydrogen resonance lines Lyα and Lyβ, which are sources of pumping excited states in other atoms and ions that give rise to an inverse population and a laser effect.

2.3.3 Number of resonance scattering events – escaping of photons

In an astrophysical rarefied plasma, which is the medium of astrophysical lasers, the elementary act of resonance scattering occurs repeatedly because the optical density τ of such a medium is very high. The probability of collisions is usually many orders less than the resonance scattering cross-section, so the main part in the plasma is played by radiation processes. The optical density of resonance scattering (and resonance absorption) at the centre of the line for an ensemble of atoms, with density N_0 and average size L, is given by the expression

$$\tau_0 = \sigma_{sc}(\omega_0) N_0 L = \sigma_{abs}(\omega_0) N_0 L. \qquad (2.62)$$

The value of τ_0 for resonance lines is usually very high ($lg\tau_0 \gg 1$).

Elements of radiative quantum transitions 31

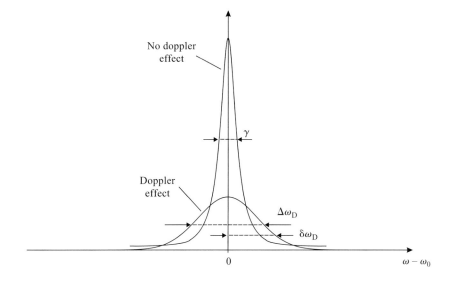

Fig. 2.6 Spectral dependence of resonance scattering cross-sections for slow atoms without Doppler effect (narrow Lorentzian profile) and for moving atoms with Doppler effect (wide Gaussian profile) illustrating effect of radiative decay on wings.

The cross-section of resonance coherent scattering determines the free path length of a photon, ℓ_{sc}, with the frequency ω in an ensemble of scattering atoms with their density N_0:

$$\ell_{sc}(\omega) = \frac{1}{\sigma_{sc}(\omega) N_0}. \tag{2.63}$$

In a region with the dimensions $L \gg \ell_{sc}$, a photon will perform random walks until it escapes this region. The number of such scattering acts, N_{sc}, in a spherical area with its diameter L is given by the simple expression

$$N_{sc} \simeq \left(\frac{L}{\ell_{sc}}\right)^2 = \tau_0^2, \tag{2.64}$$

i.e. N_{sc} seems to be huge.

In reality under astrophysical conditions (nebulae, gas condensations, outer layers of stellar atmospheres) the expression (2.64) is not valid because of the Doppler redistribution of frequency at scattering. In this case after several scattering acts a photon may find itself on a wing of the Doppler contour, where $\ell_{sc}(\omega) \gg \ell_{sc}(\omega_0)$. Such a photon is able to leave the scattering volume even when it is far from a surface layer with a thickness of about $\ell_{sc}(\omega_0)$. This is illustrated by Fig. 2.7, where photons with their frequency ω_2 in the far wing have a path length much shorter than photons with ω_1 near the centre of the spectral line.

The processes of Doppler diffusion of frequency, its relation to the number of random scattering acts N_{sc}, and the probability of photons escaping the scattering area through the wings of spectral lines, are responsible for the formation of spectral lines of stellar atmospheres and nebulae. These processes were studied in many papers in the mid 20th century and results have been presented in many review papers (for

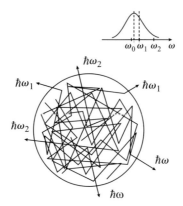

Fig. 2.7 Multiple resonance scattering of photons with various of frequencies in spherical volume.

example, Hummer and Rybicki, 1971; Mihalas and Athay, 1973) and monographs (Sobolev, 1962, 1963; Ivanov, 1971; Mihalas, 1978). In our further discussions we shall only need some key results of these fundamental studies.

If Doppler diffusion of frequency is allowed for, the number of random scattering acts N_{sc} essentially falls because of the escape of photons in the Doppler contour wings (Avrett and Hummer, 1965):

$$N_{sc} \simeq \tau_0 \left[\ell n \left(\frac{\tau_0}{2\sqrt{\pi}} \right) \right]^{1/2} \qquad (2.65)$$

This expression is valid when $a\tau_0 \lesssim 1$. With a very high optical density the scattering becomes essential in a Lorentz contour wing which gives rise to the following expression for N_{sc}

$$N_{sc} \simeq \left(\frac{\tau_0}{a} \right)^{1/2} \qquad (2.66)$$

The variation of the spectral distribution of radiation at multiple scattering is illustrated in Fig. 2.8a, where the spectrum deformation near the Doppler nucleus, i.e. for $a\tau_0 \ll 1$, is shown. The spectral distribution of the absorption line shows the same deformation (Fig. 2.8b). The flattening of the radiation spectrum reflects its approach of a radiation spectrum to the black-body radiation intensity at the corresponding temperature. This situation agrees with the definition of the effective (spectral) temperature of radiation line. In the case of the absorption spectrum this flattening corresponds to 100% absorption with $\tau_0 \gg 1$ at the centre of the line and to the manifestation of absorption in the far-removed wings of the spectral line.

To describe how the profile of an observed absorption line depends on the optical density τ_0, the so-called curve-of-growth is used in astrophysics. Since the details of this curve used for quantitative analysis of stellar spectral lines (Greenstein, 1960; Emerson, 1996) are not essential for us, we shall restrict ourselves to the qualitative growth pattern shown in Fig. 2.9. The vertical axis in Fig. 2.9 gives the area under the spectral line contour (integral intensity), which is proportional to the product of the

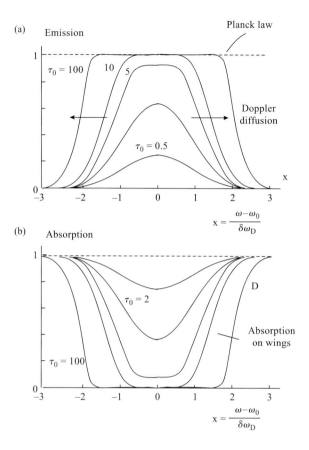

Fig. 2.8 Symmetrical evolution of spectral lines in optically thick media: (*a*) emission line in resonance scattering media; (*b*) absorption line in absorbing media. τ_0 is the optical thickness for centre of spectral lines.

line intensity and its width. For simplicity this area is normalized to 1, when $\tau_0 = 1$ for $a = 1$. The whole plot is given in log–log scale as is common in astrophysical literature. The initial linear section of this curve, where $a \lesssim 1$, corresponds to the growth of the spectral line height up to the beginning of its flattening. Below this limit the scattering (absorbing) medium is optically thin and there is no repeated scattering. Then, when $1 \lesssim \tau_0 \lesssim 1/a$, the growth becomes saturated in an optically dense medium. This results in a spectral line flattening and a very slow increase of the width caused by a sharp fall of the Gaussian wing in the Doppler profile. Finally, in an optically very dense medium, where $\tau_0 \gg 1/a$, the spectral width continues to grow at the expense of a rise of the Lorentz wing caused by homogeneous (radiative in our case) broadening.

The dependence of the spectral width of the scattered radiation $\Delta\omega_{\text{diff}}$ on the number of scattering events N_{sc} can be approximately estimated in a diffusion approximation

$$\Delta\omega_{\text{diff}} \simeq \delta\omega_D \sqrt{N_{\text{sc}}}, \qquad (2.67)$$

34 Astrophysical lasers

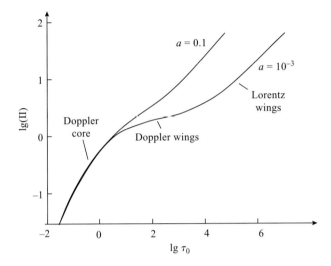

Fig. 2.9 Curves of growth of integral spectral intensity (II) of line with Doppler and radiative broadening (τ_0 is optical density, $a = \gamma/\Delta\omega_D$ – damping factor).

where $\delta\omega_D$ is the mean step of frequency variation at scattering which is approximately equal to the Doppler half-width $\delta\omega_D$.

The diffusion of the scattered radiation spectrum inside a scattering ensemble of atoms is most important for understanding the optical resonance excitation of other atoms whose absorption frequency is near the frequency ω_0 of scattered light. The probability of photon escape β is determined by the expression

$$\beta = \frac{1}{\tau_0}\left(1 - e^{-\tau_0}\right). \tag{2.68}$$

The exact expression for the Doppler contour within the range $1 \ll \tau_0 < 1/a$ has the form (Osterbrock, 1962)

$$\beta = \frac{1}{\tau_0}\left[\ln\frac{\tau_0}{\sqrt{\pi}}\right]^{1/2}. \tag{2.69}$$

The main effect consists of a decrease of β, which is proportional to $1/\tau_0$. The photons are scattered and redistributed in frequency mainly within the Doppler nucleus having the width $|\omega - \omega_0| \lesssim \Delta\omega_D$ and, as they approach this limit, they escape the scattering area. Therefore, the spectrum of the escaping radiation has two peaks at frequencies $|\omega - \omega_0| \simeq 3\delta\omega_D$ and a dip at the centre of the Doppler contour. The qualitative form of the spectrum of this radiation is shown in Fig. 2.10 for various values of the optical density $\tau_0 = \sigma_{sc}(\omega_0) N_0 (L/2)$, where L is the diameter of a spherical scattering volume.

Multiple resonance scattering leads to a radiation trapping effect when a photon moves inside the scattering area during many scatterings (Holstein, 1947). For a resonance transition to the ground state this causes an effective increase in the radiative

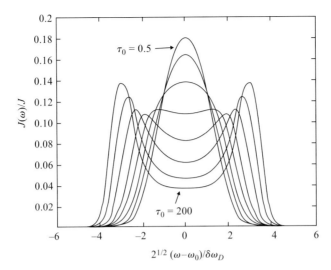

Fig. 2.10 Spectra of radiation escaping from spherical volume with optical density $\tau_0 = \sigma_{sc}(\omega_0) N_0 L/2$ (L is diameter of scattering volume obtained by numerical solution of radiation transfer equation (from Klimov et al., 2002).

lifetime of an excited atom τ_n up to the value

$$\tau_n^{\text{eff}} = \tau_n \frac{1}{\beta}, \qquad (2.70)$$

where β is the escape factor determined from (2.68) and (2.69). The radiation trapping effect in optically dense atomic media is considered comprehensively in a monograph by Molish and Oehry (1998).

In the treatment above, the scattering area has everywhere been assumed to be free from non resonance absorption ($\chi_{nr} = 0$). But, if $\chi_{nr} \neq 0$, the presence of non resonance absorption somewhat changes the ratio of line intensity with different τ_0. This can be explained by the fact that the average physical path of a photon in a scattering medium L_{ph} is larger than the medium itself L. Indeed, $L_{ph} \simeq N_{sc} \ell_{sc} \simeq L(\ell n \tau_0)^2$, i.e. the effect can be observed but with $\tau_0 \gg 1$. In this case even slight non resonance absorption, with $\chi_{nr} \simeq 1$, considerably affects the radiation intensity. The variation of the non resonance absorption factor for various spectral lines may cause the ratio between their intensities to change. This effect was numerically estimated to explain the abnormal ratio between the intensities of two very bright spectral Fe II UV lines (2507/2509 Å) and their satellites, the ion Fe II in gas condensations near the blue star Eta Carinae (Klimov et al., 2002).

In the general case, the radiation intensity inside a medium and its observable value at the boundary of the scattering area are described by transfer equations including frequency redistribution at scattering, death of photons because of absorption, etc. Radiation transfer equations and their solutions are considered in many special monographs (Sobolev, 1962; Ivanov, 1971; Mihalas, 1978) to which we refer our readers.

3
Elements of atomic spectroscopy

One of the prerequisites for obtaining stimulated emission of radiation and natural laser action in cosmic plasmas is the presence of atoms or ions having an atomic structure that allows for a build-up of population inversion. The atomic structure is equivalent to the set of stationary energy states that result from the various arrangements of electrons, as regards orbit, angular momentum and spin, predicted by quantum mechanics. A detailed knowledge of the atomic structure is, in general, essential for the understanding of the processes behind the formation of emission lines that appear in spectra of low-density astrophysical plasmas. In this Chapter, atomic structures of increasing complexity are illustrated by energy level diagrams of atoms or ions, for which fluorescence and natural laser radiation have been proposed.

In this Chapter we will also define the terminology used in connection with atomic structure, transitions and various processes because the spectroscopic language differs between atomic and stellar spectroscopy and this may constitute a barrier in scientific communication between the two fields. Stellar spectroscopists are trained in the use of Charlotte Moore's atomic data tables from the 1940s and 1950s (see e.g. Moore, 1945), which introduced the "multiplet numbers" for complete and incomplete transition arrays between LS terms (see below) of opposite parity. This notation system has not been pursued in extensions of data tables or databases that are caused by extended analyses of atomic spectra. It can thus not be used to give a complete notation of all lines observed in a stellar spectrum. Laboratory spectroscopists use standard spectroscopic notations, which are dependent on the coupling model used to describe the atomic structure. There could be more than each for each particular atom.

To address the two different systems of labeling of spectral lines we will make a compromise and use LS notation for energy levels and transitions except for those lines which are better known as members of well-known line series, e.g. Lyα, Lyβ, Hα (Balmer α) etc. in hydrogen. We will avoid multiplet numbers as no physics can be read out from them.

3.1 Basic concepts

We give here some introductory terminology and a brief explanation behind the various atomic structure concepts used in this book.

3.1.1 Structure and interactions

The atomic structure of an atom (ion) is determined by the hierarchy (amount of energy) of the various interactions. The dominating contributions come from the

Coulumb interactions: the electron–nucleus and the electron–electron interactions. The electron–nucleus interaction depends on the principal quantum number n and gives the *gross structure*. The general structure includes electron–electron interaction, which splits up the gross structure with a dependence on the orbital quantum number l. The collection of energy states associated with given values of n and l is called a *configuration*.

The smaller interactions, called relativistic effects, are associated with the electron's spin (s) and a relativistic treatment of the electron. The interaction between the spin and the electron's orbit is called *spin–orbit interaction*, which contributes energy and causes the *fine-structure splitting*. The resulting energy state is called a *level*, and is independent of atomic model characterized by its parity, energy, and total angular momentum J. The parity is odd if the sum of the individual j-values for the valence electrons is odd, and even for an even sum of j-values.

Depending on the hierarchy (relative strengths) of the various interactions different atomic models are applied when making a theoretical treatment of the atom. One basic consideration is the relative strength between the spin–ban and electron–electron interactions. When the spin–ban interaction (or the relativistic effects) is much smaller than the electron–electron interaction the LS coupling model is valid, whereas the opposite case is described by jj-coupling. (The latter applies when two valence electrons are well separated.)

3.1.2 LS coupling

As mentioned above, the LS coupling model of atoms is the one mostly referred to in astronomical literature, and the old atomic data tables of multiplets are based on that concept. Since the spin–orbit interaction is small in LS coupling the fine-structure splitting is small. The coupling conditions are:

(a) The orbital angular momenta of the electrons are coupled to give a total orbital angular momentum $L = \Sigma_i l_i$.
(b) The spins of the electrons are coupled to give a total spin $S = \Sigma_i s_i$.

The combination of a particular S value with a particular L value comprises a spectroscopic *term*, the notation for which is ^{2S+1}L. The quantum number $2S+1$ is the *multiplicity* of the term. The S and L vectors are coupled to obtain the total angular momentum, $\boldsymbol{J} = \boldsymbol{S} + \boldsymbol{L}$, for a *level* of the term; the level is denoted as $^{2S+1}L_J$.

For two electrons the total spin is $S = 0$ or 1, i.e., the terms are *singlets* ($S = 0$) or *triplets* ($S = 1$). For strict LS-coupling, transitions between levels in the singlet and levels in the triplet systems are forbidden. However, such LS-forbidden transitions occur in practice and they are called *inter-combination lines* (see two-electron systems below). Their strength in a spectrum indicates the validity of LS coupling in the specific case.

3.1.3 Terms, levels, and transitions in LS coupling – equivalent electrons

Simple rules govern the derivation of possible LS terms within a given configuration: $l_1 + l_2 = L$, where $L = l_1 + l_2, l_1 + l_2 - 1, \ldots, |l_1 - l_2|$ and $s_1 + s_2 = S$, where $S = s_1 + s_2, s_1 - s_2$.

The LS term is written as ^{2S+1}L. The J-value for the individual fine-structure levels within the term is derived in a similar way:

$\mathbf{L}+\mathbf{S}=\mathbf{J}$, where $J = L+S, L+S-1,\ldots,|L-S|$. The level is written as $^{2S+1}L_J$.

As an example, for the 3d4p configuration 3d gives $l_1 = 2$ and 4p gives $l_2 = 1$. This results in $L = 3, 2, 1$, which correspond to F, D and P terms. The total spin is $S = 1$ or $S = 0$, yielding triplets and singlets. The six LS terms of 3d4p are ^{3}F, ^{3}D, ^{3}P, ^{1}F, ^{1}D and ^{1}P. The J-values for ^{3}F are $4, 3, 2$ and $J = 2$ for ^{1}D. The 3d4s configuration has only two terms, one ^{1}D and one ^{3}D term.

The possible transitions between the levels in two configurations, e.g. 3d4s–3d4p are called a *transition array*, whereas the transitions between two LS terms of the array form a *multiplet*. The 3d4s ^{3}D$_2$–3d4p ^{3}F$_3$ *line* belongs to the corresponding ^{3}D – ^{3}F *multiplet*. The order of the two terms in the transitions as written above, with the lower-energy term on the left, is standard in atomic spectroscopy. The transitions allowed follow certain selection rules for electric dipole radiation: Change of parity and $\Delta J = \pm 1, 0$ (excluding $0 \to 0$). These selection rules are model independent, but in LS coupling also the selection rules $\Delta S = 0$ and $\Delta L = \pm 1, 0$ (excluding $0 \to 0$) are valid.

When a configuration contains equivalent electrons, e.g. 3p^2, the derivation of LS terms is more complicated. In two-electron systems half of the LS terms obtained for non-equivalent electrons, e.g. 3p4p, are omitted because of the Pauli principle. The np^2 configuration contains one ^{3}P, one ^{1}D and one ^{1}S term.

3.1.4 Complex spectra

The atomic systems discussed above have the property that all spectral series converge on one series limit, the *ionization limit*. In contrast to this we define *complex spectra* as atomic systems having *multiple series limits*. These can arise when

(a) the ground configuration of the next higher ion, the *parent configuration*, consists of more than one LS term, e.g. p^2, d^3 etc.
(b) there is more than one configuration at very low excitation energy, as is for example the case with the transition elements
(c) the fine structure splitting of the *parent term* is large (occurs for heavy elements) in systems with one series limit.

The first case applies to spectra of the p-shell elements, e.g. N I. The next higher ion, N$^+$, is iso-electronic with C I having 2p^2 as the ground configuration. Thus, 2p^2 is the *parent configuration* of N I and it has three LS terms, ^{3}P, ^{1}D, and ^{1}S. These form the *parent terms* or series limits of N I. The structure of N I is built up by adding an electron to each of the parent terms in the parent configuration. For example, the LS terms of the 2p^{2}3p configuration in N I are derived by applying the so-called *branching rule* in the following way: add the l of the outer electron to the $\mathbf{L}_p$ of the parent term to get the total $\mathbf{L}$ value, and do the same with s and $\mathbf{S}_p$. We start with the ^{3}P parent term, which is given in parentheses in the LS description of the final level:

2p^2(^{3}P)3p: $L_c = 1$, $l = 1 \Rightarrow L = 2, 1, 0$ give D, P and S terms, and $S_c = 1$, $s = 1/2 \Rightarrow$ $g = 3/2, 1/2$ give quartets and doublets. The result is ^{4}D, ^{4}P, ^{4}S, ^{2}D, ^{2}P, ^{2}S. The full LS notation is 2p^2(^{3}P)3p ^{4}D, 2p^2(^{3}P)3p ^{4}P etc.

$2p^2(^1D)3p$: $L_c = 2$, $l = 1 \Rightarrow L = 3,2,1$ give F, D and P terms and $S_c = 0$, $s = 1/2 \Rightarrow g = 1/2$ gives doublets. The result is ^2F, ^2D, ^2P. The full LS notation is $2p^2(^1D)3p\,^2F$ etc.

$2p^2(^1S)3p$: $L_c=0$, $l=1 \Rightarrow L=1$ gives one P term and $S_c = 0$, $s = 1/2 \Rightarrow g = 1/2$ gives doublets. The result is ^2P and the full LS notation is $2p^2(^1S)3p\,^2P$.

The configuration $2p^2\,3p$ is split up in three *subconfigurations* $2p^2(^ML)3p$, based on three different parent terms. Each subconfiguration is a one-series limit representation. The strongest lines in a complex spectrum correspond in most cases to transitions between subconfigurations having the same parent term.

3.1.5 Metastable states, pseudo-metastable states – forbidden lines

The ground configuration of a complex spectrum contains several LS terms that cannot decay by obeying the selection rules for electric dipole radiation (E1). Hence, they have long radiative lifetimes (order of ms–s) compared to other excited states (order of ns–μs). Such levels are called *metastable states*. In laboratory sources the metastable states decay by collisional deexcitation, which is also the case in dense astrophysical plasma. However, in low-density astrophysical plasma, such as a nebula, the metastable states may decay radiatively in electric quadrupole (E2) or magnetic dipole (M1) radiation. Such transitions are in astrophysics called *forbidden lines*.

In complex spectra with many valence electrons and many parent terms the ground configuration may extend above the next higher excited configuration. For example, in spectra of singly ionized iron group elements that may result in transitions from high 4s states *down* to lower 4p states. Such high 4s states show metastability with radiative lifetimes of the order of milliseconds even if they can decay in what seems to be regular LS transitions, and they are called *pseudo-metastable* states.

3.2 One-electron systems

3.2.1 One-electron atoms and ions

The simplest of all atomic systems is the hydrogen atom, as it only contains one electron and one proton. In the special case of hydrogen – a one-electron-atom – the energy of the stationary states only depends on the principal quantum number, n, and is thus well described by the Bohr atomic model. This is the so-called *gross structure* of the atom, shown in Fig. 3.1a, only regarding the attraction between the electron and the nucleus. If the energy level system is displayed according to the orbital quantum number (as an ordinary one-electron system, such as Li or Na) the spectrum of radiative transitions between even- and odd-parity levels is illustrated in Fig. 3.1b which also indicates the fine-structure levels of the various terms and assigns them in LS notation. This is only for convenience as no coupling of electrons occurs. It should be noted that in most practical cases the gross structure of hydrogen is enough for illustration, as the various fine-structure components are not resolved in astrophysical spectra.

Owing to the high cosmic abundance of hydrogen its spectrum lines are dominant in astrophysical spectra. The principal excitation process of hydrogen emission lines

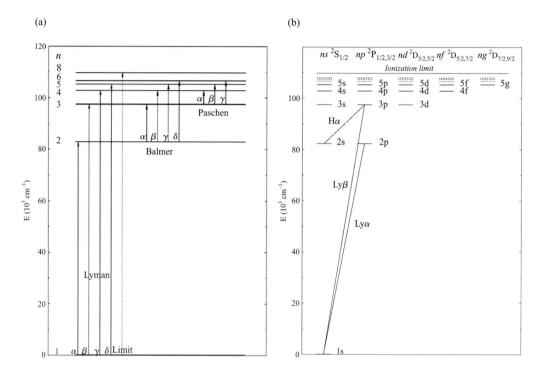

Fig. 3.1 The gross energy level structure of hydrogen (atom energy level diagram) (a). An extended diagram including the spectral lines Lyα and Lyβ, which are important for astrophysical lasers (b).

in low-density media (e.g. the interstellar medium (ISM)) around stars is radiative capture or recombination, as discussed in Chapter 4. The recombination spectrum occurs all the way from 912 Å to far out into the microwave region. For direct radiative line pumping by hydrogen the most important lines are the Lyα and Lyβ transitions, indicated in Fig. 3.1b. During the cascading of H I downwards in the energy level diagram after recombination, a substantial proportion of the H atoms are populating the $n = 2$ and $n = 3$ states, producing a substantial flux in Lyα and Lyβ. We will see later in Chapter 4 applications of resonant line pumping of transitions in other elements by Lyα and Lyβ, resulting in inverted population and stimulated emission of radiation. We have also inserted Balmer alpha (Hα) to make the reader familiar with the diagram.

3.2.2 Alkali and alkali-like spectra

The influence on the level structure of the orbital angular momentum is evident when looking at the alkali atoms, where there is one valence electron outside a closed rare gas shell. The shielding of the nucleus by the core electrons is the second largest effect in the level diagram as it splits up the gross structure for a given n. It is sometimes referred to as the penetration effect, where the valence electron penetrates

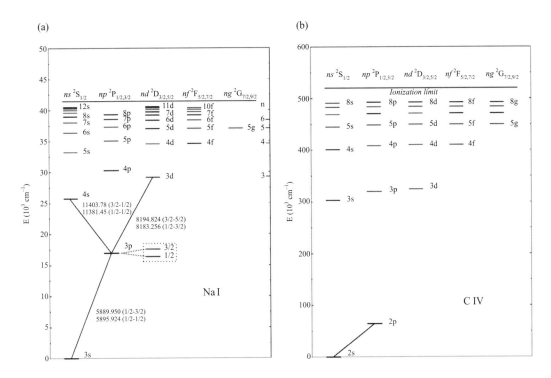

Fig. 3.2 Energy level diagram of neutral sodium, Na I (a), and Li-like carbon C IV (b).

the atomic core and experiences an effective net charge that is strongly dependent on l, the geometric eccentricity of the orbit. In Fig. 3.2a we show the structure of the sodium atom, Na I, with its famous Na D lines around 5990 Å, where the difference in binding energy between the 3s and the 3p electrons corresponds to the photon energy of the Na D lines. However, at such a long optical wavelength the Na D lines have very little possibility of having any influence on line pumping in astrophysical media. This possibility increases, however, when looking at ions of elements that are iso-electronic with the alkali spectrum of Li I, such as C IV and N V (Fig. 3.2b).

Owing to the much shorter wavelength of the resonance lines and the high cosmic abundance these species are involved in a number of fluorescence processes in various astrophysical objects, e.g. symbiotic stars, which are binary systems with a hot and a cool component. The resonance line of triply ionized carbon, C IV at 1550 Å, is formed close to the hot star, and it photo excites resonantly singly ionized iron, Fe II. This interaction results in a selective population of certain Fe II levels, which decay by emission of radiation in fluorescence lines. In the symbiotic star RR Tel the wavelength coincidence between one of the C IV resonance lines and quite a moderate Fe II line results in the selective photo excitation of a specific Fe II level at about 8 eV, which decays in ten observed transitions. These transitions become significant fluorescence lines in the ultraviolet spectrum of RR Tel contributing to the cooling of the nebula (Johansson, 1983; Hartman and Johansson, 2000).

3.3 Two-electron systems

3.3.1 Alkaline earth elements and iso-electronic spectra

The alkaline earth elements are two-electron systems having two valence electrons outside a closed rare gas shell, i.e. having two ns valence electrons. Among these, magnesium and calcium have high cosmic abundances and contribute strong lines in astrophysical spectra. Among two electron systems with similar atomic structure one should also include iso-electronic ions of other abundant elements, such as Be-like C III, Mg-like Al II, and Si III. The two possibilities of parallel and anti-parallel spins of the two valence electrons are reflected in two types of LS terms: triplets and singlets. The connection between the two spin systems introduces the concept of inter-combination lines or spin forbidden transitions, which play a major role in astrophysical emission line spectra.

The strong LS-coupling in Mg I is demonstrated by the atomic parameters describing the inter-combination line to the ground state, $3s^2\ ^1S_0 - 3s3p\ ^3P_1$ (see Fig. 3.3). The resonance transition at 2852 Å within the singlet system $3s^{2\,1}S_0 - 3s3p\ ^1P_1$ has a transition probability (A-value) of $5 \cdot 10^8$ s^{-1}, whereas the inter-combination transition at 4571 Å has an A-value of $2 \cdot 10^2$ s^{-1}, i.e. six orders of magnitude smaller. In general, the intensity ratio between the resonance line and the inter-combination line get considerably smaller for heavy elements.

The presence of inter-combination lines in astrophysical emission line spectra indicates low densities, and the intensity ratio between spin-forbidden and spin-allowed lines is used as a density diagnostics in planetary nebulae (Osterbrock and Ferland,

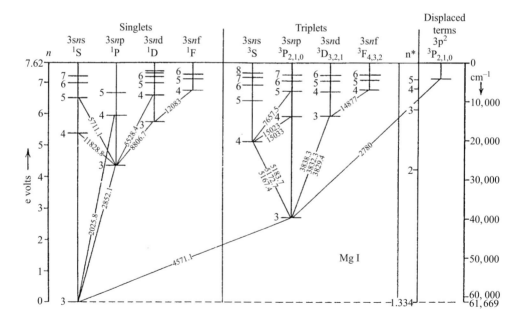

Fig. 3.3 Energy level diagram of neutral magnesium Mg I.

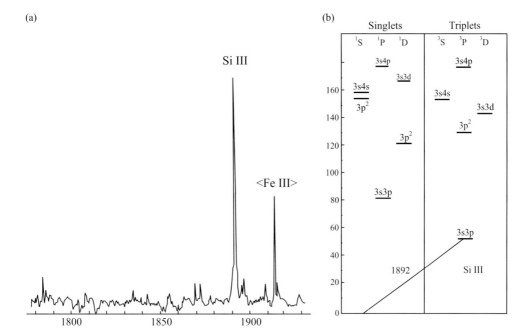

Fig. 3.4 The intercombination line of Si III in the spectrum of Eta Carinae (*a*) and the lower part of the energy level diagram of the Mg-like spectrum of Si III (*b*).

2006). To illustrate the significance of inter-combination lines in low-density media we show in Fig. 3.4 the region 1780–1930 Å of the blob spectrum of Eta Carinae, as recorded with the High Resolution Spectrograph (HRS) onboard the Hubble Space Telescope (HST). Besides a Lyα pumped fluorescence Fe III line, the intercombination line of Si III at 1898 Å is very dominant in this wavelength region. An extract of the energy level diagram of the Na-like Si III system is also displayed in Fig. 3.4 to indicate the location of the intercombination line. The Si III line might be enhanced in the Eta Carinae spectrum owing to resonance-enhanced two-photon ionization (RETPI) (Johansson *et al.*, 2006).

3.3.2 Elements with two electrons or two holes in the p-shell

The elements having an atomic structure with two valence electrons in an open p-shell show similar energy level diagrams to elements with four p-electrons, which is equivalent to two electron holes in the p-shell. Some of the most important elements in astrophysics meet this description – carbon (C I) with the ground configuration $2p^2$, and oxygen (O I) with the ground configuration $2p^4$. The corresponding elements in the 3p-shell are silicon (Si I) and sulphur (S I) with the ground configurations $3p^2$ and $3p^4$, respectively. All these elements have their resonance lines far down in the ultraviolet wavelength region. In Fig. 3.5 we show the energy level diagrams of C I and O I. The diagrams of Si I and S I will be very similar to the diagrams of the homologous elements C I and O I, respectively.

44 *Astrophysical lasers*

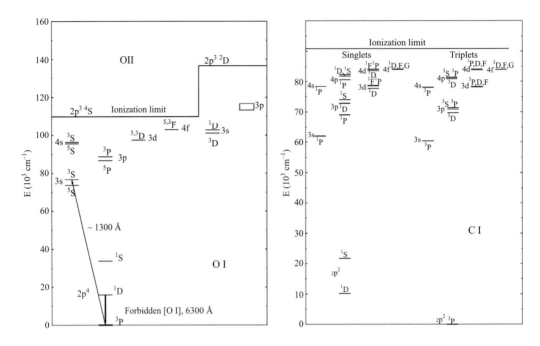

Fig. 3.5 Energy level diagrams of O I and C I.

The ground configuration of the four p-shell elements mentioned above consists of three LS terms, ^{3}P, ^{1}D and ^{1}S, whereas in all groups of elements discussed above there has only been one LS term, viz. ^{1}S in the alkaline earth elements and ^{2}S in the alkali elements. In these latter two groups the resonance line corresponds to a transition where $\Delta n = 0$, i.e. the transition occurs within the same electron shell. (The only exceptions are hydrogen and helium where the resonance transitions require $\Delta n = 1$.) The consequence of $\Delta n = 0$ transitions is that the resonance lines of the alkali atoms appear in the optical wavelength region (e.g. Na I at 5890 Å), which is also the case for the inter-combination lines of the alkaline earth elements (e.g. Mg I at 4571 Å). The spin-allowed resonance lines of the alkaline earth elements occur at somewhat shorter wavelength. In the p-shell atoms the situation is different. The excitation of the ground configuration requires at least a $\Delta n = 1$ transition, which for C I and O I means a 2p–3s or a 2p–3d jump. The energy involved in a jump to an outer shell is often larger than the transition energy within a shell, which means that the resonance lines in C I and O I appear in the far-ultraviolet region, at 1600 Å and 1300 Å, respectively. Thus, to trace the behaviour of two of the most important elements in the galactic evolution of stars there is in many cases a need for satellite observations. The forbidden [O I] line at 6300 Å (indicated in Fig 3.5) has been observed in spectra of cool stars recorded with ground-based telescopes, and it has been used for abundance studies (Zoccali et al., 2006). In hotter stars the O I 3s–3p transitions in the near-infrared region are observed and used for abundance.

3.4 Complex spectra – with emphasis on Fe II

3.4.1 Definition and properties of complex spectra

In Section 3.1.4 we defined complex spectra as energy level systems that are built on more than one low series limit. We gave three different cases behind the formation of complex spectra:

1) the ground configuration of the next higher ion, the *parent configuration*, consists of more than one LS term, e.g. p^2, d^3 etc.
2) there is more than one configuration at very low excitation energy, as is the case, for example, with the transition elements
3) the fine structure splitting of the *parent term* is large (occurs for heavy elements) in systems with one series limit.

We gave as an example of case 1 the structure of N I, which is built on three different LS terms, ^{3}P, ^{1}D, and ^{1}S of the parent configuration $2p^2$ in N II. The same is true for P I, as phosphorous is homologous with nitrogen.

In this Subsection we will focus on the complex spectra of transition group elements, often referred to as the iron group filling the 3d shell, the palladium group filling the 4d shell, and the platinum group filling the 5d shell, respectively. However, in the neutral atoms the binding energy (BE) of the nd electron is often smaller than the BE for the $(n+1)$s electron, i.e. in the iron group the 4s electron is more tightly bound than the 3d electron for neutral atoms. In spectra of the first ions (the second spectra) the BE's of the nd and $(n+1)$s electrons are very similar, and the combination of cases 1 and 2 above results in very complicated level structures. This is illustrated in Fig. 3.6 by the level structure of Zr II, for which the BEs of the 4d and 5s electrons are very similar (case 1), and the parent configurations $4d^2$, $4d5s$, and $5s^2$ contain eight different LS terms (case 2). This means that the structure of Zr II is built on eight different parent terms, and it represents a simple complex spectrum.

3.4.2 Astrophysical importance of Fe II

The structure of singly ionized iron, Fe II, deserves a special Section as it is involved in, and many times dominates, spectra of nearly all astrophysical sources. The high cosmic abundance of iron, the level structure of Fe II with its line-rich spectrum, and the match between stellar temperature and the ionization energy of Fe I explain the richness of Fe II lines in stellar spectra. It is interesting to note that this wealth of Fe II lines had already been noticed at the time when only optical spectra were available, but the real wealth became more evident with the satellite-ultraviolet observations as well as the near-IR observations with sensitized photographic plates.

The large number of Fe II lines in stellar spectra increases the probability of accidental wavelength coincidences with lines of other elements. In normal stellar spectra, like the solar spectrum, where absorption lines appear in the background Planck radiation, such coincidences are notified as blends. Such features are in general useless in, for example, abundance analyses, where the total area of the absorption line profile is a measure of the relative abundance of an element. However, in astrophysical emission line spectra the discrete line radiation is a reemission of energy supplied by a

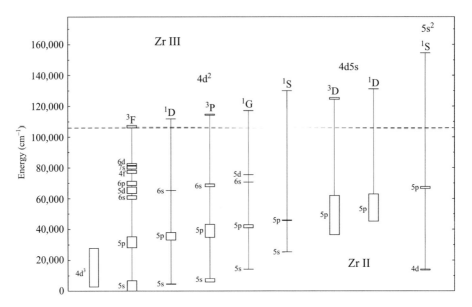

Fig. 3.6 Term diagram of a rather simple, complex spectrum, Zr II, having three parent configurations with a total of eight parent terms.

nearby star, and such a spectrum may contain information about surrounding nebular regions. As mentioned in the introduction and discussed in detail in Chapters 9–12 various cases of the fluorescent Bowen mechanism occur in low-density media outside stars and in most of these cases Fe II and H Lyα are involved. Eta Carinae is probably the best example of a star that has been thoroughly studied as regards fluorescent emission lines originating from surrounding clouds or blobs. The most frequent ion participating in these fluorescence processes is Fe$^+$, and that is the reason why Fe II is selected as a representative of a really complex spectrum. We will emphasize the importance of considering the level structure of Fe II when we treat the fluorescence and laser processes in detail in this book.

3.4.3 The atomic structure of Fe II

Fe II has seven valence electrons, and the level system is based on two low parent configurations in Fe III: $3d^6$ and $3d^54s$. We present the energy levels in two separate diagrams: the singly excited (SE) level system $3d^6 nl$ in Fig. 3.7, and the doubly excited (DE) system $3d^54snl$ in Fig. 3.8. The SE system is thus built on the 16 parent LS terms in the $3d^6$ ground configuration of Fe III, having an energy span of about 98000 cm^{-1} (12.1 eV). The 16 parent terms are located above the ionization limit of Fe II, which is defined by the ^{5}D term in Fe III. The vertical line, starting on the ^{5}D term in Fe III and terminating on the (^{5}D)4s subconfiguration in Fe II, contain several boxes, which represent the location of the $3d^6(^5D)nl$ subconfigurations in Fe II. The lowest dashed box includes 62 metastable levels, and the smaller box 12 pseudo-metastable levels. The level mixing occurring between levels in the left-most 5p box and two high 4p boxes,

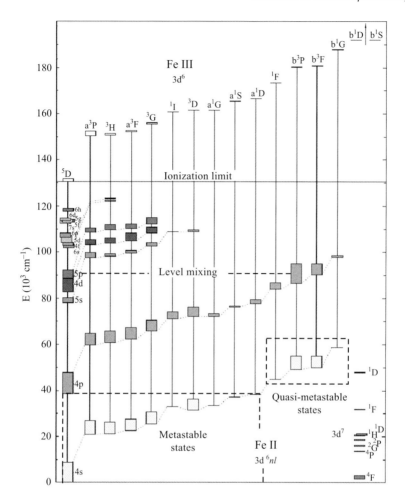

Fig. 3.7 The singly excited (SE) system of the complex spectrum of Fe II, built on the lowest parent configuration of Fe III, $3d^6$, having 16 parent terms. The level mixing, metastable and pseudo-metastable states are indicated by dashed line and boxes.

all of odd parity, is also indicated. To the lower right in Fig 3.7 we have included the $3d^7$ configuration, for which the Pauli principle for equivalent electrons excludes some terms. Thus, the LS terms of $3d^7$ cannot be assigned to a particular parent term.

As shown in Fig 3.9, all these subconfigurations can now be displayed as a one-electron system with one series limit, the ^5D term of Fe III. This procedure can now be done for each of the parent terms in Fe III representing a one-electron system, where one valence electron ($nl = 4s$, 4p, 5s, 4d, etc) is added to the parent term ^ML to form the subconfigurations $3d^6(^ML)nl$ in Fe II. The energy difference between the parent term and a valence electron represents the binding energy of that particular electron.

48 *Astrophysical lasers*

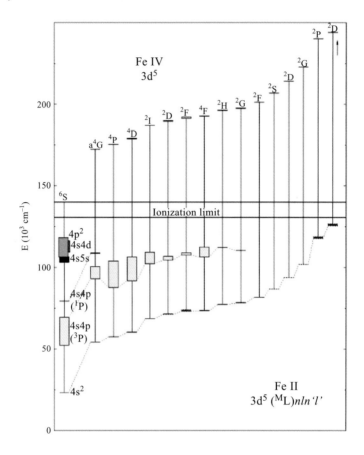

Fig. 3.8 The doubly excited (DE) system of the complex spectrum of Fe II, built on the lowest grand parent configuration of Fe IV, $3d^5$, having 16 grand parent terms.

It is clear from Fig 3.7 that the pattern of the location of the parent terms in Fe III is similar to that of the $(^ML)nl$ boxes for a given valence electron. This means that the binding energy for a valence electron is nearly independent of the parent term. The recipe, the so called branching rule, for deriving the LS terms within the various boxes is given in the beginning of this Chapter.

The spectroscopic notation of configurations, subconfigurations, LS terms etc. used so far in this chapter is the unambiguous standard notation defined in atomic spectroscopy. However, in the astronomical literature a simplified notation is used, which is developed from the extensive work on the solar spectrum by C. Moore-Sitterly. She tabulated all the energy levels known in the 1950s in three volumes: the Atomic Energy Levels (AEL), where all levels are presented as LS terms (Moore, 1949, 1952, 1958). From these tables she arranged all observed and some predicted spectral lines in LS multiplets in the Multiplet Tables – one for the optical region ($\lambda > 3000$ Å) (Moore, 1945) and one for the satellite-ultraviolet region ($\lambda < 3000$ Å) (Moore, 1962),

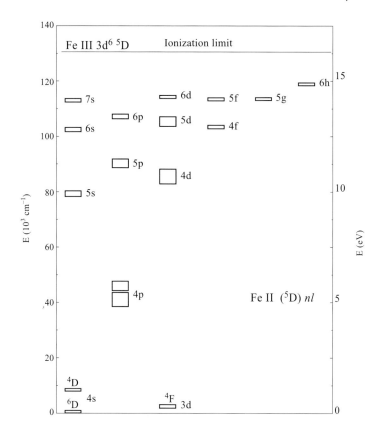

Fig. 3.9 One set of subconfigurations, $3d^6(^5D)nl$ (the leftmost one in Fig. 3.7), displayed as a one-electron system (see e.g. Fig. 3.2).

the limit set by the opacity of the earth's atmosphere. The extensive use of these outdated tables by astronomers creates communication problems between laboratory and stellar spectroscopists. Moore-Sitterly introduced a non-physical "quantum number" by giving the LS terms preceding letters in the following way: Starting from below in the level diagram the LS terms of the same multiplicity (doublet, quartet, sextet, etc.) and of even parity were labeled a, b, c in order of appearance, and in the same way the LS terms of odd parity were labeled z, y, x, etc. These notations are also used in the Multiplet Tables. In the extracted level diagrams in Chapters 9–12 we have also used these simplified notations in the figures, but we have written out the full spectroscopic notation in the text. It should be mentioned that the Moore-Sitterly system breaks down in complex spectra, since the LS coupling is only valid for the low energy levels and the concept LS multiplet has in most cases no physical meaning.

The doubly excited (DE) system is displayed in Fig 3.8. For atomic physics reasons concerning the strength of the electrostatic interactions between the electrons it is more reasonable to couple the pair of outer valence electrons, 4s4s, 4s4p, and then couple these pairs to the grand parent configuration in Fe IV. This will result in the configurations $3d^5(^ML)4s^2$ and $3d^5(^ML)4s4p$. In the examples of lasing transitions in Chapters 9–12 only a few LS terms of the DE system are involved, and we will not go into more detail of the properties of this system. We will notify the reader about the fact that since the two 4s electrons result in a closed shell, they will not contribute any angular or spin momenta, so the resulting LS term will be the same as the corresponding grand parent term.

4
Elementary excitation processes in rarefied plasmas

An atom (ion) in a rarefied astrophysical plasma near a hot (blue) star takes part in a chain of radiative and collisional processes initiated by the UV radiation from the star. Fig. 4.1 illustrates in a simplified way the sequence of these processes. The UV blackbody radiation $P(\omega)$ from the star photoionizes atoms A and thus creates a low-density plasma. Collisions between the formed photoions A^+ and the electrons recombine with a certain probability to make neutral atoms in the ground state A or in excited states A*. The excited atoms A* decay to the ground state A by emitting photons in allowed radiative transitions. A very small fraction of the atoms are kept in metastable states for a longer time. Collisions between electrons themselves and with ions result in a thermalization at a definite temperature T_e determined by the intensity of the stellar UV radiation. The collisions of electrons with atoms also lead to their excitation and ionization. The emitted recombination lines of atom A can resonantly excite atoms (ions) of a different transition or even a different element in the plasma if one of their absorption lines coincides in wavelength with one of the recombination lines.

The main constituents of an astrophysical plasma are hydrogen (90% of the atoms) and helium (10%) with the ionization potentials $I(\mathrm{H}) = 13.6\,\mathrm{eV}$ and $I(\mathrm{He}) = 24.6\,\mathrm{eV}$, respectively. Therefore, the processes mentioned occur mainly with hydrogen as well as helium but in the vicinity of hotter stars and at a short distance from them. Here we shall restrict ourselves to processes that involve the hydrogen atom. More detailed information on elementary processes can be obtained from the monographs by Massey and Burhop (1952) and by Sobel'man *et al.* (1981).

4.1 Photoionization of atoms

The dependence of the photoionization cross-section on the frequency ν for a hydrogen atom in state n ($n = 1$ is the ground state), where $\nu > \nu_c$ and ν_c is the ionization limit (Lyman limit $\lambda_c = 912$ Å, $\nu_c = 3.3 \cdot 10^{15}$ Hz), is determined with a good accuracy by the Kramers' formula

$$\sigma_i(\nu) = \frac{64}{3\sqrt{3}} \alpha \left(\frac{\nu_c}{\nu}\right)^3 \frac{\pi a_0^2}{n^5}, \qquad (4.1)$$

where $\alpha = e^2/\hbar c$ is the fine structure constant, πa_0^2 is the geometrical cross-section of an atom, and $a_0 = \hbar^2/me^2 = 0.529 \cdot 10^{-8}$ cm is the Bohr radius. This relation is plotted in Fig. 4.2, where we can see that the photoionization cross-section has its

52 Astrophysical lasers

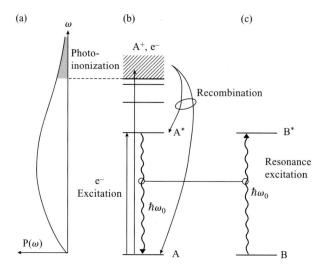

Fig. 4.1 Basic processes under irradiation of atom A by black-body radiation with UV short wavelength wing (a), including resonance excitation of impurity atom B (b) by recombination spectral line of atom A (c).

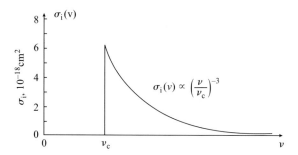

Fig. 4.2 Frequency dependence of photoionization cross-section of hydrogen $\sigma_i(\nu)$ near ionization threshold (Lyman limit) $\nu_c = 3.3 \cdot 10^{15}$ Hz ($\lambda = 912$ Å).

maximum at the ionization limit and then, just above the limit, rapidly decreases according to the law $(\nu_c/\nu)^3$. For complex atoms, however, the frequency dependence $\sigma_i(\nu)$ is more complicated, with several so-called autoionization resonances of different widths. The absolute values of $\sigma_i = \sigma_{ph}$ near the ionization limit lie within 10^{-18}–10^{-19} cm^2 (Ditchburn and Opic, 1962).

Figure 4.3 presents the ionization potentials for the neutral atoms of the first 40 elements in the periodic table. All of these and even heavier elements can be found in astrophysical plasmas according to their average abundance in the Universe. The diagram shows that, even when the entire stellar UV radiation is absorbed because of photoionization of hydrogen, radiation with the wavelength $\lambda > \lambda_c$ is able to photoionize

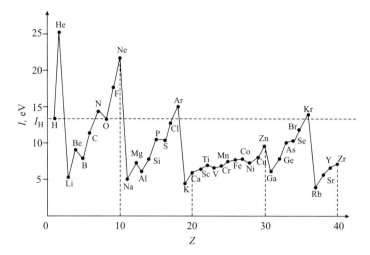

Fig. 4.3 Ionization potential I of elements with atomic number Z. I_H – ionization potential of hydrogen.

most of the other elements with lower ionization potentials $I < I_H$. This is essential for neutral regions, where the electron density and energy are too small to ionize other elements by collisions.

Under irradiation by a spectral brightness of $P(\nu, \Omega)$ (in photons/cm² · s · st · Hz) the photoionization rate $W_i = W_{ph}$ (in s⁻¹) is determined through the cross-section $\sigma_i(\nu)$:

$$W_i = \int_{\nu_c}^{\infty} \int \sigma_i(\nu) P(\nu, \Omega) \, d\nu d\Omega. \tag{4.2}$$

For isotropic equilibrium radiation this expression is reduced to

$$W_i = 4\pi \int_{\nu_c}^{\infty} \sigma_i(\nu) P(\nu, \tau) \, d\nu. \tag{4.3}$$

The presence of a UV wing in the radiation spectrum is essential for photoionization of hydrogen (Fig. 4.2a). Therefore, UV radiation often means radiation with $\lambda < \lambda_c = 912$ Å.

4.2 Recombination

Radiative recombination is the reverse process of photoionization. For hydrogen, for example,

$$H^+ + e^-(\varepsilon) = H(n, e) + h\nu, \tag{4.4}$$

where $\varepsilon = mv^2/2$ is the electron's kinetic energy, i.e. the energy $(h\nu - I_c)$ transferred to a photoelectron by radiation with the photon energy $h\nu > I_c = I_H$. The cross-section of radiative recombination to the level $n\sigma_r(n)$ depends on the electron velocity in the same manner as the photoionization cross-section depends on the excess of photon energy $h\nu$ over the ionization potential I_c. The cross-sections of photoionization $\sigma_i^n(\nu)$ and radiative recombination $\sigma_r^n(v)$ are related by the Milne relation

$$\sigma_r^n(v) = \frac{g_i(n)}{g_{i+1}(1)} \left(\frac{h\nu}{mcv}\right)^2 \sigma_i(\nu) \tag{4.5}$$

where $g_i(n)$ is the statistical weight of level n of the atom, and $g_{i+1}(1)$ is the statistical weight of the ion in the ground state. In the temperature range $T_e = (5-10) \cdot 10^3$ K, typical for a low-density astrophysical plasma, the recombination cross-section is about 10^{-20} cm^2, i.e. much smaller than the geometrical cross-section of the hydrogen atom πa_0^2.

The photo electrons appearing in the astrophysical plasma will reach equilibrium with the rest of the charged particles at a certain temperature T_e. Thus, the recombination rate (in cm^3/s) at the temperature T_e can be expressed by the factor of recombination to a bound state n as:

$$\alpha_n(T_e) = \int_0^\infty \sigma_r(n, v) \, v f(v) dv, \tag{4.6}$$

where $f(v)$ is the velocity distribution of the electrons (Maxwell–Boltzmann distribution)

$$f(v) = 4\pi \left(\frac{m}{2\pi k T_e}\right)^{3/2} v^2 \exp\left[-\left(\frac{mv^2}{2kT_e}\right)\right]. \tag{4.7}$$

The dependence of α_n on the final quantum state n for hydrogen is given in Table 4.1. The most probable recombination is to the ground state $n = 1$. But, in astrophysical plasma with high optical density, a photon L_c in the Lyman continuum, i.e., with energy $h\nu > I_c$, is absorbed again to photoionize another H atom. Through recombination and successive radiative decays from high-lying states 70 per cent of all recombination to excited states populates the 2p levels and gives rise to radiation in the Lyα line. The total factor of recombination to all the states is equal to the sum over all final states

$$\alpha(T_e) = \sum_n \alpha_n(T_e). \tag{4.8}$$

Table 4.1 Recombination coefficient α_n for hydrogen at $T_e = 10^4$ K (in units 10^{-15} cm^3/s)

$\alpha_{4s} = 3.6$	$\alpha_{4p} = 9.7$	$\alpha_{4d} = 11$	$\alpha_{nf} = 5.5$
$\alpha_{3s} = 7.8$	$\alpha_{3p} = 20$	$\alpha_{3d} = 17$	
$\alpha_{2s} = 23$	$\alpha_{2p} = 54$		
$\alpha_{1s} = 160$			
$\alpha_A = \sum_{n=1}^\infty \alpha_n = 420;$		$\alpha_B = \alpha - \alpha_{1s} = 260$	

The dependence of $\alpha(T_e)$ on the temperature is well described by the approximate expression

$$\alpha(T_e) \simeq 3 \cdot 10^{-11} \frac{(Z-1)^2}{\sqrt{T_e}} \left[\text{cm}^3 \cdot \text{s}^{-1}\right], \qquad (4.9)$$

where, for generality, the parameter Z is introduced as the degree of ionization of the recombining ion, i.e. for a singly ionized atom, e.g. H^+, $Z = 2$, for a doubly ionized atom $Z = 3$, etc.

The probability of recombination per time unit W_{rec} depends on the electron density n_e and the recombination factor

$$W_{\text{rec}} = n_e \alpha(T_e). \qquad (4.10)$$

The expressions above for the rates of photoionization and radiative recombination allow us to consider the ionization balance in a hydrogen gas cloud (nebula, gas condensation) where astrophysical lasers seem to operate.

4.3 Electron excitation and ionization

A collision between an atom and an electron with such a high kinetic energy $\mathcal{E} = mv^2/2$ that it exceeds the excitation energy E_n of the n-th state, or the ionization energy I, can excite or ionize the atom, respectively. The cross-sections of such collisional processes have values within a wide range around the gas-kinetic cross-section πa_0^2. Different methods of estimating these cross-sections for atoms and ions are considered in detail in the exhaustive monograph by Sobel'man et al., (1981). Here we restrict ourselves to illustrative data for the hydrogen atom.

Fig. 4.4 shows the excitation cross-section(s) for the allowed 1s-2p and forbidden 1s-2s transitions of H I as a function of energy $\mathcal{E}$. The experimental results and the calculations in a simplest Born approximation show a satisfactory agreement. According to the dipole character of this approximation, the calculated cross-section for allowed transitions (Fig. 4.4a) is above the experimental curve whereas the agreement is fairly good for forbidden transitions (Fig. 4.4b). But these cross-sections are many orders less than for the cross-section of resonance radiative transitions. Therefore, the metastable states are excited either by a cascade of radiative decay from high-lying states or through collisional electron excitation from the ground state.

Fig. 4.5 shows the cross-section of ionization from the ground state for hydrogen as a function of electron energy $\mathcal{E}$. It can be seen that this cross-section is much higher than that of radiative recombination in a rarefied plasma.

4.4 Total and local thermodynamic equilibrium (TE and LTE) in plasma

In the case of a gaseous (plasma) medium hosting astrophysical lasers it is very important that at least one pair of levels of at least one component (atom, ion) should be under nonequilibrium in this medium. The main mass of particles in this case may still be under thermodynamical equilibrium. In a total equilibrium state the atoms

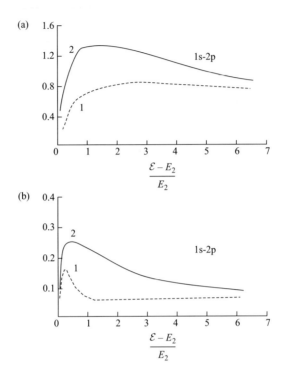

Fig. 4.4 Effective cross-sections (in units πa_0^2) for the allowed 1s-2p (a) and forbidden 1s-2s (b) transitions of the H atom: (1) experiment and (2) Born approximation (modified from Sobel'man et al., 1981).

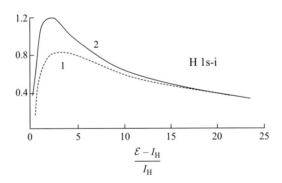

Fig. 4.5 Effective cross-section (in units πa_0^2) for ionization of the H atom from ground state: (1) experiment and (2) Born approximation (modified from Sobel'man et al., (1981).

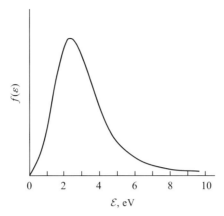

Fig. 4.6 Maxwell distribution of electron energy in electron ensemble with temperature $T_e = 10^4$ K.

(ions) and the radiation are under thermodynamical equilibrium, where the frequency distribution of radiation is described by the Planck distribution (2.1), and the distribution of atoms among the stationary energy levels by the Boltzmann distribution (2.2). In the plasma they must be supplemented by the energy distribution of electrons and the distribution of atoms among the various ionization stages (the ionization balance).

The distribution of electrons with energy $\mathcal{E}$ is determined by the Maxwellian distribution derived from the velocity distribution of atoms (4.7)

$$f(\mathcal{E}) = \frac{2}{\sqrt{\pi}} T^{-3/2} \frac{1}{\sqrt{\mathcal{E}}} e^{-\varepsilon/T}, \qquad (4.11)$$

which, for convenience, is written in unit energy (1 eV = 11605 K) for $\mathcal{E}$ and the temperature T. Fig. 4.6 illustrates this distribution for an electron gas with $T_e = 10^4$ K and $kT_e \simeq 1$ eV. The electrons in the tail of this distribution can excite energy levels in most atoms, except for hydrogen and helium which require an energy from 10 to 20 eV to be excited by electrons.

The distribution of atoms over different ionization stages (Z) is described by the Saha distribution

$$\frac{N^{(Z+1)}}{N^{(Z)}} = \frac{g_{Z+1}}{g_Z} \frac{2}{n_e} \left(\frac{mkT_e}{2\pi\hbar^2}\right)^{3/2} e^{-I_Z/kT_e}, \qquad (4.12)$$

where I_Z is the ionizing potential of atom A_Z, n_e is the density of free electrons with temperature T_e, g_Z is the statistical weight of the ground state of atom A_Z ($Z = 1$ corresponds to a neutral atom). At high temperatures and in complex atoms the statistical weights are replaced by the partition functions to account for excited states. In cold plasmas ($kT_e \ll I_1$) the degree of ionization is low, but as the temperature increases up to the levels $T_{Z-1} \lesssim T \lesssim T_Z$, where T_Z is the temperature for which $N^{(Z+1)} = N^{(Z)}$, the degree of ionization becomes high.

58 Astrophysical lasers

The properties of astrophysical media (stellar atmospheres, nebulae, etc.) are determined from their radiation, which implies that they are open systems in terms of the thermodynamics. But the degree of their "openness" is usually low as the radiation flow in the characteristic time for establishing thermodynamic equilibrium is much less than the total radiation energy in the system. Therefore, the concept of local thermodynamic equilibrium (LTE) has been introduced for such media. It is assumed that at LTE the distribution of kinetic and internal (excitation, ionization) degrees of freedom is identical to that at thermodynamical equilibrium. But, under LTE conditions these distributions depend on the local temperature. Thus, the intensity distribution is not necessarily determined by the Planck distribution for a given local temperature of the atoms, electrons, and ions since the radiation may come from areas having different local temperatures. In such a situation the population distribution over the energy levels may differ from the Boltzmann distribution and even include population inversion of some level pairs. These are the cases that are of interest for astrophysical masers and lasers.

4.5 Non-LTE astrophysical media

Deviations from LTE are observed in stellar atmospheres as the plasma density decreases. LTE is achieved in deep layers of the atmosphere where the ionization of atoms takes place mainly by collisions. In these deep layers the population distribution over levels in atoms (ions) and adjacent ionization states is described by the Boltzmann–Saha distribution, and the frequency distribution of radiation by the Planck distribution. Towards the upper layers of the atmosphere the role of collisions decreases and, finally, in the uppermost layers the excitation of atoms by collisions is small as compared to excitation by radiation processes. Thus, in the uppermost layers of a rarefied stellar atmosphere radiation processes are dominant, and, therefore, the population of levels well below the ionization limit is not described by the electron temperature T_e but by the effective radiation temperature at the frequency of the photo excitation. This change from LTE to non-LTE is illustrated qualitatively in Fig. 4.7, where E_b is the conventional boundary in excitation energy between the LTE and non-LTE regimes corresponding to upper and lower energy levels, respectively. This boundary depends on the plasma density, and it is rising when the density decreases. Thus, the

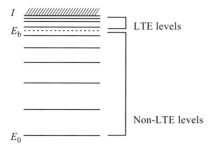

Fig. 4.7 Transition of LTE populations of excited levels to non-LTE populations of lower levels depending on density of plasma.

variations from LTE in astrophysical media are conditioned by the fact that the energy distribution of free particles (atoms, electrons, and ions) can sometimes not be characterized by a single temperature.

The population N_n(non-LTE) of an atomic level is expressed in terms of the population of the same level N_n(LTE) at LTE using the coefficient b_n

$$N_n(\text{non-LTE}) = b_n N_n(\text{LTE}), \qquad (4.13)$$

where the population N_n(LTE) is determined by the Boltzmann and Saha distributions. The factor b_n characterizes the degree of deviation of the level population N_n from local thermodynamical equilibrium. For astrophysical lasers it is essential that b_n for some energy levels can be much larger than 1.

The situation with non-LTE and $b_n > 1$ for a certain pair of levels may occur around the energy $E \simeq E_b$ (Fig. 4.7), where the upper level is collisionally excited, whereas the lower level is depleted by fast radiative transitions to lower levels. The probability for this process to occur forms the basis for the so-called collisional astrophysical lasers considered in Chapter 8.

4.6 Non-LTE: photoselective excitation

The probability of pronounced deviation from LTE exists in multicomponent media. This situation is rather characteristic of astrophysical plasmas, which may contain a mixture of nearly all the elements in the periodic table, normally in accordance with their cosmic abundance. Table 4.2 gives us an idea of the mean abundance of some elements, which are of importance for astrophysical lasers. In a typical and real situation a relatively low number of B atoms are surrounded by A atoms (usually hydrogen) whose concentration is several orders higher ($N_B \simeq (10^{-3}-10^{-5})N_A$). The A atoms control the whole process of LTE, and the influence of the fewer B atoms can, in principle, be neglected. There are, however, rare but important exceptions to this general statement. For example, one of the intense absorption lines of the B atoms may accidentally coincide with a broad emission line of the A atoms as shown in Fig. 4.8. As a result, a considerable fraction of radiation energy of the A atom may escape through radiation of the excited B atom. The radiation of the B atom occurs at longer wavelengths and can be much more transparent in an astrophysical medium than the radiation of the A atom.

For astrophysical lasers the most important fact is that, in the process of accidental photoselective absorption, the level population of the B atom may differ greatly from

Table 4.2 Relative abundance (in number of atoms) of some chemical elements (in %)

H – 92.5	Ne – $7.4 \cdot 10^{-3}$
He – 7.3	Si – $2.9 \cdot 10^{-3}$
C – $3.7 \cdot 10^{-2}$	Ca – $2.2 \cdot 10^{-4}$
N – $0.93 \cdot 10^{-2}$	Mg – $1.5 \cdot 10^{-5}$
O – $6.8 \cdot 10^{-2}$	Fe – $1.85 \cdot 10^{-3}$

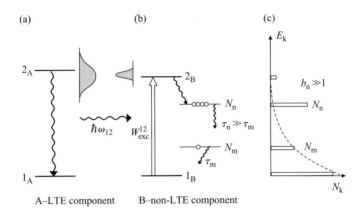

Fig. 4.8 Non-LTE in two-component media with emission line ω_{21} of main LTE-component A (*a*) and coinciding absorption line of minor component B (*b*) with strong non-LTE population of long-lived level n (*c*).

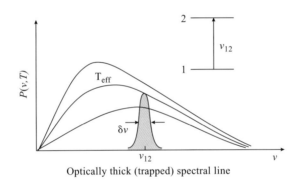

Optically thick (trapped) spectral line

Fig. 4.9 Definition of effective (spectral) temperature of spectral line (gray contour) by comparison with spectral brightness of black-body radiation with equivalent T_{eff}.

LTE, i.e. $b_n \gg 1$ for certain levels. A cascading radiative decay to lower levels may bring an excited atom to a long-lived state. In the absence of collisions in a rarefied medium this may result in an abnormally high population of this level, very far from LTE (Fig. 4.8). This situation leads to the inverse population discussed in Chapters 10–12 generated by hydrogen atoms (A) and applied to Fe^+ ions and O I atoms (B).

The probability of photoselective excitation of the B atom can be expressed in terms of an effective (spectral) temperature T_{eff} of the radiation from the A atom by using (2.29) and (2.34)

$$W_{\text{exc}}^{1',2'} = \frac{g_2^B}{g_1^B} \frac{A_{21}}{\exp\left(\frac{h\nu}{kT_{\text{eff}}}\right) - 1}. \tag{4.14}$$

The effective temperature of a monochromatic line is determined by comparing it with the spectral brightness of nonpolarized black-body radiation, which, at T_{eff}, has the same spectral brightness at a line frequency determined by the Planck distribution (2.1)

$$P_\nu(T) = \frac{c}{h\nu_{12} 4\pi} U_\nu(T) = \frac{2}{\lambda_{12}^2} \frac{1}{\exp\left(\frac{h\nu_{12}}{kT_{\text{eff}}}\right) - 1}, \quad (4.15)$$

where $\lambda_{12} = c/\nu_{12}$. This is illustrated in Fig. 4.9. In an optically thick medium a captured spectral line may have T_{eff} much higher than the temperature of the continuous radiation. This fact is a prime consideration in the treatment of optically pumped astrophysical lasers.

A situation similar to photoselective excitation may occur owing to resonance transfer of energy in collisions between atoms (molecules) with close energy levels. Such resonance collisions have cross-sections materially exceeding the gas-kinetic cross-sections πa_0^2. This case can be realized for vibration-excited N_2 molecules colliding with CO_2 molecules in planetary atmospheres (Chapter 8). But, it is the radiation of the nearest star that always serves as a primary source of energy.

5
Astrophysical rarefied gas/plasma

The ranges of gas (plasma) density, energy density and temperature in astrophysical media are rather wide. For astrophysical lasers the most interesting media are those where significant deviations from LTE occur. Then, it is necessary that the collisions of particles do not dominate over radiative transitions, i.e. collisions should play a limited part in establishing steady state, which allows non-LTE in the plasma. Since the probability of radiative (allowed and forbidden) transitions of atoms (ions) in the optical spectrum lies typically within $A_{nm} \simeq 1\text{--}10^8$ s^{-1}, this gives an upper estimate of the particle concentration (hydrogen atoms) in the gas

$$N_H \lesssim \frac{A_{nm}}{\langle \sigma_{coll} v_e \rangle} \simeq 10^7\text{--}10^{15} \text{ cm}^{-3}.$$

This density range covers the upper layers of stellar and planetary atmospheres, gaseous nebulae and gas condensations. Exciting and ionizing the atoms require that the kinetic energy of the electrons or the energy density of the photons must be sufficiently large. This criterion is probably met where nearby stars are of spectral class B and O known to be rich in UV radiation. The deviation from LTE caused by photoselective excitation requires that the effective temperature of the exciting spectral line exceeds $T_{eff} \gtrsim 10^4$ K. Therefore, the gas condensations near to blue stars and the upper layers of stellar atmospheres are suitable astrophysical media for the laser effect to be active. They are placed in the lower part of the diagram in Fig. 5.1. It is very difficult to observe the laser effect in a stellar atmosphere because of the photospheric background radiation. The observations of the upper atmosphere or chromosphere of a single star require a very high angular resolution and will perhaps become possible with future instruments. However, spectral information about the outer parts of a star can be obtained for eclipsing binaries. Here we shall restrict ourselves to gaseous media with low densities nearby a hot blue star (blobs, shells, etc.). Regarding density they take up an intermediate position between planetary nebulae and stellar atmospheres, and it is in these gas condensations that laser effects and even nonlinear optical effects, the subject of this book, may take place.

5.1 Low-density gas nearby a hot star – Strömgren sphere

As far back as the classical work of Strömgren (1939) the ionization state of the hydrogen gas in the vicinity of a star was modelled (Fig. 5.2). The simple model is very convenient to describe processes in nebulae, shells, and gas condensations because

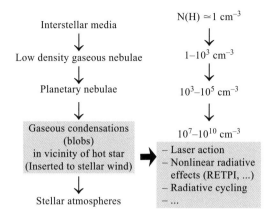

Fig. 5.1 Density scale of various gas/plasma astrophysical media. Gas condensations in the vicinity of hot blue stars are places of discovered astrophysical lasers.

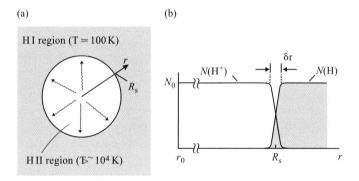

Fig. 5.2 Strömgren sphere (*a*) separating ionized (H II) region and neutral region (H I) gas (*b*).

hydrogen is approximately 90% of the nearby stellar gas. It is usually convenient for describing a real multicomponent astrophysical medium as He is only photoionized by far UV radiation ($I_{He} = 24.6$ eV), and the remaining elements contribute only impurity concentration.

The formation rate of photoionized H^+ atoms in a unit volume is expressed as

$$\frac{d}{dt}N(H^+) = W_i(r)N(H), \qquad (5.1)$$

where the photoionization rate W_i determined by (4.3) depends on the spectral brightness of the radiation $P(\nu, r)$ in the actual volume at the distance r from the star. This radiation occurs in the UV spectrum with the Lyman continuum $\nu > \nu_c$ at a narrow solid angle Ω. The rate of disappearance of ions as a result of recombination of H^+

ions in a unit volume is given by the expression

$$-\frac{d}{dt}N(\mathrm{H}^+) = -\alpha N(\mathrm{H}^+)n_e = \alpha N^2(\mathrm{H}^+), \tag{5.2}$$

where the recombination probability is given by (4.10), and the electric neutrality of the medium is taken into account: $N(\mathrm{H}^+) = n_e$. Under steady conditions characteristic of astrophysical media these rates must be the same, i.e.

$$W_i(r)N(\mathrm{H}) = \alpha N^2(\mathrm{H}^+). \tag{5.3}$$

This condition dictates the concentration ratio between ions and neutral atoms

$$N(\mathrm{H}^+) = \left[\frac{W_i(r)N(\mathrm{H})}{\alpha}\right]^{1/2}. \tag{5.4}$$

In the vicinity of the star, at distances $r \lesssim R_s$, the H atoms are completely ionized owing to a high radiation intensity, while that is not the case at distances $r \gtrsim R_s$. At distances $r > R_s$ the radiation intensity, proportional to $1/r^2$, is not sufficient for complete ionization of H, as shown in Fig. 5.2. The distance R_s, at which this transition occurs, i.e. the distance where $N(\mathrm{H}^+) = N(\mathrm{H})$, is referred to as the Strömgren radius and its corresponding sphere with $r = R_s$ as the Strömgren sphere. Let us estimate a value of R_s in terms convenient for a reader not familiar with this concept.

The photoionization probability $W_i(r)$ depends on the radiation dilution factor w as the radiation propagates at a distance $r > R_s$ from the stellar surface

$$w = \frac{\Omega}{4\pi}\left\{1 - \left[1 - \left(\frac{r_0}{r}\right)^2\right]^{1/2}\right\}, \tag{5.5}$$

where Ω is the solid angle at which a star with its radius r_0 can be seen at the distance r from the centre of the star. For distances $r \gg r_0$ the dilution factor has a simple form

$$w = \frac{1}{4}\left(\frac{r_0}{r}\right)^2. \tag{5.6}$$

Thus, the photoionization probability W_i is determined by the expression (4.3) reduced because of the dilution factor w

$$W_i = w \int_{\nu_c}^{\infty} \sigma_i(\nu) P(\nu, T) d\nu, \tag{5.7}$$

where the frequency dependence of the cross-section $\sigma_i(\nu)$ is shown in Fig. 4.2.

The cross-section rapidly decreases above the ionization limit. For example, for $h\nu - I_\mathrm{H} \simeq 3$ eV, the cross-section for hydrogen is only half of its value at ν_c. So we can simplify expression (5.7) by using the averaged cross-section $\langle \sigma_i \rangle \simeq \sigma_i(I_\mathrm{H} + 3 \text{ eV})$ and the average number of photons in the Lyman continuum emitted by a star with temperature T per unit time in all directions

$$L_c = 4\pi(4\pi r_0^2)\int_{\nu_c}^{\infty} P(\nu, T) d\nu. \tag{5.8}$$

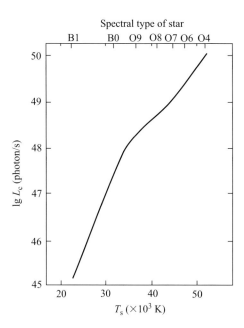

Fig. 5.3 Number of Lyman continuum photons emitted by star per sec as a function of star temperature according Harvard spectral classification of stars (adapted from Avrett, 1976).

Fig. 5.3 shows the dependence of L_c on the temperature T of the stellar photosphere of different spectral classes. In this approximation the photoionization of H at a distance r from the star is determined by a simple expression

$$W_i(r) \simeq w \langle \sigma_i \rangle \frac{L_c}{(4\pi)^2 r_0^2} = \frac{\langle \sigma_i \rangle L_c}{(8\pi)^2 r^2}. \tag{5.9}$$

The Strömgren radius R_s is determined from its definition that the densities of H^+ ions and H atoms are the same at the point $r = R_s$ as shown in Fig. 5.2

$$N(H^+) = N(H) = \frac{N_0}{2}, \tag{5.10}$$

where N_0 is the total density of hydrogen. From (5.4), (5.9), and (5.10) we can derive a simple estimate for R_s

$$R_s \simeq \frac{1}{8\pi} \left[\frac{2 \langle \sigma_i \rangle L_c}{\alpha N_0} \right]^{1/2}. \tag{5.11}$$

For example, for the hydrogen density $N_0 = 10^2$ cm^{-3} nearby a B-class star with $T = 30{,}000$ K $L_c = 10^{47}$ s^{-1} (Fig. 5.3), $\langle \sigma_i \rangle \simeq 3 \cdot 10^{-18}$ cm^2 (Fig. 4.2) and, according to (4.9), the recombination factor α with $T_e = 10^4$ K, we have $R_s \simeq 0.56 \cdot 10^{18}$ cm $= 0.18$ pc (1 pc $= 3.09 \cdot 10^{18}$ cm). This value agrees with the results of more precise calculations.

The thickness of the transition layer δr between the H II and H I regions is very small as compared to R_s. It depends on the free path of a photon in the Lyman continuum near the ionization boundary

$$\delta r \simeq \ell_c = \frac{1}{\chi_c} = \frac{1}{N_0 \langle \sigma_i \rangle}, \qquad (5.12)$$

where $\chi_c = \langle \sigma_i \rangle N_0$ is the photoionization absorption factor (cm^{-1}) of this photon. For the above numerical instance, $\delta r \simeq 3 \cdot 10^{15}$ cm $= 10^{-3}$ pc. The value of δr corresponds to the unit optical density of a photon in the Lyman continuum $\tau_c = \delta r \chi_c = 1$. At the distance $r \gtrsim R_s + \delta r$ the radiation of the Lyman continuum drastically falls off, and a low degree of ionization is maintained by photons at frequencies far from the ionization limit.

The transition boundary between the H II and H I regions is characterized by a sharp change in the parameters of the medium. In the transition region the fraction of H I increases drastically and at the same time the ion density in the Lyman continuum decreases sharply. Thus, in this narrow region $\tau_c \simeq 1$ and in the H I region $\tau_c \gg 1$. The free path length of a Lyα photon in the scattering region ℓ_{sc} is much shorter than in the Lyman continnum ℓ_c because of a great difference between the cross-sections of resonance scattering σ_{sc}^{12} and photoionization $\langle \sigma_i \rangle$

$$\ell_{sc} \simeq \ell_c \frac{\langle \sigma_i \rangle}{\sigma_{sc}^{12}}, \qquad (5.13)$$

where $\sigma_{sc}(\text{Ly}\alpha)$ is derived from (2.61) to be $\sigma_{sc} \simeq 1.4 \cdot 10^{-14}$ cm^2. In the transition region $\ell_c \simeq \delta r$, and $\ell_{sc} \simeq 10^{-4} \delta r$. Hence, according to (2.66), a Lyα photon can here be subjected to 10^4–10^5 events of resonance scattering and only thereafter escape the transition zone through the far wings of the spectral line (Fig. 2.19). In the H I region the escaped photons undergo strong resonance scattering in the radiative wings because of a high H I density and the very high optical density in the line wings determined by (2.63) to be: $a\tau_0(\text{Ly}\alpha) > 1$, where a is the damping factor (2.57) (Adams, 1972).

A drastic change in the parameters of the hydrogen gas in a thin transition layer can be observed at sufficient angular resolution studying the fluorescence radiation from other atoms obtained by resonant excitation by Lyα (Chapter 9 and following). It should be kept in mind, however, that the simple plasma model used here does not allow, for example, for the "microturbulence" of the real astrophysical plasma. This effect causes additional Doppler broadening, which somewhat smoothes out the sharpness of the transition boundary.

The effective conversion of photons of the Lyman continuum to resonant Lyα photons and their repeated scattering (capture) must cause an increase in the effective spectral temperature of Lyα. This effect is considered further in Section 5.3 by an example: gas blobs of hydrogen near to a hot blue star.

5.2 Planetary nebulae

Among gaseous diffuse nebulae the so-called *planetary nebulae* (PNe) are close to the subject of this book as they have a comparatively high density of particles, $N_0 \simeq$

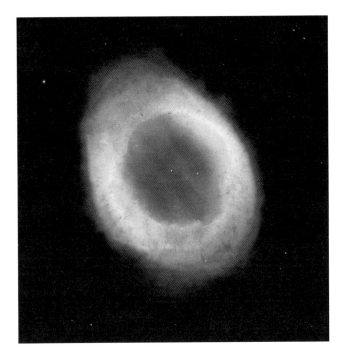

Fig. 5.4 Picture of M57 Ring Nebula taken by Hubble Space Telescope (HST) (Credit HST team and NASA). (See Plate 1)

10^3–10^5 cm^{-3}. Owing to their extension on the sky they resemble the planets in the solar system, when observed with a telescope, from which they got their name. Planetary nebulae often look like small discs and ovals and sometimes like rings. For illustration, a photo of the Ring Nebula (PN M57) in the constellation Lyra is shown in Fig. 5.4 with a shape reminiscent of the Saturn rings. It is located at a distance of $2 \cdot 10^3$ light years from Earth and its diameter is around one light year. On the average, the angular dimensions of planetary nebulae are less than 10 arcsec and the mean radius is $R_N \simeq 10^{17}$ cm. However, the dimensions of various PNe may differ by orders of magnitude. They can be very peculiar in shape but 20% of them are ring-shaped (shown in Fig. 5.4), which makes their intepretation simpler.

PNe represent an important stage of stellar evolution in the Galaxy as they are expanding shells of gas expelled by a star. In the centre of a planetary nebula there is a star (its nucleus) with a very high photospheric temperature of $T = 10^5$ K. The main part of the radiation energy of such stars lies in the far UV spectrum. However, the emission of a nebula in the visible range is caused by the absorption of unobservable UV photons, which are converted to longer wavelengths owing to recombination, and are reradiated in the visible spectrum. But, because of the high temperature of the stellar radiation, this conversion occurs not only for H and He atoms but also for other elements (O, N, Ne, S, etc.) and their multiply charged ions.

68 *Astrophysical lasers*

The ionization/recombination equilibrium in PNe for hydrogen is described by the Strömgren model (1939), which was developed for hydrogen-rich diffuse nebulae in the vicinity of hot stars of spectral O- and B-class. The results obtained by Strömgren and briefly described in Section 5.1 are applicable to PNe. The Strömgren boundary R_s lies inside a planetary nebula. A high effective temperature of the radiation of the central star results in photoionization of atoms, He I ($I_{\text{He}} = 24.6$ eV, $\lambda_c(\text{He I}) = 504$ Å), and ions, He II ($I_{\text{He II}} = 54.6$ eV, $\lambda_c(\text{He II}) = 228$ Å). The Strömgren radius for He II is closer to the central star, i.e. inside the PN, since the photoionization of He II demands short-wavelength UV radiation, R_s can be estimated from (5.11), where the recombination factors α for He I and He II should be used, and L_c means the total stellar radiation in the continua of He II or He I.

In the case of elements (atoms) with a larger number of electrons (O, N, Ne, S, etc.) several Strömgren zones can arise for each element in successive ionization stages. For oxygen atoms, for example, ionization zones from O V to O I (O^{4+} to neutral O) can occur, if the radiation temperature of the star $T \simeq 1.5 \cdot 10^5$ K. This extended Strömgren model is illustrated in Fig. 5.5 for oxygen in a PN. Close to the central star oxygen exists as four times ionized, i.e. as O V, which requires that UV radiation at $\lambda < 160$ Å is absorbed in this zone. Photons with $\lambda < 226$ Å will be absorbed in the next zone, and so on.

The information on planetary nebulae implies that the radiation temperature T of the central star (nucleus) lies within the range $4 \cdot 10^4$–10^5 K, i.e. it is very high. The electron temperature(s) of the planetary nebulae, however, is much lower and lies within the range $T_e \simeq (10\text{–}15) \cdot 10^3$ K. This is due to the fact that, firstly, the stellar radiation reaching the nebular part is subjected to strong dilution ($w \simeq 10^{-13}$). Secondly, it can also be explained by recombination radiation and subsequent cascade radiation (emission) of excited hydrogen atoms and impurity atoms. The chemical composition of the PN agrees with that of the central star. Therefore, besides

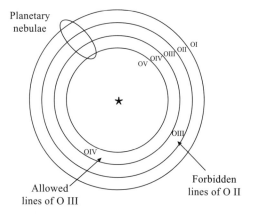

Fig. 5.5 Distribution of ranges with various degrees of photoionization and corresponding emission lines for oxygen.

hydrogen and helium, planetary nebulae contain heavier elements O, N, Ne, S, Fe ..., etc. with a total concentration of about 0.1% of the hydrogen abundance. The radiation from the heavy elements easily escapes the PN, which leads to thermostatting the electron temperature. Owing to the low density, excited atoms can irradiate not only in allowed but also in forbidden transitions, which is a characteristic feature of PN.

The distribution of line radiation over the ionization zones is shown in Fig. 5.5 with oxygen as an example. Spectral lines of O II caused by recombination are evidently emitted in the O III zone and the lines of O III must be emitted in the O IV zone, and so on. But, besides allowed lines, there are forbidden lines in the emission line spectra. In the O III zone, for example, very intense lines of unknown origin as well as abnormally intense emission lines from highly excited states of O III were observed early Bowen (1928, 1947). Based on his analysis of the laboratory spectrum and level system of O III, Bowen managed to explain both these puzzles. The first one was attributed to the existence of low-lying, long-lived (metastable) levels of O III decaying by radiative transitions that are forbidden in a dipole approximation (see Section 3.1.5). The second puzzle was explained in rather an exotic way – a high-lying state of O III is populated by photoselective excitation by (a) the strong resonance Lyα line of He II. The frequency of the He II line coincides accidentally with the frequency of an absorption line of O III, originating from two fine structure levels in the ground term. Both these mechanisms are illustrated in Fig. 5.6. Since they are closely related to the subject of book we consider them in more detail later.

5.2.1 Forbidden transitions

Many atoms and their iso-electronic ions have long-lived (metastable) excited states (Chapter 3). A classical example is the 2s state of the hydrogen atom, which undergoes radiative decay to the ground state 1s, only by two-photon transitions with the probability $A_{21} \simeq 8$ s^{-1}. Hydrogen atoms fall into this state as a result of recombination and radiative decay of high-lying states. Two-photon radiation of hydrogen has a broad spectrum in the UV range and makes a contribution to the continuous radiation of interstellar media. Other atoms (ions) have long-lying metastable states whose radiative decay is forbidden in a dipole approximation, too. Such a case for O III is shown in Fig. 5.6a, where the $^1D_2$ and $^1S_0$ states have radiative lifetimes $\tau_m = 38$ s and 0.6 s, respectively. In laboratory light sources, forbidden transitions are usually not observed because of quenching collisions. And when observed in nebular spectra, they were for several years even attributed to the unknown element "nebulium" until Bowen (1928) explained them as forbidden radiative transitions of O III, O II, N II etc. from low-lying metastable states.

In the photoionized rarefied plasma of a PN the density of hydrogen and electrons usually lies within the range of N_H, $n_e \gtrsim 10^4$ cm^{-3}. Therefore, the collision frequency γ_{coll} (2.47) and the probability of excitation of metastable levels by electrons W_e satisfy the condition

$$W_e, \gamma_{\text{coll}} \ll A_{m0}, \qquad (5.14)$$

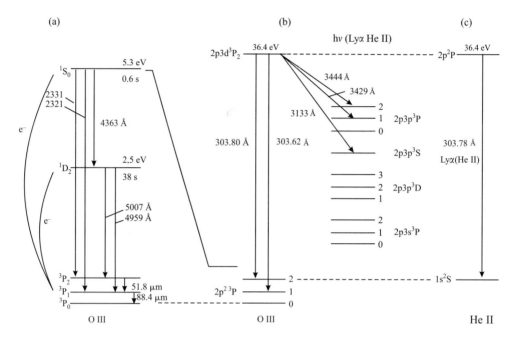

Fig. 5.6 Schematic partial diagram of O III, illustrating the electron excitation of low-lying metastable states (a) and the Bowen mechanism of optical excitation and following fluorescence (b, c).

where A_{m0} is the radiative transition probability for a forbidden line connecting the metastable state with the ground state. The electron temperature in a PN $T_e \gtrsim 10^4$ K, provided through photoionization of hydrogen by central star radiation, is sufficient to excite both metastable levels in O III by collisions with electrons with an energy of $E > 2.5$ eV (and 5.3 eV). The O III ions are produced by photoionization owing to a very high effective radiation temperature T_s of the central star. Since the probability of excitation by electronic collisions W_e satisfies the condition (5.14) such collisions do not deactivate the metastable states. The population of two metastable levels with different energies depends on the electron temperature, so the electron temperature T_e can be measured using the intensity ratio of two forbidden lines (Ambartzumian, 1933). It is reasonable that in nebulae with a higher hydrogen density and, hence, electron density n_e (for example, $n_e > 10^7$ cm^{-3}) the depletion of metastable levels increases because of the more frequent collisions between electrons and atoms.

Another potential channel of depletion of a metastable level m may be, in principle, induced allowed transitions to an upper level n under continuous UV stellar radiation at a radiation temperature $T_s \gg T_e$. Transitions to the ground level are insufficient because they are compensated for by the same transitions from the metastable ground level according to the detailed balancing principle (2.6). The rate of stimulated absorption is determined by Eq. (2.26), which in our case can be presented as

$$W_{mn} = A_{mn}\left(\frac{\Omega}{4\pi}\right)\langle n\rangle = A_{nm}\frac{w}{\exp(h\nu_{nm}/kT_s) - 1}, \qquad (5.15)$$

where Ω is the solid angle of radiation of a star in the location of the PN, w is the dilution factor (5.6), $\langle n\rangle$ is the occupation number for the equilibrium Planck radiation of a star with temperature T_s at the transition frequency ν_{mn} and $kT_s \gg \hbar\nu_{nm}$. In this case the ratio between rates of absorption depletion and radiative decay of a metastable level is

$$\frac{W_{mn}}{A_{mo}} \simeq \frac{A_{mn}}{A_{m0}} w \exp\left(-\frac{h\nu_{mn}}{kT_s}\right). \qquad (5.16)$$

Even with high values of $A_{mn}/A_{m0} \simeq 10^8$–$10^9$, because of strong dilution of the stellar radiation, $W_{mn} \ll A_{m0}$, i.e. upward transitions do not reduce the lifetime of the metastable state m.

The situation is approximately the same for photoionizing transitions in the continuum. The photoionization rate W_i, according to (4.2), is

$$W_i \simeq \langle\sigma_{ph}\rangle\frac{L(\nu_c - \nu_{m0})}{R_N^2}, \qquad (5.17)$$

where $\langle\sigma_{ph}\rangle$ is the photoionization cross-section, $L(\nu_c - \nu_{m0})$ is the total luminosity of the central star (photons/s) in the continuum $\nu > \nu_c$–ν_{m0} for a metastable state in the atom (ion), and R_N is the distance from the central star to the nebula. Even with a high effective temperature of the star $T_s \simeq 10^5$ K, $W_i \lesssim 1$ s^{-1} owing to a long distance R_N. However, for levels with a long lifetime ($\gtrsim 10^2$ s^{-1}) and for nebulae near to the star, the effect of photoionization depletion of the metastable level must be taken into account. This, in particular, is true for metastable hydrogen atoms appearing in nebular clouds near to the central source in blue star Eta Carinae (Johansson et al., 2004).

Complex atoms (ions), such as Fe II, have metastable high-lying states, which are essential for the formation of inverse population and a laser effect in gas blobs near to blue stars, more dense than PNe (Chapter 11).

5.2.2 Accidental coincidences of spectral emission and absorption lines

Planetary nebulae with a high degree of ionization are characterized by intense emission lines observed in the UV spectrum, often originating from high-lying states. Most of them result from recombination, but some lines, e.g. O III (3444 Å, 3429 Å, 3133 Å, etc.), have abnormally high intensities. Bowen (1934, 1947) explained the anomalous intensities in an original manner directly related to astrophysical lasers. He showed that the anomalous O III lines are caused by radiative decay of one and the same upper level of O III (Fig. 5.6b). This level is excited by the optically strong He II line at 303.78 Å, because of an accidental wavelength coincidence with an O III line at 303.80 Å, as shown in Fig. 5.6b, c. The 303.80 Å line feeds the upper level of the

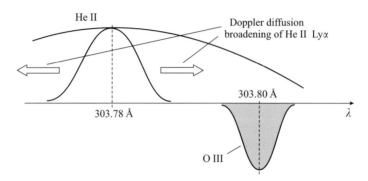

Fig. 5.7 Coincidence of wavelength of He II Lyα emission line and O III absorption line. Detuning $\Delta\lambda \simeq 0.02$ Å is compensated for by Doppler diffusion of optically thick He II Lyα emission line.

strong optical O III lines at 3444 Å, 3429 Å, and 3133 Å, known as fluorescence (see also Chapter 3).

The excitation efficiency of a fluorescence line depends on the overlap between the emission and absorption line profiles as indicated by the wavelength difference or detuning. In case of the radiative interaction between the pair He II–O III discussed in the previous paragraph the value of the detuning $\Delta\lambda = 0.02$ Å is four times the Doppler half-width of the He II Lyα line ($\frac{1}{2}\Delta\nu_D = 0.55 \cdot 10^{-2}$ Å). However, the optical density τ_0 for the He II line at 303.78 Å in PNe is very high because it is the hydrogen-like resonance line He II Lyα. It plays the same role as the HI Lyα line at 1215 Å in the hydrogen spectrum. The cross-section of the radiative transition for this line $\sigma_{2p-1s} = 0.6 \cdot 10^{-14}$ cm^2 (for $T = 10^4$ K and $A_{2p-2s} = 10^{10}$ s^{-1}). With a He II density of $N(\text{He II}) \simeq 10^2$ cm^{-3} in the photoionization area of the nebula the free path of the He II Lyα photon according to (2.64) is $\ell_{sc} \simeq 1.6 \cdot 10^{12}$ cm. For diffusive broadening of the line up to the value of the wavelength detuning (Fig. 5.7) less than 10^2 scattering events are enough. This value is much less than the number of scattering events $N_{sc} = (\Delta R_N/\ell_{sc}) \simeq 10^4$ in a nebula with a thickness of the He II shell of $\Delta R_N \simeq 10^{16}$ cm.

The efficiency of the Bowen fluorescence process is rather high. It can be defined as a fraction of all the He II Lyα photons that are formed through recombination and converted into a fluorescence line of O III (except the direct transitions 2p3d-2p^2). The values of the observed efficiencies are typically about 0.5 (Pottasch, 1984). Such high values can be understood from Fig. 5.7. In view of the Doppler diffusion about half of the long-wavelength wing of the He II emission line passes through the Doppler contour of the O III absorption line. Hence, a continuous "run off" of all the long-wave photons takes place. The He II Lyα photons, which have not been absorbed by the O III ions and have crossed the absorption line contour, can cross it again by scattering with a short-wave jump of the scattered photon frequency. This short-wave jump is very important since the average frequency jump of the Lyα photon at scattering is approximately equal to half the Doppler width of the He II line $\Delta\lambda_D(\text{He II})/2$, which is twice as much as the Doppler width of the O III absorption line.

Owing to another accidental wavelength coincidence, a different O III line at $\lambda = 374$ Å is able to excite N III in PNe. But, a large number of absorption lines in spectra of many atoms and ions coincide with a restricted number of strong emission lines. Conclusively, the Bowen mechanism itself and similar fluorescence processes for other pairs of elements have proved to be extremely important for astrophysical optically pumped lasers, e.g. the pairs H I Lyα and Fe II, H I Lyβ and O I considered in Chapters 9–12.

5.3 Gas condensations (blobs) in the vicinity of a hot star

Gas condensations (GC) near to hot stars are the most interesting nebular medium of astrophysical lasers owing to their being in the relatively high density and their being in the vicinity of the central star. It became possible to discover and study such compact GCs within our Galaxy thanks to the advance of modern methods of optical astronomy, e.g. the Hubble Space Telescope and large adaptive telescopes, and, particularly, high-spatial resolution imaging interferometry (Saha, 2002).

The most detailed information on compact gas condensations has been gained from HST data of the close surroundings of Eta Carinae. These data have become the starting point for our studies of astrophysical lasers. Eta Carinae is one of the most studied but still not well understood massive stars in our Galaxy (Davidson and Humphreys, 1997). A critical question is whether the central object is a binary or a single star. The interpretation of recent observations favours the binary star model (Damineli, 1996). The luminosity of the central object is $5 \cdot 10^6$ times higher than that of the Sun $L_\odot$, i.e. it is one of the brightest stars in our Galaxy. Eta Carinae is classified as a luminous blue variable (LBV) star, and it has experienced a vigorous life for two centuries. After a strong eruption in the 1840s a huge bipolar nebula, known as the "Homunculus" because of its shape, has grown around it. In the 1840s, Eta Carinae was after Sirius, the second brightest object in the sky and then its apparent magnitude dropped because of the surrounding, obscuring gas. About 160 years after the eruption, the nebula occupies about $20''$ in the sky, which is 45,000 astronomical units at a distance of 2.3 kpc ($\sim$7500 l.y.). The high angular resolution of the Hubble Space Telescope has made it possible to obtain wonderful images of the Eta Carinae Homunculus, one of which is displayed on the cover of this book. The physical process behind The Great Eruption, as it is called, or behind later and smaller eruptions, is not fully understood. The periodical events appearing every 5.5 years (Damineli, 1996) may indicate that the central object is a binary star. During these events, having a duration of a few months, spectra of the GCs change drastically with time. The information embedded in all spectra recorded with ground-based and space-borne (e.g. the HST) instruments during the event around midsummer 2003 is still far from being extracted and interpreted.

The near vicinity of the central star was investigated by speckle interferometric techniques which resulted in the discovery of a complex nodal structure. In the equatorial region of the star three bright compact condensations were discovered (Weigelt and Ebersberger, 1986), referred to as the Weigelt blobs B, C, and D. At the distance of $R_\mathrm{b} \simeq 10^{16}$ cm (several light days) from the star the projected spatial extension of a blob is of the order of 10^{15} cm, and its density $N_\mathrm{H} \simeq 10^7$–10^{10} cm^{-3} (Davidson and

74 Astrophysical lasers

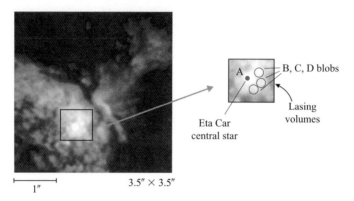

Fig. 5.8 Eta Carinae's near surrounding with compact blobs B, C, D (reprinted from van Boekel et al., 2003 with courtesy and permission Elsevier). (See Plate 2)

Humphreys, 1997). Fig. 5.8 shows images of these Weigelt blobs taken with the Very Large Telescope Interferometer (VLTI) in Chile by van Boekel et al. (2003).

The surprising fact with these strange compact ejecta nearby the star is that they move at low velocities, about 50 km/s. The high angular resolution of HST and speckle interferometers, as well as the low velocities of the blobs, allows the studying of the spectra of these objects separately from the central star and high-velocity ejecta. These observational facts made it possible to conclude that the blobs emit simultaneously both allowed and forbidden spectral lines, which had previously been associated with the central star. Moreover, many lines are narrow, i.e. emitted from comparatively cold areas of the blobs in spite of the fact that these are very close to the very bright central star (Davidson, 2001a, b). The two-temperature structure of the GC in the form of a hot and a cold region dictates the position of the Strömgren boundary inside the blob. For this purpose, the volume-average density of hydrogen $N_0 (\simeq N_H)$ must satisfy the condition (see Chapter 10):

$$N_0 > N_0^{cr} = \left(\frac{W_i}{\langle \sigma_i \rangle D \alpha} \right)^{1/2}, \tag{5.18}$$

where N_0^{cr} is the so-called critical density of hydrogen (Johansson and Letokhov, 2004a), at which the photoionizing radiation of the Lyman continuum of a hot star is absorbed in the region of a gas condensation with diameter D. The photoionization model of a GC is discussed in more detail in Chapter 10 by the example of blob B in the vicinity of Eta Carinae. This simple model qualitatively explains the origin of the complex spectra registered with the Hubble Space Telescope including their periodic changes (Hartman et al., 2005). It was the observation of strange spectral lines in the blob spectra that led us to the nature of abnormal Fe II lines and to detect those Fe II

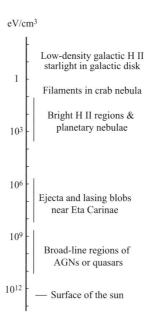

Fig. 5.9 Typical thermal (near-UV, visible, near-IR) radiation densities (eV/cm^3) in various astrophysical emission-line objects. HLyα emission is not included here, but most other 0.1–1.5 µm emission is included (adapted and modified from Davidson, 2001b).

and O I transitions that are involved in the astrophysical lasers and are described in Chapters 10–12.

The radiative processes occurring in compact gas condensations near a star are similar to those in a planetary nebula, and the GCs can thus be studied by using the extensive research of planetary nebulae (Seaton, 1960; Pottasch, 1984; Kwon, 2000). But, there is an essential difference related to the much shorter distance between the blob and the star and, hence, by a much smaller (ten orders of magnitude) flux reduction due to radiative dilution. Therefore, a gas condensation like a Weigelt blob has a high internal radiation density comparable to the free energy density of charged particles. Fig. 5.9 illustrates typical thermal radiation (near-UV, visual, near-IR) densities in various astrophysical emission-line objects including the ejecta (the GCs) near Eta Carinae (Davidson, 2001b). These ejecta have a relatively high energy density almost comparable with broad-line regions (BLR) of active galactic nuclei (AGNs) or quasars. Hence, such gas condensations near to a hot star can be referred to as *radiation-rich* regions.

As a conclusive result, a photoionization–recombination cycle in GCs near hot stars produces very high intensities of the hydrogen resonance lines Lyα and Lyβ (see Fig. 3.1) formed through recombination. Because of this fact the effective spectral temperature of such lines is many orders of magnitude higher in the GCs than in planetary nebulae. Moreover, this temperature can be comparable to the surface temperature of the star in spite of the fact that the distance of the GC from the star exceeds the stellar radius by about 10^2 times. This effective spectral temperature is, however, that high only for some spectral lines and not for the entire continuum. It is quite sufficient for the formation of abnormally bright spectral lines of Fe II, for example, and for the occurrence of the laser effect (Johansson and Letokhov, 2002, 2007).

The effect of a high spectral temperature of H I lines in the Weigelt blobs (Johansson and Letokhov, 2004a) can exist in other compact gas condensations as well, but that is still beyond the reach of detailed investigations.

5.4 Surroundings of symbiotic stars

Like the gas condensations near to hot stars the surrounding nebulae of symbiotic stars are also remarkable astrophysical media, in which astrophysical lasers can operate. Symbiotic stars are long-period (a few years) variable binary systems, whose unresolved composite spectra have various origins (Kenyon, 1986). Boyarchuk (1969) proposed that a symbiotic star was composed of three sources. Firstly, the spectrum contains characteristics usually associated with a red giant. Secondly, the spectrum has a feature typical for a compact hot source, such as a white dwarf with an effective temperature of $T \sim 10^4-10^5$ K. Finally, the two stars are embedded in an ionized nebula with an electron temperature of about $T_e \simeq 15 \cdot 10^3$ cm and an electron density of $n_e > 10^6$ cm^{-3}. The presence of a surrounding nebula has been confirmed by the observation of a Raman process by Schmid (1989). The cool component, i.e. the red giant, produces many emission lines in the optical region, which are observable owing to the low continuum level. The gas dynamics and gas flow between the cool and hot components is complicated in symbiotic stars. Most symbiotic stars show LS-allowed as well as LS-forbidden, (intercombination) lines from higher ions, which are associated with the hot regions of the symbiotic star.

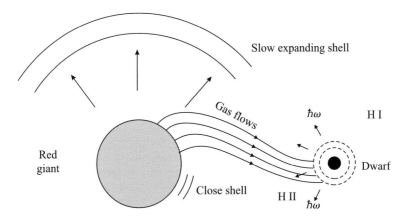

Fig. 5.10 Mutual position of large cold star and small very hot star. Intensive UV irradiation of extended cold gas surrounding the large star is provided by the close bright companion (dwarf) in this symbiotic star system.

The symbiotic stars represent a late stage of stellar evolution including a brief span in the life of a binary star. Because this late stage has a very short timescale, symbiotic stars are very rare stellar objects. An interesting feature of a symbiotic star is the presence of a gas flow from the cool component irradiated by bright UV light from the hot component. A schematic representation of a symbiotic binary star containing a red giant, a white dwarf and accreting material from the giant's stellar wind is shown in Fig. 5.10. High energy photons emitted by the white dwarf ionized a part of the wind (H II region), although some of the circumstellar nebula remains neutral. This unusual situation is favorable for building up non-LTE conditions and, therefore, for the operation of astrophysical lasers. The symbiotic stars, that are studied in detail, are located in our Galaxy. However, even if they are extended objects they are still not possible to resolve spatially, but belong to the challenges for future high-resolution experiments with ground-based or even orbital instruments.

6
Basic elements of laser physics

The basics of laser physics have been described in many classical monographs (e.g., Sargent et al., 1974; Siegman, 1986; Svelto, 1998; Yariv, 1976). The readers can use these books for details. The purpose of this Chapter is to present the basic elements of laser science. Knowledge of these elements will allow the readers of later chapters to more easily distinguish the specific properties of astrophysical lasers from properties of laboratory lasers. The latter knowledge is essential for the development of methods to search for astrophysical lasers. It goes without saying that brief description of laser physics elements is restricted only by continuous wave (CW) gas lasers with nonhomogeneous amplification spectral line, which are closest to astrophysical lasers.

6.1 Amplification coefficient

The usual way to obtain light amplification in laser media is to use population inversion of quantum levels of a particle (atom, ion, molecule). Other possibilities of amplification, which are due to the presence of noncoherent or coherent radiation from another source in the medium, have been considered briefly in Section 1.3. In this book we restrict ourselves to the case of inverted populations, as other possibilities are not realistic in natural masers and lasers. Particles with an inverted population, $N_2 > N_1$ of the levels E_2 and E_1, are located far from each other because of the low particle density N_0 ($N_0^{-3} \gg \lambda_{12}$). Thus, the density of the inverted population $\Delta N = \Delta N_{21}$ is connected with the average densities of population of levels N_1 and N_2:

$$\Delta N = \left(N_2 - \frac{g_2}{g_1} N_1\right). \tag{6.1}$$

The amplification coefficient $\alpha_{21}(\omega)$ and the shape of the amplification spectral line due to the $2 \rightarrow 1$ transition are determined by the expression:

$$\alpha_{21}(\omega) = \Delta N \sigma_{21}(\omega). \tag{6.2}$$

The cross-section $\sigma_{21}(\omega)$ of the radiative transition $2 \rightarrow 1$ is determined by expressions (2.37) and (2.38), which might be presented in a more convenient form:

$$\sigma_{21}(\omega) = \frac{\lambda_{12}^2}{4} a(\omega) = \frac{\lambda_{12}^2}{4} A_{21} g(\omega). \tag{6.3}$$

The function $a(\omega)$ describes the amplification of a spectral line with normalization (2.36), $g(\omega)$ is the form-factor amplification line with normalization $\int g(\omega) \, d\omega = 1$. In case of a Lorentz contour with the FWHM $\Delta\omega$, expression (6.3) can be written as

$$\sigma_{21}(\omega) = \sigma_{21}(\omega_0) \left[1 + \left(2\frac{\omega - \omega_0}{\Delta\omega}\right)^2\right]^{-1}, \qquad (6.4)$$

where the cross-section of the radiative transition for the center of the line is:

$$\sigma_{21}(\omega_0) = \frac{\lambda_{12}^2}{2\pi} \frac{A_{21}}{\Delta\omega}. \qquad (6.5)$$

Expressions (6.2) and (6.5) are convenient for estimations.

However, in the case of small values of the radiation and collision widths in comparison with the Doppler width $\Delta\omega_D$, which corresponds to a low pressure of the media active in astrophysical lasers, it is necessary to use the following expression for $\sigma_{21}(\omega)$ according to (2.54),

$$\sigma_{21}(\omega) = \sigma_{21}(\omega_0) \exp\left[-4\ell n 2 \left(\frac{\omega - \omega_0}{\Delta\omega_D}\right)^2\right] \qquad (6.6)$$

instead of expressions (6.4) and (6.5). Expression (6.6) differs from (6.5) for $\Delta\omega = \Delta\omega_D$ by the coefficient $(\pi \ell n 2)^{1/2} = 1.5$. It emanates from a difference in the amplitudes between the Gaussian and Lorentzian profiles for equal area A_{21} (2.36). For a simple estimation of the amplification coefficient this difference is insignificant, since the main coefficient in $\sigma_{21}(\omega_0)$ is $(A_{21}/\Delta\omega) \simeq 10^{-6} - 10^{-2} \ll 1$.

So far, it has been assumed that the amplification is independent of the light polarization. The real atoms and molecules have an angular momentum J_n, which has $2J_{n+1}$ projections on a chosen direction corresponding to $2J_{n+1}$ magnetic sublevels. The contribution of the magnetic sublevels to the amplification depends on their population. An unequal population of the various sublevels because of selective pumping leads to a polarization dependence on the amplification. The presence of a magnetic field eliminates the orientation degeneracy of the angular momentum $\vec{J}$ due to Zeeman splitting of the magnetic sublevels. In this case the frequency splitting of the σ^+ and σ^- components of amplification connects to the strength of magnetic field. This effect was studied in details for astrophysical OH masers (Elitzur, 1991, 1992; Vlemmings, 2007). The observation of polarization properties of astrophysical lasers will be the subject of future research.

6.2 Saturation of amplification

The amplification for low intensity radiation does not depend on the intensity and it is called *linear* amplification. The rate of stimulated transitions in the linear regime is small as compared to the rates of population relaxations of levels 1 and 2 for an amplified transition. Let us estimate the *saturation intensity* I_{sat} of amplification at which the inverted population ΔN and, correspondingly, the amplification coefficient will be reduced by a factor of 2. We consider the simplified level scheme shown in Fig. 6.1. To avoid cumbersome expressions and still reveal an effect, the upper level 2

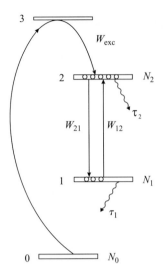

Fig. 6.1 Simplified model of quantum transitions of four-level laser: 2-1 is the transition with inversion of level populations N_1 and N_2, 1 is the fast decaying level ($\tau_1 \ll \tau_2$), W_{exc} is the rate of level 2 excitation (pumping rate).

is populated with the rate $W_{\text{exc}}(\text{s}^{-1})$ by excitation of particles with density N_0. The excited particles in levels 1 and 2 relax during the times τ_1 and τ_2, respectively. Let us set $\tau_1 \ll \tau_2$, so that there is population inversion in levels 1 and 2, i.e. $\Delta N > 0$. In this approximation, which in laser physics is known as a "four-level scheme" and which is typical for an astrophysical CW laser, the coefficient of linear amplification α_{21}^0 equals:

$$\alpha_{21}^0 = \sigma_{21}\Delta N \simeq \sigma_{21} N_2 = \sigma_{21}\left(W_{\text{exc}}\tau_2\right) N_0 \qquad (6.7)$$

For simplicity, assume that for the portion of excited particles N_2 in expression (6.7) $N_2 \ll N_0$ because $W_{\text{exc}} \ll 1/\tau_2$.

6.2.1 Homogeneous broadening

To account for the outgoing of particles from levels 2 and 1 due to stimulated transition with rates $W_{21} = \sigma_{21}I$ and $W_{12} = \sigma_{12}I$, the expression for the amplification coefficient is:

$$\alpha_{21} = \frac{\alpha_{21}^0}{1 + W_{21}\tau_2} = \alpha_{21}^0 \Big/ \left(1 + \frac{I}{I_{\text{sat}}}\right), \qquad (6.8)$$

where I is the radiation intensity within the amplification spectral line (in photons/cm$^2 \cdot$ s). This simple expression can be derived easily for the steady-state regime, considering the balance of populations in levels 1 and 2 and accounting for the smallness of stimulated transitions $1 \to 2$ because $\tau_1 \ll \tau_2$ and, hence $N_1 \ll N_2$. So, if the rate of stimulated transitions W_{21} is comparable with the relaxation rate $1/\tau_2$ the amplification coefficient is halved. It takes place at $I = I_{\text{sat}}$, where

$$I_{\text{sat}} = \frac{1}{\sigma_{21}\tau_2}\left[\text{photon/cm}^2 \cdot \text{s}\right]. \tag{6.9}$$

The expressions (6.8) and (6.9) can be generalized for the case of arbitrary times τ_1 and τ_2, for the value W_{exc} being comparable with $1/\tau_2$. However, it leads to more cumbersome expressions, which do not change the physical meaning of the saturation effect.

Strictly speaking, in the estimations of the level populations made above, we implied that the rate of induced transitions W_{21} is less than the rate of the dephasing of the phase difference of the wave functions of states 1 and 2. In other words, $W_{21} \ll \Delta\omega$, where $\Delta\omega$ is the amplification line width. This assumption is always valid for astrophysical lasers as long as the so-called saturation parameter G is satisfied by the conditions:

$$G = I/I_{\text{sat}} \ll \tau_2 \Delta\omega. \tag{6.10}$$

In this approximation we neglect the coherent effects such as Rabi oscillations and field splitting of energy levels (Sargent et al., 1974) which are improbable in astrophysical lasers. We refer the reader to the papers on saturation of a two-level atom in chaotic monochromatic fields (Zoller, 1979; Ficek et al., 2000).

The most essential assumption is that all particles in levels 1 and 2 interact with resonant directed radiation, which can have a spectral width much narrower than the spectral width of the amplification line $2 \rightarrow 1$ (Fig. 6.2b). It corresponds to *homogeneous* broadening of the amplification spectral line. The shape of this amplification line is unchangeable during saturation as shown in Fig. 6.2b. A Doppler-broadened spectral line is, in essence, a set of a large number of much narrower spectral lines from particles with different velocities (Section 2.2). Therefore the Doppler broadening is often described as *inhomogeneous* broadening.

6.2.2 Doppler broadening

Let us for clarity introduce the following spectral widths ($\Delta\omega = 2\pi\Delta\nu$):

(1) $\Delta\nu_{\text{hom}}$ is the homogeneous spectral width determined by the sum of radiative and collisional widths (Fig. 6.3a);

(2) $\Delta\nu_{\text{geom}}$ is the geometrical broadening of the spectral interval for the interaction between the radiation and the gas due to a finite divergence of radiation (Fig. 6.3b):

$$\Delta\nu_{\text{geom}} = \left(\frac{\Theta}{\pi}\right)\Delta\nu_D, \tag{6.11}$$

where $\Delta\nu_D$ is the Doppler width. For isotropic radiation ($\Theta = \pi$) and $\Delta\nu_{\text{geom}} = \Delta\nu_D$ (Fig. 6.3c);

(3) $\Delta\nu_{\text{int}}$ is the spectral interval of the Doppler profile involved in the interaction with radiation (Fig. 6.3b):

$$\Delta\nu_{\text{int}} = \Delta\nu_{\text{hom}} + \Delta\nu_{\text{geom}}. \tag{6.12}$$

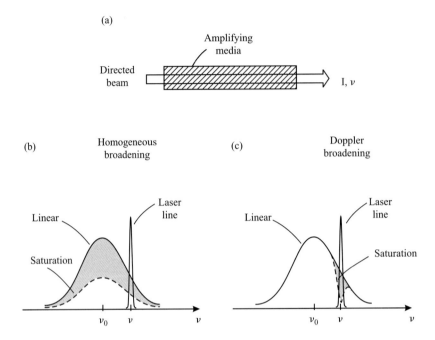

Fig. 6.2 Saturation of amplification by the directed beam with frequency ν (a) and evolution of spectral line shape in the case of homogeneous (b) and Doppler broadened (c) lines. Solid lines are linear amplification and dashed lines are saturated amplification.

Thus, when spectral-line broadening is inhomogeneous, the light wave interacts only with the particles with which it is at resonance. The portion of particles that interact with the field depends on the ratio between the homogeneous width and the Doppler width. Strictly speaking, it depends also on the spatial wave configuration. A plane travelling wave described by $E \cos(\omega t - kz)$ interacts only with particles located within the spectral range of the homogeneous width $\Delta \omega_{\text{hom}}$ at the resonance frequency $\omega = \omega_0 + k\text{v}$ (Fig. 6.2c). In other words, the travelling light-wave interacts only with the particles that have a definite velocity projection on the travelling wave propagation direction,

$$|\omega - \omega_0 - k\text{v}| < \Delta \nu_{\text{hom}}. \tag{6.13}$$

Let us underline the fact that the resonance width depends also on the divergence of a light beam according to geometrical broadening (6.11), which increases the spectral interval $\Delta \nu_{\text{int}}$ of the light beam interaction with the Doppler profile up to the value given in (6.12) as shown in Fig. 6.3b. If a monochromatic field is isotropic, i.e. comprises a series of waves that propagate in different directions, *all the particles* can interact with the field, whatever the velocity may be (Fig. 6.3c). In this limiting case of isotropic radiation the geometrical broadening (6.11) coincides with the Doppler width $\Delta \nu_D$.

In the saturation regime the radiation interacts and saturates the Doppler profile in the spectral interval $\Delta \nu_{\text{int}}$ causing "hole burning" inside the Doppler profile (Fig. 6.2c).

Basic elements of laser physics

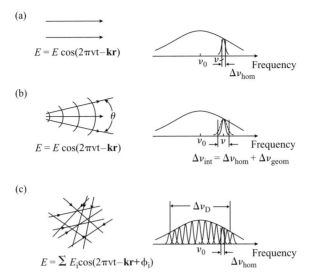

Fig. 6.3 Illustration of the effect of the divergence of the monochromatic field on the resonant interaction width with inhomogeneously (Doppler) broadened spectral line for: (*a*) plane light wave; (*b*) light field with angular divergence Θ; (*c*) isotropic monochromatic field.

Strictly speaking, it is necessary to account for the effect of an intensity broadening of $\Delta\nu_{\text{int}}$ (Letokhov and Chebotayev, 1977):

$$\Delta\nu_{\text{int}}(I) = \Delta\nu_{\text{hom}}\sqrt{1 + \frac{I}{I_{\text{sat}}}} = \Delta\nu_{\text{hom}}\sqrt{1+G}, \quad (6.14)$$

The intensity broadening takes place until $\Delta\nu_{\text{int}} \lesssim \Delta\nu_{\text{Dopp}}$, i.e. in the wide intensity range:

$$I_{\text{sat}} \lesssim I < I_{\text{sat}}\left(\frac{\Delta\nu_D}{\Delta\nu_{\text{int}}}\right)^2. \quad (6.15)$$

For higher intensities the whole Doppler profile interacts with the directed stimulated radiation.

In the case of Doppler-broadened amplification line, the dependence for a homogeneously broadened line (6.8) is replaced by a gradual dependence of the intensity saturation (Letokhov and Chebotayev, 1977):

$$\alpha_{21} = \alpha_{21}^0 \bigg/ \left(1 + \frac{I}{I_{\text{sat}}}\right)^{1/2} = \frac{\alpha_{21}^0}{\sqrt{1+G}}. \quad (6.16)$$

This dependence is valid until the intensity broadening of the interaction interval $\Delta\nu_{\text{int}}$, determined by expression (6.14), reaches the Doppler width $\Delta\nu_D$. It takes place at high intensities determined by the interval (6.15). Thereafter, all Doppler profiles interact with the directed radiation.

84 *Astrophysical lasers*

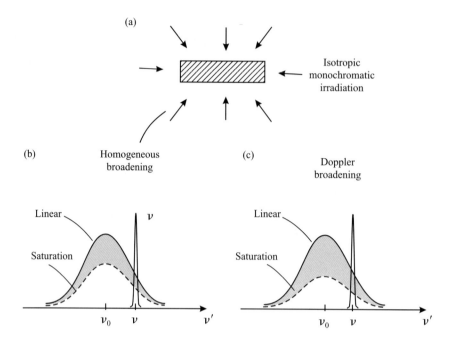

Fig. 6.4 Saturation of amplification by isotropic monochromatic (a) of homogeneously (b) and Doppler broadened amplification line (c). Solid lines are linear amplification and dashed lines are saturated amplification.

The monochromatic *isotropic* radiation (Fig. 6.4) interacts with all Doppler profiles in the linear and saturated amplification regimes (see expression (6.8) for the homogeneous broadening). In other words, the isotropic radiation provides the "homogenization" of the Doppler profile. As a result, the saturation of amplification occurs without distortion of the line shape (Fig. 6.4b, c). This is essential for understanding the influence of the amplifying volume geometry on the spectral line of astrophysical lasers.

6.3 Laser with resonant optical feedback (cavity)

The idea of light amplification by stimulated emission in a medium with inverted population has been mentioned in a number of publications over a 20–30-year period of the 20th century. However, this idea had slightly opened the way to finding how to produce a beam of identical photons, but subject to one condition: identical photons to be fed to the input of the amplifying medium. Otherwise, any imperfection of the light beam at the input of the amplifying medium would be repeated in an amplified form at the output. All conventional sources emit chaotic radiation, from which it is impossible to extract a flow of identical photons of any intensity for a subsequent amplification. A more revolutionary idea appeared, namely, to place the amplifying medium (molecular beam) in a *resonant cavity* (Gordon et al., 1954; 1955; Basov and

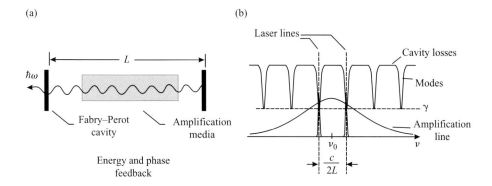

Fig. 6.5 Laser with resonant coherent optical feedback by means of the Fabry–Perot cavity (a), and longitudinal resonant modes of cavity with amplification spectral line (b).

Prokhorov, 1954; 1955), wherein the photon loss factor was lower than the gain factor, i.e. to introduce a *positive resonant optical feedback*. The amplification process could then be started with a single or a few spontaneous photons that, while repeatedly passing through the amplifying medium, give birth to a large number of identical photons. This chain process of *undamped* multiplication of identical photons is usually called *generation* or *oscillation*, and the device developed and used is known as the *maser* in the case of microwaves, and the *laser* in the case of light. However, the idea of the quantum oscillators (masers and lasers) entered the field of optics from the radio field, as illustrated by the "arrow across from "Maser (1959)" to "Laser (1960)" in Fig. 1.1.

Thus, the key idea of masers and lasers was using resonance optical feedback in the form of a closed cavity for the masers and an open cavity for the lasers. The Fabry–Perot cavity was the first simple version of the open cavity for the laser (Prokhorov, 1958; Schawlow and Townes, 1958), as shown in Fig. 6.5. The generation takes place on a limited number of resonant modes with well-defined optical field configurations if the number of resonant modes within the amplification line above threshold is limited. In the optimal case the standing waves are formed along the optical axis of a cavity with the frequency interval $\Delta\nu_m = c/2L$ (see Fig 6.5) – a Gaussian field distribution with a spherical or almost plane wavefront is formed in the lateral direction. In this case the generated light wave is spatially coherent. Such a feedback may be called *coherent* or *resonant*. This type of feedback performs two functions: (1) it returns some part of the electromagnetic energy emitted by the active (amplifying) medium and (2) it forms an electromagnetic field configuration in the cavity owing to the constructive interference between the incident and reflected waves.

6.3.1 Simple consideration

It is convenient to characterize the laser medium by an amplification coefficient distributed per unit length at a given frequency $\alpha(\omega)$. Energy losses in the medium and on mirrors, including useful output light through the partially transparent mirror, can for simplicity be described by a loss coefficient γ at the resonant frequencies of the modes. It is important that the amplification coefficient α exceeds the loss coefficient γ for some number of resonant modes within the amplification line only (Fig. 6.5b).

The light wave of a single-mode laser in the classical case, illustrated in a simplified way in Fig. 6.5, has the form:

$$E(t, x, y, z) = E(x, y) \sin(\omega t - kz), \qquad (6.17)$$

where $E(x, y)$ is the transverse electric field amplitude distribution. The sinusoidal expression describes the longitudinal distribution of the standing wave field inside the laser cavity, part of which exists in the form of a coherent traveling wave. The intensity of radiation, with no regard to frequency and phase, is

$$I = \frac{1}{\hbar\omega} \frac{c}{8\pi} E^2 \; [\text{photons/cm}^2 \cdot \text{s}]. \qquad (6.18)$$

A change of intensity I of single-mode laser radiation inside the amplifying non-coherent medium, in the case when the phases of the independent dipole emitters (two-level systems) are not correlated, is given by the simple kinetic equation:

$$\frac{1}{c} \frac{dI}{dt} = [\alpha(I) - \gamma] I. \qquad (6.19)$$

When the initial (linear) amplification coefficient $\alpha_0 > \gamma$, the laser is above the threshold. Then, the intensity of radiation grows exponentially up to the steady-state value I_s, at which the saturated amplification equals the losses:

$$\alpha(I_{\text{st}}) = \frac{\alpha_0}{1 + \frac{I_s}{I_{\text{sat}}}} = \gamma \qquad (6.20)$$

It is convenient to use the coefficient of amplification excess above the threshold, $\eta = \alpha_0 \gamma$, because the steady-state intensity is

$$I_s = I_{\text{sat}}(\eta - 1). \qquad (6.21)$$

Simple dependences of amplification of intensity and of the steady-state intensity at an excess of the pumping rate above threshold are shown in Fig. 6.6. Let us now leave out the considerations of interaction of laser modes with different frequencies, a synchronization of modes etc., because astrophysical lasers do not require these details.

6.3.2 Quantum considerations

Laser oscillations start from the level of spontaneous photons inside the resonant mode of a cavity, for which $\alpha_0 > \gamma$. Quasi-classical equations (6.19) do not account for the spontaneous emission and, hence, do not describe the initial process of oscillation. For correct description of the initial stage, the right part of (6.19) must contain a term with the number of spontaneous photons. These are emitted at a spontaneous

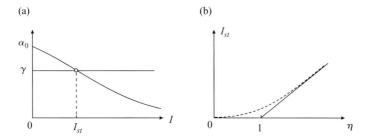

Fig. 6.6 Intensity dependence of amplification providing the steady-state operation of laser (a), and steady-state intensity I_{st} dependence of the excess of the pumping rate above threshold in the quasiclassical limit (solid line) and the quantum regime (dashed line) nearly threshold.

decay rate A_{21} by excited particles of density N_2 inside a cavity, which is occupied by radiation (mode volume). The quantum consideration of laser action is presented in the monographs of Sargent et al. (1974), Meystre and Sargent (1990).

In quantum language Eq. (6.19) can be presented in the form

$$2\frac{d\langle n\rangle}{dt} = c\left[\alpha\left(\langle n\rangle\right) - \gamma\right]\langle n\rangle + A_{21}\int N_2 dN, \qquad (6.22)$$

where $\langle n \rangle$ is the mean number of photons in mode n, and the last term describes the addition of spontaneous photons to the cavity mode. This simplified equation allows us to trace the change of the mean number of photons in the mode, $\langle n \rangle = \sum nP(n)$, starting from spontaneous emission, as well as to understand the gradual change of the line intensity or $\langle n \rangle$ near the threshold of oscillation, as shown in Fig. 6.6b. However, Eq. (6.22) does not allow us to understand the dynamics of the evolution of the distribution number of photons $P(n)$ in the mode. That requires the presentation of the field in the form of quantum harmonic oscillators with corresponding flows of transition probabilities up and down, as shown in Fig. 6.7.

The quantum theory of lasers has been developed in works of the Lamb and Scully (Sargent et al., 1974) and Haken (1970) schools. Below we elucidate the elements of quantum theory following the works of Lamb and Scully (1967).

The evolution of the photon number distribution $P(n)$ in the mode n during the transition through the threshold can be studied by means of the so-called master equation for the distribution $P(n)$. This master equation defines the photon number in mode n with energy $n\hbar\omega$, coupled with the levels having the energies $(n-1)\hbar\omega$ and $(n+1)\hbar\omega$ by amplifying and dissipative quantum transitions. The amplification rate (s^{-1}) can be described by the coefficient αc and the loss rate by the coefficient γc:

$$\frac{1}{c}\frac{d}{dt}P(n) = -(n+1)\alpha P(n) + n\alpha P(n-1)$$
$$- \gamma n P(n) + \gamma(n+1)P(n+1). \qquad (6.23)$$

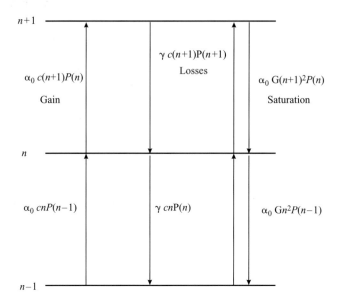

Fig. 6.7 Flows of probabilities of up and down transitions between quantum states of quantized harmonic oscillators (adapted from Sargent et al., 1974).

The coefficient α in this equation describes the saturation amplification, which for weak saturation $(G \ll 1)$ can be presented in the form

$$\alpha = \frac{\alpha_0}{1 + (n+1)G} \simeq \alpha \left[1 - (n+1)G\right]. \qquad (6.24)$$

The corresponding quantum transition rates are shown also in Fig. 6.7.

It is clear that the value $\alpha c(n+1)P(n)$ in equation (6.23) is the flow of the probability of transitions from state $|n\rangle$ of the photonic field to state $|n+1\rangle$, caused by photon emission of the excited atoms. The similar interpretation of the value $\alpha c(n+1)P(n)$ follows directly from the origin of the $\alpha_0 c$ coefficients corresponding to the rate of induced transitions.

So, the product of the value $\alpha_0 cP(n)$ and the number of final states $(n+1)$ defines the total rate of photon output from the field states $|n\rangle$ to the state $|n+1\rangle$. By analogy, the term $\alpha_0 cnP(n-1)$ is the flow of probabilities from state $|n-1\rangle$ of the photonic field to state $|n\rangle$. The term $\gamma c(n+1)P(n+1)$ describes the probability flow from state $|n-1\rangle$ to state $|n\rangle$ caused by decay of $|n-1\rangle$ states due to loss of light.

Expansion of the expression (6.24) in the small parameter (degree of saturation) $G = I/I_{\text{sat}}$ in the case of the quantized field gives the interpretation of the terms responsible for the saturation of a laser media. Thus, the linear term of the expansion

$$(n+1)\alpha_0 c(n+1)GP(n) = \alpha_0 cG(n+1)^2 P(n) \qquad (6.25)$$

corresponds to the process in which the photon is absorbed at first and emitted later. An analogous explanation can be given for the expansion terms of higher orders. The higher order terms are not essential for us, because the steady-state distribution $P(n)$, governing the laser photon statistics, exists already at the weak saturation.

The distribution $P(n)$ of the number of laser mode photons n depends on the rates of stimulated transitions and cavity losses. Let us consider the statistics of the laser photons for the steady-state oscillation regime below ($\eta < 1$) and above ($\eta > 1$) the oscillation threshold.

Equation (6.23) in the steady-state regime and a small degree of saturation ($G \ll 1$) is reduced to follow the equation for the photon distribution:

$$-\alpha_0(n+1)P(n) + \alpha_0 n P(n-1) - \gamma n P(n) + \gamma(n+1)P(n+1). \tag{6.26}$$

This second order differential equation can be written for consequent values of n. As a result of this operation the system of equations is reduced to two systems of equations. Each of these equations is the first-order difference equation

$$\alpha_0 n P(n-1) - \gamma n P(n) = 0. \tag{6.27}$$

This difference equation has the solution

$$P(n) = \left(1 - \frac{\alpha_0}{\gamma}\right)\left(\frac{\alpha_0}{\gamma}\right)^n, \tag{6.28}$$

with the standard normalization

$$\sum_{n=0}^{\infty} P(n) = 1. \tag{6.29}$$

Hence, the probability of photon absence inside the cavity of the laser below the threshold ($\alpha_0 < \gamma$) is

$$P(0) = 1 - \frac{\alpha_0}{\gamma}, \tag{6.30a}$$

and the probability of presence of a single photon is:

$$P(1) = \left(1 - \frac{\alpha_0}{\gamma}\right)\left(\frac{\alpha_0}{\gamma}\right), \tag{6.30b}$$

etc. Fig. 6.8a shows the steady-state distribution of photon numbers below the threshold and for a very small excess of the pumping rate above the threshold.

It is quite interesting to compare expression (6.28) with the distribution of the number of photons in a single mode of black-body radiation at the same frequency ω:

$$P(n) = \exp\left(-n\hbar\omega/kT\right) / \left[1 - \exp\left(-\hbar\omega/kT\right)\right]. \tag{6.31}$$

The comparison of the distributions (6.28) and (6.31) shows that the statistical properties of laser radiation below the threshold of laser operation coincides with the statistical properties of black-body radiation in a single mode at the same frequency ω and at an effective temperature determined by the expression

$$\exp\left(-\hbar\omega/kT\right) = \frac{\alpha_0}{\gamma}. \tag{6.32}$$

90 Astrophysical lasers

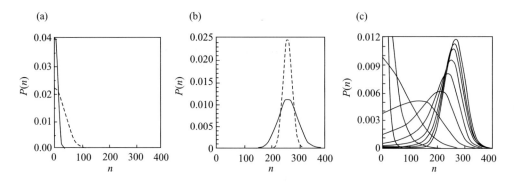

Fig. 6.8 Graphs of photon number probabilities (steady-state distributions) $P(n)$: (a) below the generation threshold (solid line) and slightly above the threshold (dashed line); (b) above threshold (20%) for the mean number of photons $\langle n \rangle = 400$ (solid line) and (c) Poisson distribution for the same $\langle n \rangle$ (dashed line) (adapted and modified from Sargent et al., 1974).

This temperature can be called the effective temperature of laser radiation *below* the threshold of oscillation.

Equation (6.23) above the threshold of oscillation ($\alpha_0/\gamma \gg 1$) for the steady-state regime can be written in the form:

$$-\left(\frac{\alpha_0^2}{\gamma}G\right)P(n) + \left(\frac{\alpha_0^2}{\gamma}G\right)P(n-1) - \gamma n P(n) + \gamma(n+1)P(n+1) = 0. \quad (6.33)$$

This difference equation of the second order is reduced to two difference equations of the first order:

$$\left(\frac{\alpha_0^2}{\gamma}G\right)P(n-1) - \gamma n P(n) = 0. \quad (6.34)$$

This equation has the normalized solution in the form of the Poisson distribution

$$P(n) = \frac{\langle n \rangle^n}{n!}e^{-\langle n \rangle} \quad (6.35)$$

with the average number of photons in mode $\langle n \rangle$:

$$\langle n \rangle = \frac{\alpha_0^2}{\gamma}G. \quad (6.36)$$

Thus, in the case of a significant excess of the threshold of laser oscillation the statistics of the number of laser photons in single mode is governed by the Poisson distribution (6.35) (Fig. 6.8b). This distribution corresponds to the coherent state of a laser field. Fig. 6.8c illustrates the gradual passage through the initial noncoherent distribution to the steady-state Poisson distribution during laser generation. This evolution has been studied in experiments by Arecchi (1965).

The descriptive presentation about the relative value of the fluctuations of the photon number of a laser field gives the Mandel parameter, Q, which determines the normalized deviation of the mean-quadratic photon number from the square of the mean value:

$$Q = \frac{\langle n^2 \rangle - \langle n \rangle^2}{\langle n \rangle} - 1. \tag{6.37}$$

In the case of Poisson statistics for the photon number: $Q = 1$. For $Q > 1$ the photon statistics is super-Poisson, and for $Q < 1$ the statistics is sub-Poisson. For single-mode laser radiation the Mandel parameter equals:

$$Q = \frac{\gamma}{\alpha_0 - \gamma}. \tag{6.38}$$

Correspondingly, the distribution of the photon number below the threshold oscillator ($\alpha_0 < \gamma$) is sub-Poisson, and above the threshold ($\alpha_0 > \gamma$) the distribution $P(n)$ is always super-Poisson. In the case of a significant excess of the threshold ($\alpha_0 \gg \gamma$) the distribution $P(n)$ is close to Poisson distribution.

6.4 Coherent properties of laser light

An optical cavity, i.e. an optical system forming the light field with a stable spatial configuration at frequency ω close to the resonant frequency ω_0 of the optical amplifying transition (Fig. 6.5), provides the coherence of laser light. The amplification $\alpha_0 > \gamma$ compensates for the damping of the light oscillations caused by unavoidable losses. For lasers with a large active volume $V = a^2 L \gg \lambda^3$, where a is the transverse size and the length of the active laser media, an open cavity is most convenient. The number of modes with a given polarization and with the propagation vectors within the solid angle Ω along the optical cavity axis and within the spectral width $\Delta\nu$ of the amplification lines equals:

$$N_\nu = 2\Omega \frac{V}{\lambda^3} \frac{\Delta\nu}{\nu}. \tag{6.39}$$

The minimal solid angle is determined by diffraction

$$\Omega_{\min} \simeq \Omega_{\text{dif}} \simeq \left(\frac{\lambda}{a}\right)^2. \tag{6.40}$$

In this case the number of modes with wave vectors within the solid angle Ω_{dif} with given polarization equals

$$N_m = \frac{2L}{\lambda} \frac{\Delta\nu}{\nu} = \frac{2L}{c} \Delta\nu. \tag{6.41}$$

Thus, the frequency interval between neighbouring longitudinal modes is $\Delta\nu/N_m = c/2L$ [Hz], as shown in Fig. 6.5.

6.4.1 Spatial coherence

A Fabry–Perot open cavity is the optical feedback system, which provides the drastic reduction of the density of high-quality modes. Between two parallel mirrors the two opposite light waves can propagate along the optical axis. Some part of the radiation will be lost because of diffraction at the mirror edges. The off-axis waves have big losses.

92 Astrophysical lasers

The radiative (diffractional) losses cause the formation of weakly decaying modes with the definite stable spectral configuration of the electromagnetic field. The Fabry–Perot cavity has small diffraction losses in the case of a high Fresnel number

$$N = \frac{a^2}{\lambda L} \gg 1. \tag{6.42}$$

This is due to the angular size of one mirror, which is less than the diffraction(al) angle λ/a viewed from the point of the location of the other mirror. The Fresnel number equals approximately the number of Fresnel zones, which are observable on one of the mirrors from the position of the other mirror. The diffractional losses are very small for $N \gg 1$, but grow drastically for $N \lesssim 1$.

A laser with a Fabry–Perot cavity emits a light beam with a diffraction divergence of

$$\varphi \simeq \frac{\lambda}{a}. \tag{6.43}$$

This beam has a definite phase along the wavefront that provides full spatial coherence of the light field. Practically, an open cavity with spherical mirrors is more stable and convenient. A laser with spherical mirror(s) emits a divergent spherical wave with a definite phase along the spherical wavefront (Fig. 6.9a). Such a spherical beam is spatially coherent, too, and can be transformed easily to the directed beam with a diffractional divergence. A laser can operate inside a thin optical fiber or thin optical layer with the transverse size of the same order as the optical wavelength. The output divergence of such a laser is large (Fig. 6.9b), but again, owing to the definite phase along the wavefront, such a divergent light beam can be converted to a directed beam with a diffraction divergence. The spatial coherence (diffractional directivity) of the laser beam is one of the most impressive properties of lasers.

6.4.2 Temporal coherence and spectral bandwidth

The amplitude of laser oscillations of the ideal single-mode laser, according to (6.21), is stable to an accuracy of the quantum fluctuations, governed by the photon distribution $P(n)$. The quantum fluctuations are responsible for phase fluctuations of the laser

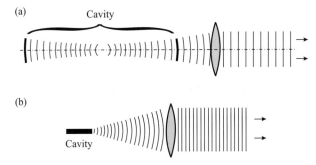

Fig. 6.9 Transformation of the divergence of the coherent light beam from a laser with a confocal Fabry–Perot cavity (a) and a very thin fiber or diode laser (b).

wave. Spontaneous photons are occasionally merged with the single mode and shift the phase of the wave by a very small random value. This effect causes a diffusion of the phase and produces the bandwidth of the laser radiation spectrum, which can be called the *natural linewidth* of the laser.

In order to study the phase fluctuations of the laser field, it is quite convenient to present the electromagnetic field in the form of a complex vector (Fig. 6.10). This vector undergoes small changes in phase and amplitude due to arriving spontaneous photons to the mode. For a laser, far above the threshold, the amplitude fluctuations are negligible. That field strength can be written in the form

$$E(t) = \sqrt{\langle n \rangle} e^{i\Theta(t)} e^{-i\omega_0 t}, \tag{6.44}$$

where $\langle n \rangle$ is the mean number of photons with frequency ω_0. The phase $\Theta(t)$ makes a one-dimensional random walk around a circle, as indicated in Fig. 6.10.

The quantum laser theory gives the distribution $P(\Theta)$, which is governed by the diffusion equation

$$\frac{\partial P(\Theta, t)}{\partial t} = D \frac{\partial^2 P(\Theta, t)}{\partial \theta^2}. \tag{6.45}$$

D is the diffusion coefficient

$$D = \frac{\alpha c}{\langle n \rangle} \text{ [rad/sec]}, \tag{6.46}$$

where amplification α gives the arriving rate of spontaneous photons to the mode. The phase diffusion causes the decay of the mean amplitude of an electric field:

$$\langle E(t) \rangle = \int d\Theta P(\Theta, t) E(\Theta) = E_0 e^{-Dt}, \tag{6.47}$$

and leads to a spectral shape with a finite bandwidth of the laser oscillations:

$$S(\nu) = \frac{\langle n \rangle}{\pi} \frac{D}{(\omega - \omega_0)^2 + D^2} \tag{6.48}$$

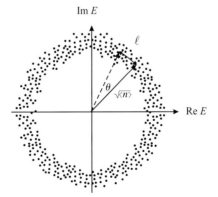

Fig. 6.10 Electromagnetic field as a vector with fluctuating amplitude and diffused phase due to a spontaneous photon, merged with the arriving to laser mode (adapted and modified from Sargent et al., 1974).

and the full bandwidth at half-weight:

$$\Delta\omega_{\text{las}} = 2D = \frac{(\alpha - \gamma)c}{2\langle n \rangle}. \tag{6.49}$$

This limiting (natural) spectral width of a laser mode was derived by Schawlow and Townes (1958). The natural bandwidth is very small for lasers, ranging from 10^{-3} Hz for He–Ne lasers to 10^5 Hz for miniature diode lasers.

The possibility of generating a light wave with a very narrow spectrum, which does not restrict the interaction time with quantum emitters, was quite nontrivial for physicists in the 1960s. It is worth giving the following citation on this subject from the Lamb article (Lamb et al., 1999):

To many physicists steeped in the uncertainty principle, the maser's performance, at first blush, made no sense at all: Molecules spend so little time in the cavity of a maser, about one-ten-thousandth of a second, that it seemed impossible for the frequency of the radiation to also be narrowly confined. Townes recalls a conversation with Niels Bohr on a sidewalk in Denmark in which Bohr emphasized this very argument. After Townes persisted, he said: "Oh, well, yes, maybe you are right". But, the impression was that he was simply trying to be polite to a younger physicist.

To conclude this brief discussion of the unique properties of coherent laser-light one of the authors (V.L.) remembers the discussions in the early 1960s of the nature of a coherent laser field and the comparison of the laser light with the radiation of an extremely hot black body. Let us assume that an ultranarrow filter and a diaphragm form a beam of black-body radiation having the same intensity, spectrum, and divergence as a laser beam. The electromagnetic wave of the perfect laser has a stable amplitude and frequency, and a slightly fluctuating phase. There is no linear spectrometer that can distinguish the single-mode laser radiation from the hypothetical black-body radiation mentioned above. However, the nonlinear correlation measurements will demonstrate a great difference. The laser wave has negligible fluctuations of the amplitude (shot noise of photons) and relatively slowly varying phase (Fig. 6.11a). In quantum terms, this means that the distribution of photon number in mode $P(n)$

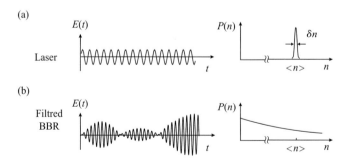

Fig. 6.11 Comparison of single-mode radiation of laser (a) and single-mode radiation of hypothetical very bright black-body radiation (BBR) (b). Classical light wave (a) and photon number probabilities $P(n)$ (b).

has a very narrow peak. At the same time, however, the light wave of a single mode of the hypothetical black-body radiation has deep fluctuations of the amplitude and the phase with the correlation time determined by the spectral width of the radiation (Fig. 6.11b):

$$\tau_{\text{corr}} \simeq \frac{1}{\Delta \nu_{\text{rad}}}. \tag{6.50}$$

In quantum terms the distribution of the photon number in single mode $P(n)$ is very wide even for $\langle n \rangle \gg 1$. This fundamental difference is a result of the resonant optical feedback in the presence of an amplifying medium.

6.4.3 Coherent state of light field

"Coherent states" of light, which have been introduced and studied by Glauber (1963a, b), may be considered as an example of one-dimensional classical harmonic oscillations. The properties of coherent states are most similar to properties of the classical oscillator. The laser coherent field is almost classical and, hence, the coherent states are relevant to the laser light.

Let us assume that the harmonic oscillator is in a quantum state with the wave function Ψ_n, which can be designated $|n\rangle$ for brevity. The mean values of the coordinate x and the momentum p in such quantum states are equal to zero:

$$x_n(t) = \langle n | \hat{x} | n \rangle = 0, \quad p_n(t) = \langle n | \hat{p} | n \rangle = 0. \tag{6.51}$$

In terms of classical mechanics

$$x(t) = x_0 \cos(\omega t + \varphi). \tag{6.52}$$

It is possible to find a state α_{cl}, for which the following relationship is valid:

$$x(t) = \langle \alpha_{\text{cl}} | \hat{x} | \alpha_{\text{cl}} \rangle. \tag{6.53}$$

Such a state was expressed by Shrödinger (1926) as:

$$|\alpha\rangle = e^{-\frac{|\alpha|^2}{2}} \sum_{n=0}^{\infty} \frac{\alpha^n}{\sqrt{n!}} |n\rangle \tag{6.54}$$

The mean number of vibrational photons (for a classical oscillator) or photons in the field oscillator in this coherent state is

$$\langle n \rangle = |\alpha|^2, \tag{6.55}$$

and the probability of occupation $P(n)$ of an oscillator in the state $|n\rangle$ is determined by the Poisson distribution

$$P(n) = e^{-\langle n \rangle} \frac{\langle n \rangle^n}{n!}. \tag{6.56}$$

For the coherent state

$$(\Delta x)^2 = \bar{x}^2 - \bar{x}^2 = \frac{\hbar}{2\pi\omega}, \tag{6.57a}$$

and
$$(\Delta p)^2 = \bar{p}^2 - p^2 = \frac{m\hbar\omega}{2}. \tag{6.57b}$$

The expressions (6.57a, b) give
$$\Delta x \Delta p = \frac{\hbar}{2}, \tag{6.58}$$

i.e. in the coherent state the product of the uncertainties of (a) the coordinate and (b) the momentum is minimal.

Let us move from the classical oscillator (6.52) to the quantum field oscillator (6.45). The quantum field states with a given number of photons n are called Fock states. Glauber (1963a, b) studied the coherent properties of laser light in terms of photon distribution $P(n)$ and Fock states. He found that the coherent states can be presented in the form of the coherent superposition of the states (6.54) with given photon number (Fock states). This result leads to a Poisson distribution for the photon number $P(n)$ in the coherent state:

$$P(n) = |\langle n| \alpha \rangle|^2 = e^{-|\alpha|^2} \frac{|\alpha|^{2n}}{n!} \tag{6.59}$$

with a width of about $|\alpha|$.

Fig. 6.12 illustrates a coherent field state of single laser mode with the mean amplitude $|\alpha| \simeq \sqrt{\langle n \rangle}$ and phase $\varphi + \Theta$. Here the phase angle φ relates to the mean measuring field, and angle Θ is the angle of the field before the measurement. The amplitude and phase in the coherent state have minimal uncertainties. The contribution of the phase uncertainty is shown in Fig. 6.12:

$$\delta\varphi = \frac{1}{2|\alpha|} = \frac{1}{2\sqrt{\langle n \rangle}} \tag{6.60}$$

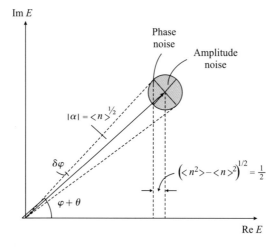

Fig. 6.12 Geometrical representation of the amplitude and phase fluctuations of the coherent state of laser field E.

The uncertainty of the photon number Δn equals

$$\Delta n = |\alpha| = \sqrt{\langle n \rangle}. \tag{6.61}$$

Thus, the product of the uncertainties from photon number and phase equals:

$$\Delta n \delta \varphi = \frac{1}{2}. \tag{6.62}$$

This result is similar to the Heisenberg relationship although it does not result in a direct consequence of a commutative relationship between the operators for the number of photons and the field phase. However, the expression (6.62) describes correctly the balance of the uncertainties of phase and amplitude. The strictly definite phase of the field corresponds to large amplitude fluctuations and vice versa. This effect leads to a concept of the squeezed states of the field, for which an ellipse of uncertainties is valid instead of a circle. For details the interested reader should consult quantum optics monographs (Loudon, 1983; Scully and Zubairy, 1987; Meystre and Sargent III, 1990; Walls and Milborn, 1995).

6.5 Laser as an amplifier

A laser medium without a resonant cavity is an amplifier of optical waves. There are two types of amplifiers, which are shown in simplified forms in Fig. 6.13. The amplifiers

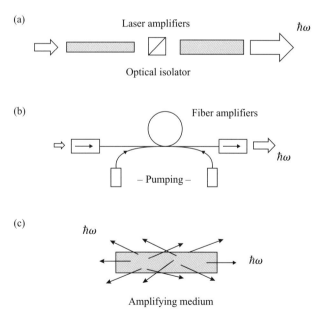

Fig. 6.13 Laser amplifier with controllable amplification of external laser radiation in a multi-stage system (*a*) a fiber system (*b*), and a noncontrollable amplification of spontaneous emission (*c*).

of the first type (Fig. 6.13a, b) have quite a large amplification $K = \exp(\alpha L)$. These amplifiers operate as *controllable* amplifiers of a relatively weak continuous wave or pulsed input of an optical signal. The suppression of the internal noise of the spontaneous photons is an important feature of such an amplifier. In the ideal case every amplified photon corresponds to a spontaneous photon. However, the amplified (stimulated) photons are emitted to the mode of the signal only, whereas the spontaneous photons are emitted in 4π steradians (to all modes within the amplification line).

Another situation occurs in the amplified medium with an *uncontrollable* amplification of its own spontaneous emission in the absence of an external optical signal (Fig. 6.13c). This amplification regime is parasitic because in it the uncontrollable self-generation of the amplified medium takes place. Various methods of optical isolation of separate cascades of the chain of amplifiers are used in order to suppress the parasitic generation. On the other hand, this is the only way to construct an amplifying laser where the input optical photons for amplification are very rare or the possibility of using the optical feedback does not exist. Particularly, it is true for the important case of X-ray free-electron lasers (X-ray FEL), operating in the regime of saturated amplified spontaneous emission (SASE regime). In the limiting case of the fiber amplifier laser, where the geometrical divergence is less than the diffractional divergence, the laser has the spatial coherence, determined by the diffraction (but without temporal coherence!). Temporal coherence is possible in the pulsed regime only, where the duration of the amplifying pulse is $\tau \simeq 1/\Delta\omega$, and $\Delta\omega$ is the spectral bandwidth of the amplifying pulse.

6.5.1 Amplification of coherent light

Laser amplifiers are used for two main purposes: for the amplification of the energy of high-power laser pulses and to compensate for the attenuation of the signal in a laser fiber communication line. The generation of a high-energy pulse inside the laser itself is a very difficult problem because of the limited volume of the laser medium inside a cavity with the required interferometric accuracy (Letokhov and Ustinov, 1983). The distortion of the optical quality of a large volume of active medium was one of the main limitations for the development of a laser weapon. High-power laser pulses are needed for triggering inertial thermonuclear reactions. A huge, multicascade laser setup has been constructed for this purpose (Basov et al., 1966). One of the main problems with the operation of a chain laser amplifier with high amplification $K = \exp(\alpha_0 L)$ was the optical isolation of cascades of the amplifier by means of Faraday cells, saturable absorbers etc. to prevent self-generation of the chain before the arrival of an amplifying optical pulse.

The laser fiber amplifiers of low-intensity optical signals (Fig. 6.13b) are used widely in optical telecommunication lines. The pumping of such doped solid-state fiber amplifiers is provided by very effective low-cost semiconductor lasers. This trend of quantum electronics has reached a mature industrial level.

When the input optical signal has a low intensity $I_{\text{in}} \ll I_{\text{sat}}$ (photons/cm$^2 \cdot$ s), the amplification is linear on the length

$$L_{\text{lin}} = \frac{1}{\alpha_0} \ln\left(\frac{I_{\text{in}}}{I_{\text{sat}}}\right) \qquad (6.63)$$

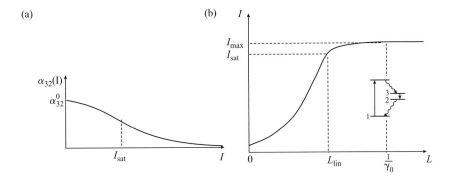

Fig. 6.14 Change in the amplification coefficient α_{32} during the transition from the linear amplification regime to the saturated amplification regime. (b) Growth in the intensity of the amplified spontaneous radiation in the linear and saturated regimes (L_{lin} is the length, over which the linear, non-saturated amplification and exponential growth of intensity occur). The insert shows a schematic of the four-level radiative cycle involving the laser transition.

The same is true for, the amplification of a laser pulse with the input energy flux $\mathcal{E}_0$ (photons/cm^2) $\ll \mathcal{E}_{\text{sat}}$, where $\mathcal{E}_{\text{sat}}$ is the energy flux at saturation $\mathcal{E}_{\text{sat}} = 1/2\sigma_{21}$ (photons/cm^2). The intensity (or energy flux) grows exponentially in the linear regime as $\exp(\alpha_0 L)$. When $I \gtrsim I_{\text{sat}}$ (or $\mathcal{E} \gtrsim \mathcal{E}_{\text{sat}}$), the amplification becomes *saturated*, and all the energy per unit length ($\frac{1}{2}\Delta N \hbar \omega$) stored in the laser medium is extracted as emission stimulated by the amplified radiation. The intensity of the amplified radiation grows linearly with length in the saturation regime. Fig. 6.14 illustrates an evolution of the intensity and amplification per unit length during the transfer from linear amplification to saturated amplification.

However, in the presence of linear losses the intensity (or energy flux) stops growing at some level because the extracted energy is only able to compensate for the linear losses (Prokhorov, 1963). This effect can be understood from simple quantitative arguments. Let us consider the length element of an amplifier $\Delta L \ll 1/\gamma_0$. The maximal energy of a pulse is achieved when the value of the energy extracted by stimulated emission from this length element $\Delta L(\Delta N_0/2)$ (photons/cm^3) equals the value of the energy loss $\mathcal{E}\Delta L \gamma_0$ in the same length element. It leads to an expression for the maximal (limiting) energy flux $\mathcal{E}_{\text{max}}$:

$$\mathcal{E}_{\text{max}} = \frac{\Delta N_0}{2\gamma_0} \quad [\text{photons/cm}^2 \cdot \text{s}] \tag{6.64}$$

In the case of amplification of continuous wave (CW) radiation, it is necessary to take into account the time of recovery τ_{rec} of the amplification α after the strong saturation, which is determined by the rate of pumping of the active laser medium. Then the expression for maximal (limiting) intensity is

$$I_{\text{max}} = \frac{\Delta N_0}{2\gamma_0 \tau_{\text{rec}}}. \tag{6.65}$$

This limitation of intensity is important for an astrophysical laser amplifier owing to the existence of linear losses of radiation (dust etc.).

Astrophysical laser amplifiers operate in the regime of *uncontrollable amplification of multimode* radiation in a low-pressure active gas medium. In this regime the transit from linear to saturated amplification is accompanied by distortion of the spectrum of the amplifying spontaneous radiation. Because of this, the laser amplifiers with Doppler-broadened amplification spectral lines are most interesting for us.

6.5.2 Amplification of spontaneous emission

The spectrum of spontaneous emission coincides with the shape of an amplification line. Thus, the spontaneous emission has maximal amplification at the centre of the line:

$$I(\nu) = I_0(\nu) e^{\alpha(\nu)L} \tag{6.66}$$

i.e. there is spectral narrowing as shown in Fig. 6.15a. Such a narrowing is determined by the parabolic shape of the top of the amplification line (Yariv and Leite, 1963):

$$\Delta \nu = \frac{\Delta \nu_0}{\sqrt{1 + \alpha_0 L}}, \tag{6.67}$$

where $L \leq L_{\text{lin}}$, and $\Delta \nu_0$ is the bandwidth of the spontaneous emission (amplification line). This expression is valid for the saturated amplification too, but αL must have a saturated value because of the reduction of α. Allen and Peters (1972b) have considered the saturated amplification regime in details.

The main contribution to the change of the spectrum of the amplified spontaneous emission is given by the distortion of the amplification line itself due to the saturation. It occurs in the case of inhomogeneous broadening, particularly for Doppler broadening of the amplification line, as shown in Fig. 6.2c. The top of the amplification line is flattened because of saturation by the minimum of the amplification ("hole burning") and, hence, the narrowing of the amplifying spectra transit to the "rebroadening"

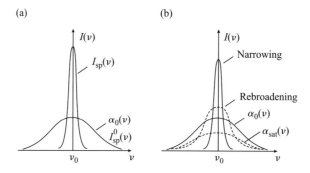

Fig. 6.15 Illustration of spectral narrowing of amplified spontaneous emission for a homogeneous broadened amplification line (a) and the rebroadening of the spectra in the case of the saturated (dashed line) inhomogeneous amplification line (b).

spectra of the amplifying spontaneous emission in the saturated regime. This is illustrated in Fig. 6.15b. Maeda and Yariv (1973) have observed this effect in laboratory experiments on the Xe amplifier tube at $\lambda = 3.51$ μm, and Casperson (1977) later published a more detailed analysis of it.

Astrophysical amplifying volumes can have various shapes (spherical, elliptical, wire-like, etc.). It means that the spectrum of amplified spontaneous emission depends on the geometry (divergence) of the emission. The equation of radiative transfer allows us to consider this effect. This task has been considered (Das Gupta and Das Gupta, 1991) for a plane-parallel cylindrical slab of an active amplifying medium with axial symmetry but not for an astrophysical laser with scattering. In principle, the equation of radiation transfer can be applied to astrophysical lasers.

7
Introduction to astrophysical lasers

Ideas of the nonequilibrium state of matter, which is out, of thermodynamical equilibrium with radiation, ideas of inverted population of quantum levels, and the possibility of an amplification (but not an oscillation!) were born in the final 30 years of the 20th century. Some of the considered schemes are surprisingly similar to the schemes for laser pumping, so they are worthy of an overview here.

7.1 Amplification under non-LTE conditions

Menzel (1969) discussed in a general form the possibility of inverted population and amplification of radiation in stellar atmospheres in non-LTE conditions. His analysis is based on the radiative transfer equation in a "microscopic" form, denoting the population of the upper (2) and lower (1) levels of the radiative transitions by N_2 and N_1, respectively,

$$\frac{dI(\nu)}{dr} = \frac{h\nu}{4\pi} N_2 A_{21} g(\nu) - \frac{h\nu}{4\pi} N_1 B_{12} \left(1 - \frac{N_2 g_1}{N_1 g_2}\right) g(\nu) I(\nu), \quad (7.1)$$

where $I(\nu)$ is the spectral intensity of radiation, $g(\nu)$ is the form-factor of the spectral line $\left(\int_0^\infty g(\nu) d\nu = 1\right)$, i.e. the mean intensity J, weighted over the line profile:

$$J = \int_0^\infty I(\nu) g(\nu) d\nu, \quad (7.2)$$

and g_1 and g_2 are the statistical weights of levels 1 and 2. The Einstein coefficients in Eq. (7.1) are: A_{21}, the transition probability for spontaneous emission from "2" to "1" per second per particle in state 2; $B_{12}J$, the transition probability for stimulated absorption (radiative excitation) from lower "1" to "2" per second per particle in state 1; and $B_{21}J$, the transition probability for stimulated emission from "2" to "1" per second per particle in state 2.

Under the condition of *radiative equilibrium* the number of radiative transitions per second per unit volume from state 1 is equal to the number of radiative transitions per second per unit volume into state 1, i.e.

$$N_1 B_{12} J = N_2 A_{21} + N_2 B_{21} J. \quad (7.3)$$

The Einstein coefficients are connected by the expressions:

$$g_1 B_{12} = g_2 B_{21} \quad \text{and} \quad A_{21} = \frac{2h\nu^3}{c^2} B_{21}. \tag{7.4}$$

The absorption coefficient $\alpha(\nu)$ is determined by

$$\alpha_\nu = \frac{h\nu_{21}}{4\pi} (N_1 B_{12} - N_2 B_{21}) g(\nu). \tag{7.5}$$

The Boltzmann relation for the relative populations, N_1 and N_2, of levels 1 and 2 with energies E_1 and E_2 is

$$\frac{N_2}{N_1} = \frac{g_2}{g_1} \exp\left(-\frac{E_2 - E_1}{kT}\right). \tag{7.6}$$

Menzel (1969) used a dimensionless parameter b (see Chapter 4) to indicate the degree of departure from thermodynamical equilibrium of a nebular gas at a temperature of T_e for the electron gas. This b-parameter (non-LTE coefficient) is equal to unity for thermodynamical equilibrium. In the general case, Eq. 7.6 can be written:

$$\frac{N_2}{N_1} = \left(\frac{b_2 g_2}{b_1 g_1}\right) \exp\left(-\frac{h\nu}{kT}\right), \quad h\nu = E_2 - E_1. \tag{7.7}$$

As long as $h\nu/kT$ is much greater than 1, we can usually neglect the term representing the stimulated emission in Eq. (7.1). If (b_2/b_1) is close to unity, i.e. for local thermodynamical equilibrium (LTE) the second term on the right side of Eq. (7.1) is approximately equal to the source function:

$$S_\nu = \frac{h\nu^3}{c^2} \left[\exp\left(\frac{h\nu}{kT}\right) - 1\right]^{-1}. \tag{7.8}$$

Menzel (1969) noted that in many cases (nebulae, highly extended stellar atmospheres) b tends to be high for the low metastable levels, i.e. the low excited levels of the same parity as the ground state tend to have b-values much less than unity. For transitions, for which $b_2/b_1 > 1$, the stimulated emission may exceed the absorption at high temperatures, i.e. α_ν determined by Eq. (7.5) will be negative. This case at non-LTE is referred to as *inverted population* and gives a net *negative absorption*.

7.2 Astrophysical predecessors of the laser

Astrophysics was a major driving force of atomic spectroscopy in the last century because astronomical spectra of stars, nebulae, and galaxies gave the main information on the structure and development of the Universe. Spectroscopy of stellar atmospheres was intensively studied (Unsöld, 1968) and modeled in parallel, with the development of computers. The emission of a star detected by an observer includes continuous Planck radiation of the photosphere, which propagates through the outer stellar atmosphere and line radiation from the enveloping regions. Therefore, depending on whether absorption or spontaneous emission dominates, the discrete absorption or emission lines of the stellar atmospheres are observed against the background of the continuous photospheric spectrum. If the continuous radiation is very intense, while the brightness of spontaneous emission of the atmosphere is low, discrete absorption lines are observed. This is the typical case for most stars, even though a discrete line spectrum can be observed in wavelength regions outside the Planck curve of continuous radiation.

However, spectra of many stars contain bright emission lines (Merrill, 1956). This is possible if the emission line intensity is comparable with the intensity of the continuous photospheric spectrum at the corresponding line wavelength. For this reason, bright lines can only be observed in spectra of extended objects or in wavelength regions where the photospheric emission is comparatively weak. Bright emission lines are very often observed in spectra of variable stars and binary stars as well as nebular regions.

Three excitation mechanisms are mainly involved in the explanation of the occurrence of bright emission lines: collisional excitation, excitation by recombination, and the Bowen mechanism (Bowen, 1934, 1935). The last of these is the first example of the coincident excitation process discussed in Chapter 9, which explains the presence of anomalously bright lines as fluorescence. These are generated by photoexcitation by accidental resonance (PAR), i.e. a wavelength coincidence of a strong emission line of one element with an absorption line of another element. The most famous case is the He II Ly α excitation of O III (Bowen, 1934), but also other cases were reported early on (Thackeray, 1935). The authors of these papers came up with ideas that are in fact the early predecessors of the concept of obtaining an amplifying laser medium by producing an inverted population of the energy levels.

It was known early on that emission lines of neutral oxygen, O I, at 8446 Å ($3p^3P^\circ \rightarrow 3s^3S^\circ$) and 7774 Å ($3p^5P^\circ \rightarrow 3s^5S^\circ$) show an anomalous behavior in spectra of Be stars (Sletteback, 1951; Merill, 1956). The 8446/7774 intensity ratio is much greater than the laboratory value. Bowen (1947) assumed that this could be explained by fluorescence in the 8446 Å line caused by a wavelength coincidence between the bright H Lyβ line (1025.72 Å) and the absorption line of O I at 1025.77 Å ($2p^3P^\circ \rightarrow 3d^3D^\circ$). A partial energy level diagram of O I showing these transitions is presented in Fig. 7.1a.

The anomalous ratio of the line intensities (8446 Å and 7774 Å) could later be explained by stimulated emission due to the population inversion for the 8446 Å transition. Since the lower $3s^3S^\circ$ level of the transition has a shorter lifetime than the upper $3p^3P^\circ$ levels, the inversion can be achieved by moderate CW pumping of the upper levels (Letokhov, 1972a; Lavrinovich and Letokhov, 1974). The 8446 Å line is further discussed in connection with laser action in astrophysical plasmas in Chapter 12.

An example of a similar intensity anomaly concerns the Mg II lines $\lambda = 3848$ Å and $\lambda = 3850$ Å of the multiplet, $5p^2P^\circ \rightarrow 3d^2D$ (Fig. 7.1b). The first line can sometimes be much brighter than the second one (Merill, 1951, 1956). Attempts were made to explain this anomaly by the wavelength coincidence between the Mg II 1025.87 Å ($3s^2S \rightarrow 5p^2P_{3/2}$) absorption line and H Lyβ at 1025.77 Å. However, the absence of an analogous anomaly for another line at 3614 Å ($5p^2P^\circ \rightarrow 4s^2S$) from the same upper level casts doubt on this mechanism (Merill, 1956).

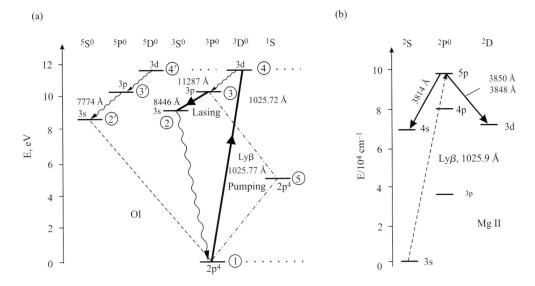

Fig. 7.1 Astrophysical schemes of pumping and fluorescence of levels in O I (a) and Mg II (b) by H Lyβ emission explaining the appearance of bright lines at 8446 Å (O I) and 3848 Å (Mg II).

The intensity anomalies in stellar spectra discussed in the 1950s were mostly observed for long-period variable stars of late spectral classes (M and S stars). Certain multiplets were observed in emission, for which the relative intensities of the lines differed from their laboratory values (Merill, 1952). One example is the anomalous behavior of the $\lambda = 4101$ Å and $\lambda = 4511$ Å resonance lines of In I corresponding to $5p^2P \to 6s^2S$ transition (Fig. 7.2a). The 4511 Å line was predominantly observed in emission rather than in absorption, which could be explained by photoexcitation of the upper ^2S level (marked "3" in Fig. 7.2) owing to the wavelength coincidence between the resonance absorption line of In I at 4101.72 Å and strong emission in the Balmer Hδ line at 4101.75 Å.

Figure 7.2a shows the scheme of transitions between the three lowest levels of In I, $5p^2P_{1/2}$ (1), $5p^2P_{3/2}$ (2) and $6s^2S_{1/2}$ (3), involved in the $3 \to 2$ fluorescence caused by Hδ pumping in the $1 \to 3$ channel. This was discussed by Thackeray as early as 1935 (Thackeray, 1935). The condition for the appearance of an emission rather than an absorption line in the $3 \to 2$ transition as found by Thackeray proved to be coincident with the population inversion condition. Let us assume that radiation of intensity I_{13} pumps the resonance channel and that radiation of intensity I_{23} is emitted in the anomalous $3 \to 2$ transition. Since the $2 \to 1$ transition is parity forbidden with an extremely low probability, a cyclic loop involving the states 1, 2, and 3 is "impossible". Since the In atom has a low ionization potential (5.76 eV), it is reasonable to include ionized atoms in state 4 into the scheme. Regarding the conditions

$$P_{14} = P_{24}, \quad P_{42} = 2P_{41}, \quad P_{32} = 2P_{31}, \tag{7.9}$$

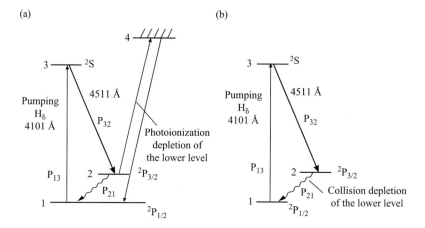

Fig. 7.2 Schemes of cyclic transitions in In I considered by Thackeray (1935) for explaining the bright emission line at 4511 Å observed owing to excitation by the 4101Å Hδ line: (a) photoionization depletion and (b) collisional depletion of the metastable $^2P_{3/2}$ level (2).

for the various transition rates, taking into account the degeneracy of the levels and the fact that levels 1 and 2 belong to the same term, Thackeray (1935) obtained the following expression for the steady-state level populations n_i:

$$N_1 \frac{P_{23}}{P_{23}+P_{24}} = N_2 \frac{P_{23}}{2(P_{23}+P_{24})} = N_3 P_{32} \left[2P_{13} + \frac{4}{3} P_{24} + \frac{2}{3} P_{13} \frac{P_{24}}{P_{23}} \right]^{-1}. \quad (7.10)$$

Therefore, the ratio of the number of emission $(3 \to 2)$ and absorption $(2 \to 3)$ events at the frequency ν_{23} is (Thackeray, 1935):

$$\gamma = \frac{N_3 P_{32}}{N_2 P_{23}} = 1 + \frac{P_{24}(P_{13}-P_{23})}{3(P_{13}+P_{24})}. \quad (7.11)$$

For $\gamma > 1$, the cyclic $1 \to 3 \to 2 \to 4 \to 1$ transition occurs, resulting in the predominant emission at the $3 \to 2$ transition.

Note that the condition of the cyclic transition $\gamma > 1$ in the $3 \to 2$ direction is simultaneously the population inversion condition for levels 2 and 3:

$$\gamma = \frac{N_3 P_{32}}{N_2 P_{23}} = \frac{N_3 \sigma_{32}}{N_2 \sigma_{23}} = \frac{N_3}{g_3} \bigg/ \frac{N_2}{g_2} > 1, \quad (7.12)$$

where σ_{nm} is the cross section for the radiative $n \to m$ transition. Therefore, the mechanism of amplification of radiation due to stimulated emission has long been used implicitly in the theory of stellar spectra for the description of the so-called emission lines. The PAR mechanism of excitation (Fig. 7.1) is in fact the astrophysical analogue of a three-level scheme with optical pumping. Astrophysicists have proposed the schemes to explain anomalies in stellar spectra, which anticipated a number of laser schemes. For example, Fig. 7.2b shows another scheme of transitions between the lower levels of In I, which was also proposed by Thackeray (1935) to explain the 4511 Å emission line. The cyclic $1 \to 3 \to 2 \to 1$ transition in this scheme and, hence, the population inversion for states 3 and 2 is achieved by fast collisional relaxation of the metastable level 2. This scheme is very close to the scheme of a collisional laser proposed by Gould (1965).

Stimulated emission lines should be narrower than spontaneous emission lines. The narrowing of the spectrum can be quite sufficient for the unambiguous determination of the origin of the emission line. This effect should be considered in more detail taking into account the existence of a large optical thickness for the amplifying (pumped) and pumping transitions and resonance scattering of radiation. In principle, two operating regimes are possible: the amplification of the photospheric emission in a stellar atmosphere and the amplification of the spontaneous line emission of the outer atmosphere. In the first case, we deal with a narrow-band nonlinear laser amplifier of an external broadband signal, and in the second case, with a resonance laser amplifier of intrinsic spontaneous noise. In the latter case, the laser amplifier can be transformed to an oscillator owing to an inevitable scattering of radiation if the rate with which the radiation returns to an amplifying region due to scattering exceeds the rate with which radiation escapes to the environment. Then, resonance scattering results in the appearance of noncoherent energy feedback into the amplifying medium (Letokhov, 1967a). This effect should divide the amplifying region into many separate regions, with lasing thresholds achieved in each of them. The effect of noncoherent feedback in the astrophysical laser is considered in more detail by Lavrinovich and Letokhov (1974) and Letokhov (1996) (see Chapter 14).

7.3 How is laser action manifested under astrophysical conditions?

The characteristic feature of laser action in laboratory lasers is a high brightness of emission, which is also inherent in astrophysical masers operating on microwave transitions in molecules. Because the brightness temperature of astrophysical masers is extremely high (10^{12}–10^{15} K), one can make an erroneous conclusion about "laser action" in the optical range from an anomalously high line intensity and an anomalous intensity ratio. This wrong conclusion is based primarily on misconceiving the fundamental difference between stimulated emission in a laboratory laser and an astrophysical laser. Stimulated emission in a laboratory laser (as distinct from spontaneous emission, Fig. 7.3) takes place in a very limited number of spatial modes of the cavity, i.e. in a very narrow solid angle. In an astrophysical laser (amplifier) both types of emission (stimulated and spontaneous) occur in all spatial modes, i.e. in a solid angle of about 4π steradians. Secondly, in the optical range the rate of spontaneous emission in allowed transitions is very high (10^8–10^9 s^{-1}) and comparable with the rates of stimulated absorption and emission, whereas in the microwave range the rate of spontaneous emission is negligible. Thus, microwave emission becomes only observable through stimulated emission.

This misconception leads to a non-optimal strategy in the search for laser action in the optical wavelength range. It is worth considering ways of detecting optical

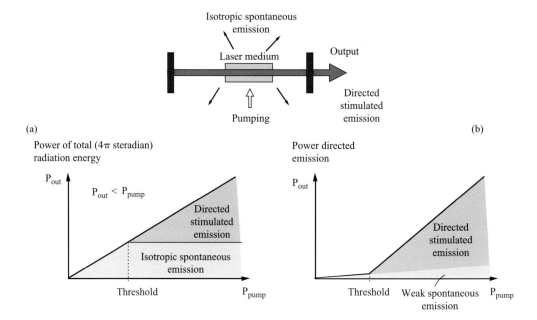

Fig. 7.3 Spontaneous and stimulated laser emission, integrated over the direction of the total emitted power (a) and over the power of directed radiation (b).

astrophysical lasers by specifically having the importance of this misconception in mind. For this purpose let us consider various possible ways to manifest laser action by investigating various properties of optical spectral lines: integrated (total) intensity, spectral width, and divergence of radiation.

7.3.1 Integral intensity of a spectral line

The active medium of a laser emits a large portion of energy, both below and above the threshold, isotropically and spontaneously in all spatial modes. Even well above the threshold, the fraction of isotropic spontaneous emission remains comparable with that of stimulated emission (Fig. 7.3a). Above the lasing threshold, the upper-level population remains constant at the threshold level (continuous regime), while the total above-threshold pump energy is transformed to stimulated emission. This emission is, however, emitted in a limited (especially over the angle) number of modes. For this reason, the intensity of the emission directed along the laser beam greatly exceeds that of the spontaneous emission within this small solid angle (Fig. 7.3b). Therefore, the high intensity of laser emission is explained by its low divergence and a small observation angle (only along the direction of the laser beam). Under astrophysical conditions, the situation is completely different. Even an anisotropic stimulated radiation of an APL is emitted within a large solid angle and should therefore not be anomalously bright against the background of spontaneous emission in allowed transitions.

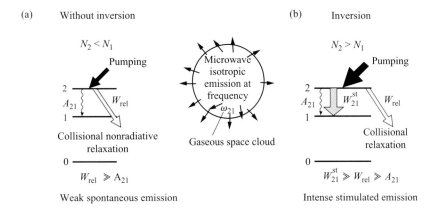

Fig. 7.4 Isotropic microwave emission of a gaseous space cloud without inversion (a) and with inversion in the $2 \to 1$ transition (b). The pump power in the absence of inversion is spent in the unobservable relaxation channel. In the presence of inversion, the pump power is transformed to the observed intense stimulated microwave emission.

This fact brings up the question: Why is the emission of an astrophysical maser so anomalously bright? The reason is completely different from the laser case and is usually not discussed in the literature. Figure 7.4 shows schematically the origin of microwave emission from a gaseous space cloud without and with inversion at the transition frequency ω_{21}. In the absence of population inversion (Fig. 7.4a), a very weak spontaneous microwave emission is observed because of an extremely small Einstein coefficient in the microwave region ($A_{21} = 10^{-7}$–10^{-9} s^{-1}), corresponding to a spontaneous radiative lifetime of 0.3–30 years (!). Obviously, collisions occur more often even in a rarefied stellar medium. This means that even the weak spontaneous emission is suppressed by collisions, and a greater part of the pump energy degrades over a non-radiative channel, i.e., an observer does not detect it (Fig. 7.4a). If the pump is intense enough to provide an inverted population in the $2 \to 1$ transition (Fig. 7.4b), and the size L of the gaseous cloud provides a gain factor $\alpha L \gg 1$ yielding a considerable gain ($K = \exp \alpha L \simeq 10^{10}$–$10^{15}$), stimulated emission develops. Since the rate W_{21}^{st} of stimulated transitions can become comparable to the rate of collisional (unobservable) relaxation and even exceed it, almost all the pump energy converts to observable stimulated emission, which can have an enormous brightness temperature. In other words, because of the maser effect, an extremely weak spontaneous emission line, which is not observed in radio astronomy, is transformed to a strong stimulated emission line acquiring energy from the unobserved channel (Fig. 7.4b). The astrophysical maser has thus a high brightness because it is not related to pumping by any microwave radiation.

There is a completely different situation with the astrophysical laser in the optical spectral region (Fig. 7.5). Let us assume that level 3 is photo excited by a spectral line of another element (directly or via a high-lying state relaxing to state 3). In the absence of population inversion (Fig. 7.5a) in the system of three levels (1, 2, 3), we

110 *Astrophysical lasers*

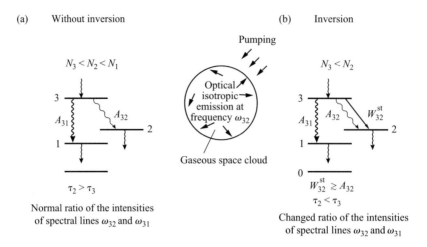

Fig. 7.5 The spontaneous isotropic emission in a medium without inversion of population (a), and spontaneous and stimulated isotropic emission in a medium with inversion of populations (b).

observe, for example, a pair of spontaneous emission lines, with an intensity ratio equal to the ratio of the Einstein coefficients A_{32} and A_{31}. We assume for clarity that the $3 \rightarrow 2$ transition in Fig. 7.5 is very weak ($A_{32} \ll A_{31}$) and, hence, almost all energy of particles excited to level 3 relaxes by the $3 \rightarrow 1$ spontaneous emission at the frequency ω_{31}.

Assume now that the spontaneous decay time τ_2 of level 2, to which particles from level 3 undergo transitions comparatively rarely, is much shorter than for level 3, so that the population inversion always exists for the $3 \rightarrow 2$ transition (Fig. 7.5b). When the size of the amplifying region is sufficiently large the gain in the $3 \rightarrow 2$ transition may exceed the ratio of the spontaneous decay rates A_{31}/A_{32}. In this case, the stimulated transition rate W_{21}^{st} can become not only equal to the spontaneous decay rate A_{31} but can also exceed it. As a result, the stimulated emission at the frequency ω_{32} has became the main relaxation channel of particles in level 3. In this case, the intensity of the weak $3 \rightarrow 2$ line or rate of stimulated radiative transition increases and approaches the pumping rate of level 3 (but does not exceed it).

It is appropriate at this point to distinguish between two different ways to pump level 3: (1) collisional excitation by electrons or recombination of ions and electrons (Chapter 8); (2) photo-excitation by an intense line of spontaneous emission from another element (the PAR process further considered in Chapter 9).

In the first case the laser line ($3 \rightarrow 2$) of stimulated emission can obviously not be brighter (in photon/cm$^2 \cdot$ sec) than a spontaneous line exciting level 3. For this reason, the criterion that a laser line in an allowed transition should have an anomalous intensity compared to bright pumping lines cannot be used in the search for laser action. This circumstance differs substantially from the case of masers, where the pumping source has a different nature (not microwave emission) and its intensity can, therefore, be very high. The conversion of the excitation energy to stimulated microwave emission makes the maser anomalously bright. It is clear from this simple qualitative reasoning that laser action should be manifested in an increasing intensity of *very weak spontaneous emission lines* (i.e., the transitions with small Einstein coefficients) up to the intensities of spontaneous bright emission lines of allowed transitions. For this reason, the search for laser action using a criterion for the integrated intensity of allowed spectral lines presents a rather difficult task. The search for laser effect in forbidden transitions with usually weak lines is much more successful since in the case of strong laser action these spectral lines might be as intense as allowed spectral lines.

The same arguments are valid for the case of collisional pumping of level 3 and population inversion in the $3 \to 2$ transition. The difference between collisional pumping and optical pumping is the limitation of the intensity of the laser line in the $3 \to 2$ transition, which is determined by the rate of collisional excitation of level 3. The fast $3 \to 2$ decay in an allowed transitions will provide a high intensity of spontaneous emission, which might mask the contribution from stimulated emission. However, the situation can be much more favorable in the case of forbidden transitions with inverted population ($N_3 > N_2$) and a sufficient gain coefficient ($\alpha_{32} L \gg 1$). The case of an intermediate rate τ_{coll}^{-1} of collisional deexcitation of level 3:

$$\tau_{\text{allow}} \ll \tau_{\text{coll}} \ll \tau_{\text{forb}}, \qquad (7.13)$$

where τ_{allow}, τ_{forb} are the decay times of allowed and forbidden transitions, respectively, is particularly favorable for detection of a laser effect. In this case the brightness of a weak spectral line can be very high owing to conversion of non-radiative deexcitation to stimulated emission just as it occurs in microwave masers (Section 7.3).

This simple qualitative consideration demonstrates that the search for laser action in the optical range must be oriented towards the observation of weak forbidden spectral lines. The search for maser action in the microwave range is more straightforward because of the absence of a significant spontaneous emission. This discussion answers to some degree the question raised by Strelnitskii (2002): why are natural masers (APM) so widespread, but natural lasers (APL) are not? Our answer looks at a first glance very similar to his explanation based on a low intensity of saturation of optical spectral transitions and a corresponding limitation of the exponential growth of intensity as compared with microwave ones.

The intensity for saturating the $n \to m$ transition is determined by expression (6.9), where $\tau_{\text{rel}} = \tau_2$ is the relaxation time of population inversion determined by collisions and radiative decay. σ_{mn} is the cross-section (cm^2) of stimulated (radiative) transitions $m \to n$, determined by expression (6.5). The Einstein coefficient A_{mn} for spontaneous radiative decay in the $m \to n$ transition (λ_{mn} or ω_{mn}) is defined by (2.32). The apparent decrease in I_{sat} at short wavelengths disappears because of an increase in A_{mn} and finally we get a simple universal expression for σ_{mn}:

$$\sigma_{\text{mn}} = 8\pi^3 \left(\frac{\omega_{\text{mn}}}{\Delta \omega_{\text{mn}}} \right) \frac{|d_{\text{mn}}|^2}{\hbar c}, \qquad (7.14)$$

where $(\Delta\omega_{mn}/\omega_{mn})$ is the relative spectral width (usually the Doppler width). The value of σ_{mn} is of more or less the same order of magnitude in the visible, IR and microwave ranges, when the spectral line width is determined by Doppler broadening. Thus, the difference of saturation intensity (in photons/cm^2·s) for visible and microwave transitions depends on the difference in relaxation times τ_{rel} of the population inversion, which does not crucially depends on wavelength.

Thus the rareness of observation of laser action in the optical range is explained most probably by the high probability of spontaneous emission, which hampers a formation of population inversion and particularly the possibility of observing a laser effect in allowed transitions.

7.3.2 Width of a spectral line

The most convincing "signature" of laser action with strong amplification in astrophysical media without cavities, i.e. an astrophysical laser amplifier of spontaneous emission, is the narrowing of the spectral line. The spectral narrowing can take place even without an enhancement of the integrated intensity of the spectral line. This problem has been studied theoretically for APMs (Litvak, 1970; Allen and Peters, 1972a,b) and has been confirmed in many observations (see, Elitzur, 1992), thanks to very high spectral resolution in radio astronomy. The APL case was considered by Lavrinovich and Letokhov (1974, 1976). The effect of spectral narrowing depends on the divergence of the stimulated emission, as the divergence of radiation affects the character of resonant interaction of radiation with the Doppler-broadened spectral line (Letokhov, 1996). This problem is considered in Chapter 13.

Unfortunately, the spectral resolution in optical astronomy today does not allow the direct measurement of the "true" spectral width of radiation or the shape of a spectral line. Thus, the method of the heterodyne intensity correlation spectroscopy with local laser heterodyne and spatially separated optical telescopes, called a Brown–Twiss–Townes heterodyne spectrometer (Johnson et al., 1974), has been proposed for the measurements of the spectral line width of APLs at sub-Doppler spectral resolution (Lavrinovich and Letokhov, 1976; Johansson and Letokhov, 2005a). The possibility of such future experiments is further discussed in Chapter 13. Thus, the most convincing and unambiguous proofs of laser action in the optical range might only be possible to obtain in the future.

7.3.3 Divergence of radiation

The most obvious evidence of laser action in a laboratory laser with a cavity is the small divergence of the stimulated emission against the background of the isotropic spontaneous radiation (Fig. 7.3 and Fig. 7.6a). It was the most convincing demonstration of laser effect in the first ruby laser (see Maiman, 2000). The natural APL without a cavity may operate in two possible regimes: (1) a regime of amplification of spontaneous emission due to stimulated radiative transitions and (2) a regime of generation (oscillation) occurring due to backscattering of radiation to the amplifying medium above the threshold, which is called a "laser with nonresonant (noncoherent) feedback" (Letokhov, 1967a). The divergence of the stimulated radiation in the first

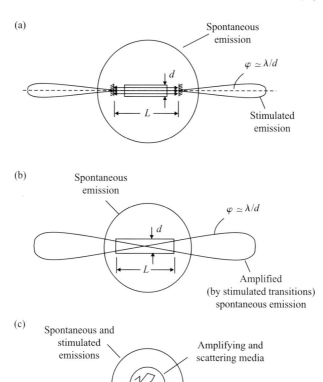

Fig. 7.6 Divergence of stimulated and spontaneous emission of three different laser sources: (*a*) laboratory laser with an open resonant cavity, which has directed stimulated emission with diffractional divergence; (*b*) laser amplifier of spontaneous emission (laboratory or natural), which has a geometrical divergence of stimulated spontaneous emission; (*c*) laser oscillator with nonresonant (noncoherent) scattering feedback, which has isotropic spontaneous and stimulated emission.

regime is determined by the geometry of the amplifying medium (Fig. 7.6*b*), i.e. by the ratio between the diameter d and the length L of the active medium. In this second regime the stimulated radiation as well as the spontaneous radiation is isotropic (Fig. 7.6*c*).

The possibility of finding a quantum "signature" of stimulated emission has been discussed by Dravins (2001). However, the quantum properties of amplified spontaneous emission and spontaneous emission are not distinguishable (Letokhov, 1967*b*). Firstly, the distinction occurs only when stimulated emission of very many identical photons takes place in a very limited number of spatial modes of the cavity. This causes a high occupation number $\langle n \rangle$ of these modes and results in a coherence of the stimulated radiation (Glauber, 1963*a*, *b*). However, this is not the case with APLs.

Secondly, the distinction can appear owing to the narrowing of spectral lines in amplified spontaneous emission. In this case the statistical properties of stimulated radiation are manifested in the form of an increased time correlation of the intensity fluctuation, which was exploited in the Brown–Twiss intensity correlation interferometry (Hanbury Brown, 1974). As mentioned above, this approach has been suggested as a diagnostic tool for APLs (Lavrinovich and Letokhov, 1976; Johansson and Letokhov, 2005a), and will be further considered in Chapter 13.

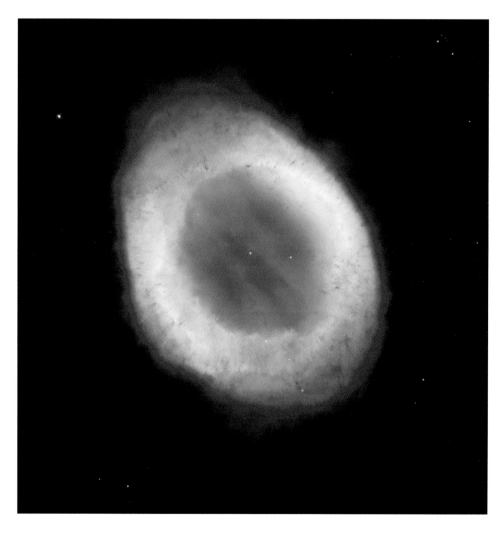

Plate 0.1 Picture of M57 Ring Nebula taken by Hubble Space Telescope (HST) (Credit HST team and NASA).

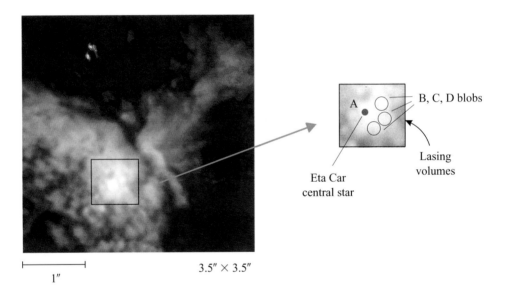

Plate 0.2 Eta Carinae's near surrounding with compact blobs B, C, D (reprinted from van Boekel *et al.*, 2003 with courtesy and permission Elsevier).

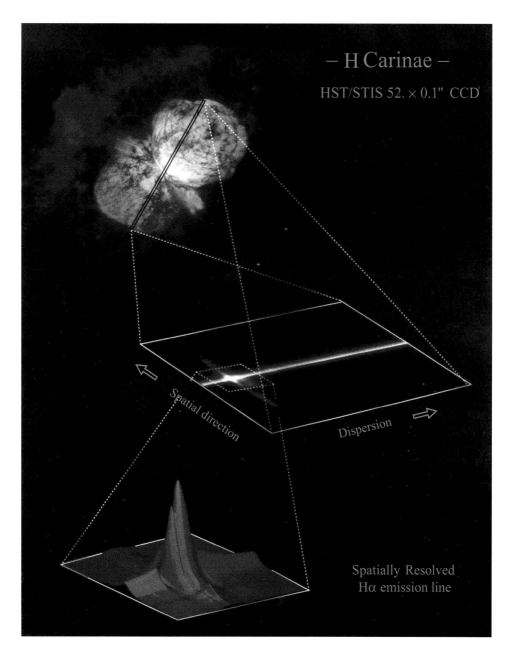

Plate 0.3 General view of a two-dimensional (angle, wavelength) observation along the slit (upper image), which is passing through the central star of Eta Carinae and (bright strip on the middle image). The blobs are resolved from the star by the STIS camera on board of HST. The wavelength range is chosen to include the strong hydrogen Balmer α line (from Ishibashi with courtesy and permission).

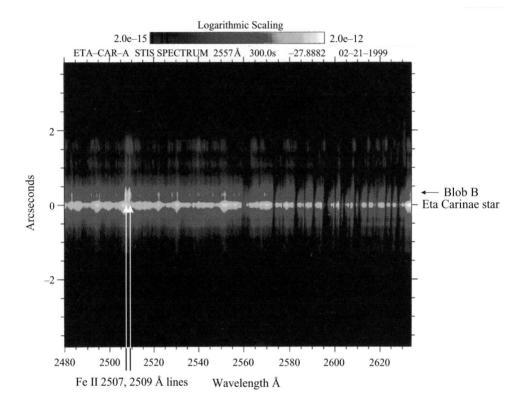

Plate 0.4 Extract from the high resolution UV spectrum of the massive star eta Carinae recorded with the space telescope imaging spectrograph (STIS) onboard the Hubble Space Telescope. The vertical scale gives the distance (in arcseconds) from the central star (located at 0) to the Weigelt blobs and other gaseous condensations of matter expelled from the star. Of special interest is Weigelt blob B at about 0.2" "above" the star, showing a distinct emission line spectrum on a very weak continuous background. Two extraordinary features around 2507.5 and 2509.1 Å (vacuum rest wavelengths) in the blob spectrum required a specific excitation mechanism to be explained.

8
Basics of collisionally pumped astrophysical lasers

The present book is focused on astrophysical lasers and nonlinear optical effects under astrophysical conditions. So far we have only explained how these effects arise due to resonant photoexcitation. Nevertheless, for completeness it is relevant to also discuss collisionally pumped mid-IR and submillimeter masers/lasers, which form a connecting link between microwave masers and optical lasers. Collisional excitation of atoms (ions) and molecules is the typical excitation mechanism in laboratory lasers of various types. In some lasers (recombination plasma lasers) electron collisions ionize the atoms, which may form a population inversion after recombination. Some lasers (CO_2–N_2–He gas-lasers) benefit from resonant energy transfer from the collisionally excited particles (N_2 molecules) to the lasing particles (CO_2 molecules). These mechanisms occur also in a long-wavelength astrophysical laser, which operates in the continuous-wave (steady-state on an atomic/molecular timescale) regime. Therefore, the formation of inverted population in astrophysical lasers (APLs) is possible according to one of the "golden rules" of quantum electronics: the lifetime of the lower level of a lasing transition should be shorter or much shorter than the lifetime of the upper level independent of the excitation mechanism.

The source of pumping energy in astrophysical lasers is radiation energy from the atmosphere of a central star, which irradiates the surrounding gas media (clouds, blobs, shells, disks, etc.) Photoionization of these media provides the storage of free energy of ions and electrons, the heating and turbulence in the medium, the recombination of ions and electrons etc., i.e. non-LTE conditions. A number of theoretical predictions of non-LTE situations with population inversions in atoms (ions) and molecules were made in the 1970s and 1980s, and they are reviewed very briefly in Section 8.1. The observations of mid-IR and submillimeter lasers were successful only for atomic hydrogen in MWC 349 (a hydrogen recombination laser) and CO_2 in the Martian and Venusian atmospheres (Sections 8.2 and 8.3).

8.1 Proposals of population inversion by collisional pumping

Elementary processes of excitation of atoms and ions were considered briefly in Chapter 4. All these processes operate in both laboratory and astrophysical lasers. Particularly these processes have been useful for proposals of population inversion by collisions. Let us summarize the numerous theoretical predictions for ions, helium, and the OH radical.

8.1.1 Fine-structure levels of ions

One of the early proposals to extend the astrophysical maser (APM) effect into a short-wavelength region was done by Smith (1969). He considered the population inversion for fine-structure levels of excited states of ions of several low ionization states obtained under specific astrophysical conditions (typical values of temperature and density in planetary nebulae) and assuming excitation by inelastic electron collisions and the ionization balance determined by the Saha equation. He examined the situation where excitation collisions from the ground state to higher levels are important, but where all other collisional rates are negligible compared with radiative excitation and deexcitation rates (Fig. 8.1a). Species with a simple atomic structure, such as the ground configurations $2p^q$ and $3p^q$ ($q = 2, 3, 4$) with long-lived metastable states, were considered. The population of the levels in the $2p^q$ and $3p^q$ configurations were studied with rate equations for ionization equilibrium in a wide range of electron densities $n_e = 10 - 10^{10}$ cm^{-3} and electron temperature $T_e \simeq 2 \cdot 10^4$ K. It was found (Smith (1969) that the typical gain coefficient is $\alpha \simeq 10^{-5}$/light year ($\approx 10^{-23}$ cm^{-1}) at an electron density of about 10^6 cm^{-3}. Only three forbidden transitions ([O III] $^3P_1 \to {}^3P_0$ at 88.2 μm; [O III] $^3P_2 \to {}^3P_1$ at 51.7 μm ($2p^2$ configuration); [S III] $^3P_1 \to {}^3P_0$ at 33.6 μm ($3p^2$ configuration) have a predicted gain length $1/\alpha$ of the order of light years. These values exceed significantly the absorption coefficient ($\sim 10^{-24}$ cm^{-1}) for free-free electron transitions.

Fig. 8.1b shows the scheme of the $2p^2$ fine-structure levels for O III and some forbidden radiative transitions between them (Smith, 1969). In this structure the larger decay rates for the $4 \to 3$ and $2 \to 1$ transitions tend to cause inverted populations between levels $3 \to 2, 3 \to 1$, and $2 \to 1$ at electron densities $n_e \approx 10^5 - 10^{6.5}$ cm^{-3}. Fig. 8.2 presents the results of the calculations of inverted population between the

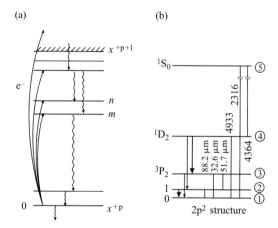

Fig. 8.1 Collisional excitation by electrons, and population inversion of fine-structure levels of ions: (a) a general model of the excitation of x^{+p} ions in an astrophysical plasma; (b) example of inversion in the $3 \to 2$ (51.7 μm), $2 \to 1$ (88.2 μm), and $3 \to 1$ (32.6 μm) transitions of the $2p^2$ configuration of O III (adapted from Smith, 1969).

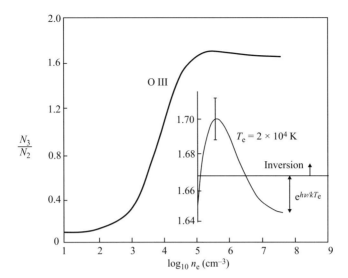

Fig. 8.2 Relative population behavior of $^3P_2$ and $^3P_1$ (O III 2p^2) in Fig. 8.1b as a function of electron density for $T_e = 2 \cdot 10^4$ K. The inset enhances the deviation from monotonic behavior. The error bars in the inset were calculated for a correlated 20% change in the collision strengths. O III ($^3P_2 \rightarrow \,^3P_1$) is not inverted at $T_e = 2 \cdot 10^4$ K (from Smith, 1969).

$^3P_2$ and $^3P_1$ levels in the ground term of O III for $T_e = 2 \cdot 10^4$ K in a wide range of electron densities. At $T_e = 10^4$ K the population inversion does not exist. Thus, this case of collisional excitation provides the formation of inverted population only in a relatively narrow range of plasma parameters (T_e, n_e, etc.).

A later detailed-balance analysis was presented by Greenhouse *et al.* (1993), who showed that population inversion exists also within the ground term and first excited terms of multiply charged atoms. The authors considered the possibility of laser radiation in infrared coronal lines because of high (>100 eV) ionization stages and higher densities in a collisional ionization balance. Inversion of population was predicted for many coronal lines of multicharged ions (from N V to Fe III) in the spectral range from 2.2 to 25 µm for electron densities of $n_e \approx 10^6$–10^9 cm^{-3} and temperatures $T_e = 10^{5.5}$–$10^{6.5}$ K. The results from these calculations of laser-gain lengths and column densities for infrared coronal lines are presented by Greenhouse *et al.* (1993).

Electron collisions in a high-temperature astrophysical plasma can provide the formation of inverted population and yield gain in many transitions of various ions. For example, it was shown (Ferland, 1993) that the [Fe VI] 6.08 µm line must have a gain at low ($n_e \ll 10^{10}$ cm^{-3}) densities. The author underlines that this is the only line of approximately 500 ionic lines incorporated in CLOUDY (Ferland, 1993), which has undergone strong maser effects during routine testing of the code.

The most promising sources for far-IR maser observations are those *highly non-thermal* ones with large fluxes in the infrared, such as quasi-stellar objects (QSO), Seyfert nuclei, AGN, and perhaps our galactic nucleus. Anyway, observation of maser action in these transitions will be difficult because of the small gain. Nevertheless, the observation of IR laser radiation in coronal lines in AGN can provide valuable new information about the structure and evolution of Seyfert nuclei (Smith, 1969; Greenhouse et al., 1993, 1996).

8.1.2 He II and He I in stellar envelopes

A similar method of computer calculations of rate equations for the level populations in collisional radiative equilibrium was used to search for inverted populations in transitions of He II (Varshni and Lam, 1976) and He I (Varshni and Nasser, 1986). The model accounts for electron excitation and deexcitation of levels, ionization and recombination with electrons, and radiative decay. Compared to Smith (1969) the authors considered even simpler atomic structures and population inversion in allowed transitions in the optical range: He II (4686 Å, $n = 4 \to n = 3$ transition) and He I (7281 Å, 3^1S-2^1P^o; 6678 Å, 3^1D-2^1P^o; 4921 Å, 4^1D-2^1P^o transition). The calculations led to a small density of population inversion $\Delta N \simeq 10^4$–10^5 cm^{-3} for helium densities of $N(He) \simeq 10^{13}$–10^{15} cm^{-3} and significant electron densities $N_0 \simeq 10^{14}$–10^{15} cm^{-3} at $T_e \simeq 10^4$–10^6 K. It corresponds to a large amplification of αL even for small lengths $L \approx 10^9$–10^{10} cm. However, the ranges of the parameter values for a positive gain are relatively narrow.

Varshni and coworkers suggested that such parameter values are achievable in surface layers of early type stars, particularly for high-temperature stars with a high mass-loss rate. Fast expansion of an astrophysical plasma (e.g. the extended atmosphere of a Wolf-Rayet star) should lead to fast recombination cooling and a corresponding non-LTE situation. Such schemes have been suggested for laboratory plasma recombination lasers (Gudzenko and Shelepin, 1963, 1965).

Observation of lasing in these transitions will be very difficult because of the high intensity of spontaneous emission in these permitted lines. The stimulated emission of radiation can yield some intensity enhancement as a result of the competition between the transition rate of stimulated emission and the rate of collisional (nonradiative) deexcitation (relaxation) of the inverted population. A significant laser effect can be expected in forbidden lines or in recombination transitions between highly excited states in the far-IR range (Section 8.2).

8.1.3 OH radical and H$_2$O molecule in star-forming regions

Modeling of OH maser emission associated with star forming regions (Gray et al., 1991, 1992) was extended to far-IR transitions of the OH radical (Doel et al., 1993). The authors predict population inversion and gain coefficients of the order of 10^{14} cm^{-1} at H$_2$ density up to $5 \cdot 10^8$ cm^{-3} for kinetic and local dust temperatures up to 125 K. The wavelengths of inverted far-IR transitions are 53.25 µm, 115.14 µm, 138.83 µm, and others. The requirement for amplification lengths of a few 10^{14} cm in comparatively dense gas suggests that such far-IR OH lasers will be pointers to the densest and most massive regions of shock-compressed gas (Doel et al., 1993).

Strong IR emission from gas phase water has been observed by the Infrared Space Observatory (ISO) in the star forming the Beclin-Neugebauer source in Orion in the wavelengths range 6–8 µm (Gonzalez-Alfonso et al., 1998). The central star in Beclin-Neugebauer object must have a Lyman continuum emission rate equivalent to that of a main sequence B0.5 star if the H II gas is photoionized (Fig. 5.3). It provides a basis for considering the possibility of laser action in vibrational transitions of the water molecule in astrophysical conditions, bearing in mind that the infrared water recombination lasers operated successfully in supersonic plasma expansion in the laboratory (Michael *et al.*, 2001). An ion-electron dissociative recombination mechanism proposed by these authors can be relevant for generating population inversion in water molecules in the astrophysical environment.

8.2 Hydrogen recombination far-IR laser in MWC 349

The hydrogen atom is a very promising quantum system for masing/lasing in an astrophysical plasma. Firstly, hydrogen is the most abundant element in Universe. Secondly, hydrogen in the vicinity of a normal star is photoionized, i.e. it has a significant amount of free energy acquired from the star. Thirdly, the recombination population of highly excited levels and their fast radiative decay have the inherent properties that contribute to the formation of population inversion, as shown in Fig. 8.3. Calculations have predicted population inversion in IR and far-IR transitions of hydrogen in H II regions (Seaton, 1959, 1964; Goldberg, 1966; Brocklehurst, 1970) with the possibility of high gain (Krolik and McKee, 1978). The lifetime τ_n of a highly excited state with principal quantum number n is proportional to n^3, i.e. lower levels decay more quickly than upper levels. Thus, it is the ideal situation for lasing in the weak transition $n+1 \to n$ ($n \gg 1$) in comparison with strong transitions according to discussions of the manifestation of the laser effect (Chapter 7).

The first experimental indications of a possible population inversion and gain $K = e^G$ with $G = \alpha L \approx 2$ for the Pfund $\beta(n = 7 \to 5)$ transition of hydrogen at $4.65\,\mu\mathrm{m}$ from the spectrum of the Orion-BN/KL source were published by Smith *et al.* (1979). This result was obtained by means of measurements of the intensity ratio of the Pfund β and Brackett $\alpha(n = 5 \to 4)$ lines. The expected intensity ratio for the optically

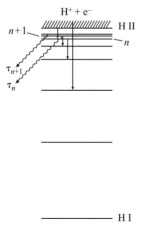

Fig. 8.3 Simplified quantum level scheme of a hydrogen recombination maser/laser, based on cascade population and fast radiative decay of highly excited states.

thin case is 0.20. The intensity of the Pfund β line observed in December 1976 is comparable to that of Brackett α. Smith et al. (1979) considered the intensity of the spectral lines in the optically thick limit case and concluded that the population in the Pfund β transition is inverted. The Brackett α transition is not inverted; a gain in Pfund β of $G = \alpha L \approx 2$ could explain the observed intensity ratio. The population inversion arises because the rates into the levels, from cascading and from radiative recombination, are comparable to the rates out of the levels by spontaneous decay with $\tau_{n+1} > \tau_n$ (Fig. 8.3). This effect occurs in small dense HII regions around the Beclin-Neugebauer (BN) source in Orion. A small size for the emitting region around BN provides a natural explanation for the observed time variations of the Pfund β flux of the order of months. If the Pfund β radiation is being amplified, the time variations may be due to effects of inhomogeneity and geometry, which change the gain or the direction of the radiation (Smith et al., 1979). Variations on this timescale are common for astronomical maser sources (Elitzur, 1992). The final conclusion of Smith et al. (1979) was as follows: "If the hydrogen lines are optically thick, and if the Pfund β radiation from BN is in fact being amplified, Pfund β is the shortest wavelength transition in which astrophysical masing has yet been observed, and the region could approximately be said to contain a hydrogen laser". However, when this observation was later repeated (Scoville et al., 1983), the Pfund β line at 4.65 µm was not as bright.

Further progress and confirmation of laser effects in HI was obtained in more long-wavelength transitions in the far-IR range by Strelnitskii et al. (1996), who chose the peculiar object MWC 349. It is a very hot, luminous (30,000 L$_\odot$) star in the constellation Cygnus at a distance of 1200 pc. It is surrounded by a massive disk of gas and dust, in which many masers have been discovered. These include submillimeter hydrogen recombination masers/lasers, particularly the H32β maser at 336.6 GHz found by Thum et al. (1995) and the H26α maser at 1775 GHz (169 µm) (Strelnitskii et al., 1995). (Hnα denotes the hydrogen transition $n+1 \to n$, i.e. between the levels with principal quantum number $n+1$ and n.) The astrophysical submillimeter maser at 169 µm has an intensity six times brighter than the non-amplified spontaneous H26α emission at the same wavelength. Thus, MWC 349 was a promising candidate source for detecting high-gain astrophysical lasers in the far-IR range (Smith et al., 1996).

The experiments with the 91 cm telescope of the Kuiper Airborne Observatory (KAO) using a cryogenic grating spectrometer enabled the observation of the hydrogen lines H10α (52.5 µm), H12α (88.8 µm) and H15α (169.4 µm) in MWC 349 (Strelnitskii et al., 1996; Smith et al., 1997). The detected far-IR lines showed radiation in excess of the predicted intensity of spontaneous emission. Fig. 8.4 compares the observed and predicted fluxes, the latter from the simplest spontaneous emission model for the optical, near-, mid- and far-IR, and submillimeter ranges. The intensitites of lines from levels having $n = 2-7$ are due to spontaneous emission. The millimeter and submillimeter lines (from levels having $n = 21-41$) show a significant excess above the level of spontaneous emission, which confirms previous observations that they are amplified by maser action. The new observed mid-IR lines ($n = 10, 12, 15$) also have an enhanced level above the spontaneous emission. The detection of these laser lines proves that population inversion exists in the recombination spectrum of hydrogen

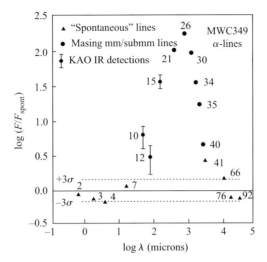

Fig. 8.4 A log-log plot of the ratio of fluxes F/F_{spont} versus λ, where F is the observed integrated flux for a hydrogen $n\alpha$ line of wavelength λ, and $F_{\text{spont}} = -7\log n - 12.24$. A triangle indicates a spontaneous line, large circles indicate masing millimeter and submillimeter lines, and small circles with error bars indicate three KAO-detected infrared lines (from Strelnitskii et al., 1996).

down to $n \simeq 10\text{--}15$ at densities higher than 10^8 cm^{-3}, which is in accordance with many previous (Storey and Hummer, 1995) calculations.

Strelnitskii et al. (1996) estimated the gain factor K and brightness temperature T_B using the angular size of the natural laser source in MWC 349. They assumed that H15α forms in the disk at an optimum density of $n_e \approx 10^9$ cm^{-3} at the radial distance $R \simeq 5$ AU from the star. Using the value of the observed integrated flux and some assumptions about the width of a single-peaked line they obtained a radiation temperature $T_B \gtrsim 10^7$ K for the laser. For the input spontaneous radiation temperature of the laser $T_{\text{sp}} \simeq 10^4$ K (the typical physical temperature of the gas in the H II region) the amplification factor is $K \geq 10^3$. Similar values of T_B and K were estimated for the H10α line. A simple modeling explains why the detectable amplification turns on near $n \gtrsim 40$ and $n \lesssim 10$ (Strelnitskii et al., 1996). These experimental results were confirmed in a study using the Infrared Space Observatory (ISO) spectrometer (Thum et al., 1998).

Thus, the observed far-IR lasers may be strong amplifiers with an amplification and effective (brightness) temperature comparable to or greater than those for the millimeter masers in the same source (Martin-Pintado et al., 1989). The luminosities ($L \simeq 0.01 - 0.1 L_\odot$) estimated by Strelnitskii et al. (1996) are comparable to those of the strongest molecular masers (Gwin et al., 1992).

8.3 IR CO$_2$ laser in the atmospheres of Mars and Venus

The first observations of nonthermal (laser) emission lines of the 10 μm CO$_2$ band in the atmospheres of Mars and Venus were performed by Johnson et al. (1976) by means of a 10 μm IR heterodyne spectrometer with sub-Doppler spectral resolution. Unexpectedly strong emission had previously been observed (Betz et al., 1977) from both planets at the line centers of the $00°1 - [10°0 - 02°0]_I$ vibrational-rotational band of $^{12}C^{16}O_2$. The emission lines had a Gaussian shape with a width corresponding to kinetic temperatures of about 170 K for Mars and 200 K for Venus. Altitudes for the 10 μm peak emission were estimated to be near 80 km for Mars and 120 km for Venus. Subsequent detailed studies confirmed the natural IR lasing in the mesospheres and tropospheres of Mars (Mumma et al., 1981) and Venus (Deming and Mumma, 1983).

The measurements of the Doppler profile of the Mars and Venus IR laser lines were made at 5 MHz spectral resolution, but search for possible narrowed emission from gain regions was carried out at 2 MHz resolution. The typical measured profile of the R8 line in Mars spectra is shown in Fig. 8.5. The three main contributions to the emitted radiation are separately identified: (a) thermal emission from the warm surface with atmospheric absorption, (b) self-emission from the atmosphere, and (c) the amplified resonant spontaneous emission within the Doppler profile. No narrowing of the Doppler profile was found, which is explained by the low value of the gain. The vertical amplification is only 0.3% (Mumma et al., 1981), and the evaluated amplification along the horizontal tangent path through the mesosphere would be ∼10% on Mars and slightly less on Venus (Deming and Mumma, 1983).

Betz et al. (1977) examined the excitation process of the 10 μm CO_2 emission of these natural lasers. They found that the emission was present only on the planet's dayside and that the intensity was approximately proportional to the local solar radiative power incident on the planet's surface. They examined several excitation mechanisms and showed that the IR emission was pumped by absorption of the combination band

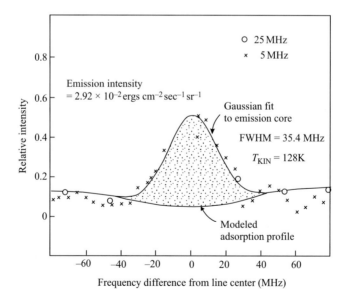

Fig. 8.5 Fully resolved shape of the R8 line of the 10.4 μm CO_2 band on Mars. The line shape corresponds to a kinetic temperature of 128 K (from Mumma et al., 1981).

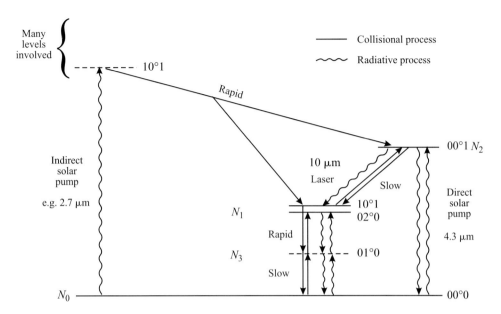

Fig. 8.6 Molecular physics of a natural CO_2 laser. Radiative and collisional processes affecting the pumping of this laser in the mesospheres of Mars and Venus are identified.

$00°0 - 10°1$ followed by resonant collisional energy transfer of a ν_3 quantum to another CO_2 molecule, which is similar to the laboratory CO_2 laser (Fig. 8.6). Radiative decay of the excited ν_3 level is fast but the optical depth of the ν_3 band is very high, and the ν_3 photon is thereby trapped, allowing a buildup of the population of the $00°1$ vibrational level. Mumma et al. (1981) studied the kinetic and excitation temperature on Mars and showed that the population of the upper vibrational level is inverted to that of the lower level $(10°0, 02°0)$. Deming and Mumma (1983) constructed detailed atmospheric models for Mars and Venus and confirmed the combination band solar pump proposed by Betz et al. (1977). They showed that about 1/30,000 of the solar constant of Mars is converted to amplified spontaneous emission at $10\,\mu m$. An independent confirmation of the $10\,\mu m$ laser action on Venus and Mars was provided by theoretical studies of Gordiets and Panchenko (1983, 1986) and Stepanova and Shved (1985).

It was suggested to make an oscillator by using the planet atmosphere as a gain medium and using huge mirrors in aerosynchronous orbit for the resonator. Generated directed output power could be used for laser communication over interstellar distances (Mumma, 1993).

9
Basics of optically pumped astrophysical lasers

From general point of view, all astrophysical lasers are optically pumped by radiation of a nearby star. In the case of the collisionally pumped APL the star radiation converts to the kinetic and free energy of colliding particles. In the case of the so-called optically pumped laser, the resonant pumping of lasing particles takes place in the strong emission line of one element (usually the intense emission lines of hydrogen Lyα, Lyβ, etc.). Such a pumping scheme requires a rare accidental coincidence of two strong emission and absorption lines and is not employed today in laboratory lasers. However, Schawlow and Townes (1958) mentioned the scheme for the first laser using cesium vapor pumped by a line of mercury. The idea was reborn in astrophysical lasers but not in the direct form.

Optical (visible and near-IR) astrophysical lasers are pumped by selective photoexcitation performed through a PAR process: the PAR acronym stands for "photoexcitation by accidental resonance" and was introduced by Kastner and Bhatia (1986). Let us consider some known examples of PAR and how inversion of population is created by the PAR effect in astrophysical conditions.

9.1 Bowen accidental resonances and fluorescence lines

The most famous case of the PAR process is the well-known Bowen mechanism (Bowen, 1934) mentioned in Section 5.2, in which a wavelength coincidence between the He II Lyα line and a ground term transition in O III results in a transfer of excitation energy from He II to O III. The subsequent decay of O III yields optical emission lines observed in spectra of gaseous nebulae. Bowen (1934) also noted the wavelength coincidence between cascade transitions in the O III decay and ground term transitions of N III, which yielded enhanced N III lines. Bowen's explanation of the origins of both the fluorescent O III and the forbidden O III lines (Bowen, 1928) laid the foundation for the interpretation of distinct emission lines in nebular spectra. In this Chapter we will briefly review the continued studies of PAR processes in medium-density regions around stars.

The probability of coincidence of two narrow spectral lines is very low. However, it is enhanced by two factors. Firstly, the most intense emission lines belong to light elements (H I, He I, H II etc.) from hot regions. These lines usually have large Doppler widths (2.55) with $\Delta\omega_D/\omega_0 \simeq 10^{-3}$–$10^{-4}$ (see example in Fig. 5.7). Secondly, the strong emission lines (Lyα, Lyβ, H I etc.) appear in astrophysical regions with very high optical depth τ_0. As a result of multiple scattering (Chapter 2), the real spectral width according to (2.67) is much larger than the Doppler width. So these rare coincidences are very useful for the existence of optically pumped astrophysical lasers.

Since the discovery of the Bowen mechanism, similar PAR processes have been proposed as the mechanism behind strong nebular emission lines. Bowen (1947) suggested that the wavelength coincidence between H Lyβ and O I caused the excitation mechanism behind the O I line at 8446 Å observed to be strong in emission in many near-IR astrophysical spectra. Some other examples are mentioned in Chapter 7 (Fig. 7.2), e.g. the Hδ pumping of an In I line, observed in variable stars (Thackeray, 1935). Cowley (1970) has summarized the PAR processes that were known at the end of the 1960s and were identified by fluorescence lines observed in astrophysical spectra in the optical region. For observational and atomic structural reasons most of the driven species concerned neutral atoms of abundant elements. A renewed interest in PAR processes appeared when satellite-ultraviolet ($\lambda > 1200$ Å) spectra became available from the International Ultraviolet Explorer (IUE) launched in 1978. Chromospheric emission lines from spectra of cool giants became visible in the absence of short wavelength Planck radiation. Chromospheric emission lines had been seen at even shorter (< 2000 Å) wavelengths in satellite solar spectra. Many lines were associated with high excitation states of Fe II and could be explained by pumping (Brown et al., 1984). Johansson and Jordan (1984) investigated the H Lyα pumped fluorescence lines of Fe II in the satellite-UV solar spectrum and the IUE spectrum of RR Tel. The strong coupling between H Lyα and Fe II occurred because the wavelengths for transitions between the low a^{4}D term (1 eV) and several 5p states in Fe II matched the Lyα wavelength. The large number of Lyα pumped Fe II fluorescent lines is due to the complex atomic structure of iron, and their strengths in spectra of many objects reflect the high cosmic abundance of iron. In a comprehensive paper on the cool giant star γ Cru by Carpenter et al. (1988) all PAR processes operating in the outer atmosphere of the star were described and tabulated with the corresponding fluorescence lines listed.

Two decay routes of the pumped Fe II levels result in two sets of combined primary and secondary fluorescence lines (see Fig. 9.1). The two decay routes occur because of level mixing between low 5p states and high 4p states at about 11 eV. Firstly, set one: the 5p states decay to 5s (primary decay) in the near-IR region and subsequently (secondary decay) to 4p at 2700–2800 Å, and secondly, set two: the high 4p states decay to high 4s states (primary decay) at about 2500 Å and subsequently to low 4p states (secondary decay) at IR wavelengths. Both sets end up in the lowest 4p states, which decay very rapidly to the ground term (a^{6}D) through resonance transitions, or to a^{4}D. A small part of the decay from the low 4p states populates quartet levels up to about 3 eV. The second set of fluorescence lines will be discussed in detail in Section 11.2.

Another group of objects showing numerous fluorescence lines are the symbiotic stars, which are binary systems consisting of one hot and one cool component. The two stars cannot be spatially resolved but the large span in ionization in the composite spectrum reveals the binarity. Spectra of symbiotic stars show many discrete emission

126 *Astrophysical lasers*

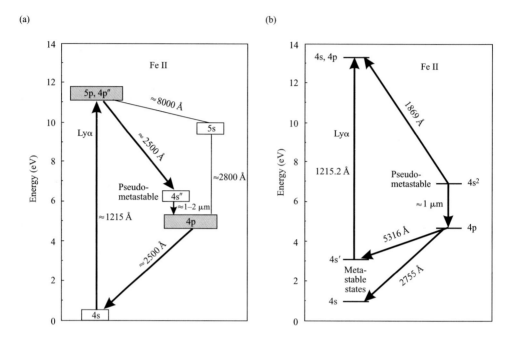

Fig. 9.1 Schematic energy level diagrams illustrating the three cases of radiative cycles (broad lines) with PAR excitation of Fe II by H Lyα.

lines from permitted (E1) s well as forbidden (E2, M1) transitions of several ionization states in the optical and UV regions, where the Planck radiation is faint between the continuous spectra of the hot and the cool components. One of the objects first studied with the IUE was RR Telescopii showing a forest of narrow emission lines in the IUE region (1200–3200 Å). The presence of secondary Fe II fluorescence lines around 2800 Å (see above) was pointed out in the pioneering paper by Penston et al. (1983). Johansson (1983) identified 10 strong lines originating from the same upper energy level of Fe II, $y^4H_{11/2}$, as being pumped by one of the IV resonance lines at 1550 Å in a PAR process. Later, the presence of O VI was indirectly detected in RR Tel through a PAR process producing fluorescence in Fe II (Johansson, 1988). The wavelengths of the resonance lines of O VI at 1030 Å are below the lower cutoff of IUE and could therefore not be directly observed from any object. At this time O VI implied the highest ionization temperature detected for RR Tel.

A systematic search in IUE spectra for Fe II fluorescence lines driven by O VI, C IV and Lyα was performed by Feibelman et al. (1991). The O VI pumped lines were of special interest for exploring the possibility of using the Fe II lines as a diagnostic for the unobservable O VI lines. Feibeiman et al. studied spectra of 13 symbiotic stars, 14 planetary nebula nuclei (PNN) and three novae. They found Fe II fluorescence to be quite common and strong in symbiotic novae as well as in normal novae. The lines were also seen, but relatively weakly, in central stars of some planetary nebulae, but were not detected in extragalactic AGNs.

A detailed study of the PAR processes in the IUE and HST regions of RR Tel has been published by Hartman and Johansson (2000), where a total of 33 pumped energy levels and 120 fluorescence lines of Fe II were identified. New PAR processes and Fe II fluorescence lines in two other objects at even shorter wavelengths have been presented by Harper *et al.* (2001) on the basis of FUSE spectra. The interaction between the radiation from high temperature species such as C^{3+} and N^{4+} and particles of low-temperature species like Fe^+ has recently been discussed by Eriksson *et al.* (2006) in terms of topology of AG Peg and 19 other symbiotic stars. The excitation mechanism, location and origin of about 600 observed lines, many of which are fluorescence lines, have been investigated.

The development of astronomical instrumentation has permitted quantitative studies of the original Bowen lines of O III and N III and cast doubts about the efficiency of the excitation process. Sternberg et al. (1988) showed that a competing process that could populate the upper levels of the O III Bowen lines is charge transfer between hydrogen and O^{3+}. On the basis of photometrically observed line ratios Kastner and Bhatia (1991) showed that the presence of N III lines in spectra of planetary nebulae is not a result of the O III pumping proposed by Bowen (1934). They found that the N III populations were nearly in statistical equilibrium. Ferland (1992) pointed out that the N III lines are produced by "continuum fluorescence", i.e. the photoexcitation is obtained by the Planck radiation. Direct measurements of the intensities of the observed N III and O III lines in AG Peg and RR Tel (Eriksson *et al.*, 2005) and in RR Tel (Selvelli *et al.*, 2007) have been compared to predicted intensities from simple models. Eriksson *et al.* found that the O III pumping contributes most of the N III intensity in RR Tel, but another process is operating in parallel in AG Peg. However, Selvelli *et al.* found that the O III pumping is solely responsible for the N III emission in RR Tel.

9.2 Inversion of population by accidental resonance pumping

The early idea of possible astrophysical lasers (Letokhov, 1972a) was associated with the explanation of intensity anomalies observed for O I in spectra of emission line objects. The O I lines at 7774 Å and 8446 Å in the spectra of Be stars were known to show strange relative intensities (see, e.g., Slettebak 1951; Merrill 1956) as the observed ratio between the $\lambda 8446$ feature (the $3s^3S^o$-$3p^3P$ triplet transition) and the $\lambda 7774$ feature (the $3s^5S^o$-$3p^5P$ transition) was much greater than the ratio derived from laboratory data. Bowen (1947) proposed that the $\lambda 8446$ line is enhanced by fluorescence caused by a wavelength coincidence between H Lyβ at 1025.72 Å and three ground-state transitions ($2p^3P_2$-$3d^3D^o$) of O I at 1025.76 Å. According to the O I-level scheme in Fig. 7.1a, the cascading of the H Lyβ pumped $3d^3D^o$ term feeds the upper levels of the enhanced $\lambda 8446$ feature. A detailed quantitative study of the "H Lyβ/OI PAR process" has been performed by Kastner and Bhatia (1995).

The anomalous ratio between the intensities of the $\lambda 8446$ and $\lambda 7774$ lines in spectra of Be stars can be naturally explained by stimulated emission of radiation (laser action) in the $\lambda 8446$ line (Letokhov, 1972a; Lavrinovich and Letokhov, 1974). An inverted population is built up in the 3s-3p transition ($3 \rightarrow 2$ in Fig. 7.1a) as a result of PAR by H Lyβ, according to Bowen's scheme (Bowen, 1947) for spontaneous and stimulated radiative pathways. Moreover, Bennett et al. (1962) obtained continuous wave (CW) lasing at 8446 Å under laboratory conditions. Population inversion was achieved in an electric discharge in a mixture of oxygen and a noble gas at a low pressure. Similar conditions can exist in some regions of stellar atmospheres. The normal intensity of the $\lambda 7774$ line can be naturally explained, since the lower level has a long radiative lifetime, and no population inversion or gain can be obtained for this transition.

For the following discussion we emphasize that the particular part of the level scheme of O I involved in the population inversion and intensity gain in the $\lambda 8446$ line is the classical four-level quantum electronics scheme (Siegman, 1986) shown in Fig. 1.2b. The specific features of this scheme are: (1) any excitation of level 4 is followed mainly by its decay to state 3; (2) $\tau_3 > \tau_2$; i.e., state 3 has a longer radiative lifetime than state 2, to which it is radiatively coupled; and (3) in the ideal case, the decay of state 2 again leads to the population of the initial state 1. This means that an atom (ion) can repeatedly participate in the radiative cycle, whose rate depends on the relationship between the excitation rate W_{exc}^{14} and the decay rate $1/\tau_3$ under the condition $\tau_2 \ll \tau_3$. All of the most effective lasers of both gas and solid-state types operate according to this scheme. The particle densities in these four-level solid-state laboratory lasers are certainly very high, and the decay processes at the intermediate stages $4 \to 1$ and $2 \to 1$ are radiationless because of the interaction between ions and phonons in crystals. Under astrophysical conditions, however, we restrict ourselves solely to radiative transitions. This is based on the fact that the transition probabilities in the optical region are, in general, much higher than the probability for collisional deexcitation, even in astrophysical plasmas with densities up to 10^{13}–10^{15} cm^{-3}. Using the four-level scheme as a prototype, one can understand the development of population inversion under isotropic and steady-state, but nonequilibrium, irradiation. Consider a four-level scheme (Fig. 9.2a), in which the selective excitation of level 4 always gives rise to population inversion in the $3 \to 2$ transition. To better understand the need for selective excitation, consider first the case in which such an atomic system interacts with a radiation field in thermal equilibrium having a temperature of T_{rad}. One should then consider both spontaneous and stimulated emission as well as absorption in the transitions of such a system, as illustrated in Fig. 9.2.

The four-level system, like any atomic system, must reach a population distribution among all four levels following the Boltzmann law for T_{rad}. The system is in radiative interaction with the matching frequencies (energy differences) of the equilibrium radiation spectrum (horizontal lines in Fig. 9.2b), and does not involve any collisions. This fact refers not only to the internal degrees of freedom of the particles, but also to the translational degrees of freedom. This statement requires that only the recoil effect should be taken into account in radiative interactions.

Even if some of the quantum transitions only interact very weakly with the radiation field, all the levels will reach a Boltzmann distribution at a rate defined by the strong allowed transitions. For example, let the $3 \to 2$ transition in Fig. 9.2a be a "semi-forbidden" line with a low transition probability, which is plausible in complex atomic systems. In such a case, the population in state 3 results from the $1 \to 4 \to 3$ sequence of allowed transitions. However, the number of particles accumulated in state

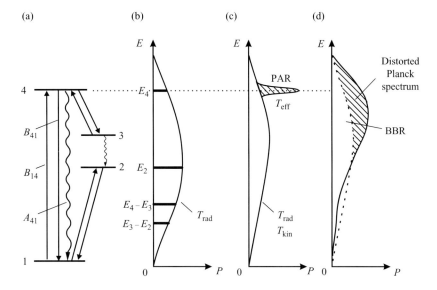

Fig. 9.2 (*a*) Scheme of a four-level atom with relevant radiative transitions, showing the radiative interaction between the four-level atom and (*b*) black-body radiation (BBR), (*c*) BBR and emission line absorbed in the $1 \to 4$ transition, and (*d*) distorted BBR with intensity enhancement at the wavelength of the $1 \to 4$ transition. (E_n is the energy of level n; A_{ik}, B_{ik} are the Einstein coefficients, T_{rad} is the radiation temperature, and T_{kin} is the kinetic temperature.)

3 will not exceed the equilibrium value, since it will be depleted in the reverse sequence $3 \to 4 \to 1$ by stimulated radiative transitions in absorption and emission. Even though state 3 is "pseudo-metastable" no population inversion can naturally be built up in the $3 \to 2$ transition. These basic facts are included here merely to show explicitly that only deviations from the Planck frequency distribution of the radiation can cause level populations to deviate from the Boltzmann distribution and specifically give rise to population inversion in an internal transition.

In a PAR process a strong spectral line of one atom (ion) can excite level 4 in another atom (ion) in a selective way at a rate that is high enough to build up a population inversion in the $3 \to 2$ transition (Fig. 9.2c). Obviously, the effective spectral temperature, T_{eff}, of this pumping radiation must exceed the equilibrium temperature of the atoms (ions) being excited. The equilibrium temperature T_{kin} in Fig. 9.2c means the kinetic temperature of particles (electrons, ions, etc.) in the astrophysical medium. The rare collisions of these charged particles with the four-level atom (ion) may lead to excitation of the lower level 2, but it is not significant if $T_{kin} \ll T_{rad}$.

Johansson and Letokhov (2003c) concluded from this general discussion that there are needs for non-equilibrium radiation in order to obtain inverted-level populations and that PAR is a most favorable process. PAR provides the optimal photoselective excitation conditions for level 4 in our four-level atom (ion). In principle, however, it is quite possible that population inversion can be formed as an effect of CW radiation even without PAR. The requirement in such a case is a nonequilibrium frequency distribution of the CW radiation (Fig. 9.2d). Mustel (1941) showed that the continuous spectrum of a stellar photosphere could strongly differ from the Planck distribution. For pumping in the four-level scheme it is necessary that the blue wing of the spectral distribution is more intense (compared to the Planck distribution) than the red wing, as shown in Fig. 9.2d. It is not difficult to obtain the general expressions for the degree of nonequilibrium required for the exciting radiation in order to obtain a population inversion in the $3 \rightarrow 2$ transition. However, these expressions are unwieldy and are of no great interest in the search for quantum transitions with population inversion. It is more expedient to obtain simple expressions for the degree of population inversion and gain in the case of PAR in a four-level system in which the formation of inversion is obvious.

We restrict ourselves to applying the four-level system (Fig. 1.2b) to atoms (ions) in a dilute astrophysical plasma in which the following nonequilibrium conditions are satisfied: (1) all collision rates are much lower than the rates of radiative processes; (2) a PAR process is operating at an effective radiation temperature T_{eff}, providing the population of state 3 by the $1 \rightarrow 4 \rightarrow 3$ excitation-deexcitation sequence; (3) the radiative lifetime τ_3 of level 3 exceeds the lifetime τ_2 for level 2; and (4) the local plasma temperature T_{kin} is substantially lower than T_{eff}. Obviously, under these conditions level 3 is excited by PAR more effectively than level 2 is excited by interactions with the astrophysical plasma, so the inverted population is maintained in the $3 \rightarrow 2$ transition.

The inverted population density $\Delta N_2 = (N_3/g_3 - N_2/g_2)$ depends on the excitation and decay rates of the levels involved. The population of level 3 is controlled by the efficiency of the PAR process

$$\frac{N_3}{g_3} = \left(\frac{N_1}{g_1}\right) \exp\left(-\frac{\Delta E_{41}}{kT_{\text{eff}}}\right), \tag{9.1}$$

where we assume a branching fraction of 1 for the $4 \rightarrow 3$ transition – i.e., all atoms (ions) excited to level 4 decay to level 3. The population of level 2 is determined by two factors: (1) the spontaneous radiative decay of level 3 in the $3 \rightarrow 2$ transition

$$\frac{N_2}{g_2} = \frac{\tau_2}{\tau_3} \frac{N_3}{g_3} = \left(\frac{N_1}{g_1} \frac{\tau_2}{\tau_3}\right) \exp\left(-\frac{\Delta E_{41}}{kT_{\text{eff}}}\right), \tag{9.2}$$

and (2) the interaction with the astrophysical plasma with temperature T_{kin},

$$\frac{N_2}{g_2} = \left(\frac{N_1}{g_1}\right) \exp\left(-\frac{\Delta E_{21}}{kT_{\text{kin}}}\right), \tag{9.3}$$

i.e.,

$$\frac{N_2}{g_2} = \frac{N_1}{g_1} \left[\exp\left(-\frac{\Delta E_{21}}{kT_{\text{kin}}}\right) + \left(\frac{\tau_2}{\tau_3}\right) \exp\left(-\frac{\Delta E_{41}}{kT_{\text{eff}}}\right)\right]. \tag{9.4}$$

The inversion condition $(N_3/g_3) > (N_2/g_2)$ is satisfied because of the following two factors: (1) the decay of state 2 is faster than the decay of state 3, and (2) $T_{\text{kin}} < T_{\text{eff}}$;

that is, the temperature of the astrophysical plasma is lower than the temperature of the PAR radiation, which is not in equilibrium with the plasma. Equations (9.1) and (9.4) yield the inverted population density.

In the outline above we have intentionally simplified the analysis in order to emphasize the basic physics and important parameters before obscuring it with relevant numbers, which differ drastically for different atoms (ions), quantum transitions, and parameters of the astrophysical plasma. Reasonable estimates for real cases have been made earlier, such as O I pumping by H Lyβ (Lavrinovich and Letokhov, 1974) and Fe II pumping by H Lyα (Johansson and Letokhov, 2002, 2003a). Following the scheme outlined here, more detailed models for complex spectra with a multitude of energy levels are still to be developed.

Let us now proceed to an ensemble of four-level atoms occupying a plasma with a geometric depth of L along the line of sight. The linear amplification coefficient in the $3 \rightarrow 2$ transition for such an ensemble is

$$K = \exp(\alpha_{32} \cdot L) \qquad (9.5)$$

where α_{32} is the amplification coefficient per unit length defined by expression (6.2) (with indexes 3, 2 instead of 2, 1). When the amplification parameter $\alpha_{32} L \gg 1$, the probability of stimulated emission in the $3 \rightarrow 2$ transition substantially exceeds that of spontaneous emission. In that sense there is a laser effect in the $3 \rightarrow 2$ transition. It is important to emphasize that the present case involves the total probability of stimulated emission of radiation in all free-space modes and not just a single mode, as is the case with the laboratory cavity laser. Thus, the stimulated emission effect is substantial, since the small occupation number of photons in a single free-space mode is compensated by contributions from all the isotropic radiation modes. This peculiarity has already been pointed out in connection with the study of laser effects with amplification in a scattering medium (Letokhov, 1967a, b).

10
Anomalous spectral effects in the Weigelt blobs of Eta Carinae

Astrophysical lasers in the visible and near-IR ranges have been discovered in the Weigelt blobs of the luminous blue variable (LBV) star Eta Carinae (η Car). This massive ($M \gtrsim 100$ M$_\odot$) and luminous ($L \gtrsim 6 \cdot 10^6$ L$_\odot$) star is a most remarkable stellar object located close enough ($\sim 7 \cdot 10^3$ light years) to be observed in great detail (Davidson and Humphreys, 1997). Using speckle interferometry Weigelt and Ebersberger (1986) discovered three remarkably compact objects at an angular distance between 0.1" and 0.3" northwest of the star, which corresponds to distances from the star of the order of 10^{16} cm, a few light days. The objects have been called "the Weigelt blobs" after their discoverer. The blobs are moving at low speeds of the order of 50 km/s (Davidson et al., 1997) and are producing many hundreds of intense, narrow emission lines unlike the spectra of any other known objects. Thus, the spectral lines from different parts of η Car have unique properties in many respects: intensity, spectral width, time and position variations (Davidson, 2001a, 2001b). It will be shown below that a number of spectral puzzles in the Weigelt blobs are conditioned by a high radiation energy density.

Spectra of η Car and its nebula have been obtained in high spectral and spatial resolution with the space telescope imaging spectrograph (STIS) onboard the Hubble Space Telescope (HST), where the slit has been positioned across and beside the star, along with and tilted against the direction of the bipolar lobes (Fig. 10.1). Distinct emission line spectra appear at particular locations close to the star, and they are very different from the spectrum of the star itself (Morse et al., 1999). A forest of narrow Fe II lines is one of the puzzles in the understanding of the object (Johansson and Zethson, 1999). At a projected position of about 0.2" above the star an emission line spectrum with narrow and bright UV lines of Fe II is related to the Weigelt blobs, which are probably located in a disk perpendicular to the expansion direction of the bipolar lobes. Fig. 10.2 shows a part of the STIS spectrum around 2500 Å, where the central star is positioned at 0 along the y-axis, giving the spatial scale in arcseconds. The blob spectrum (at 0.2" in Fig. 10.2) contains two spectacular features at vacuum wavelengths 2507.55 and 2509.10 Å, which are among the strongest single features in the whole observed spectrum of the blobs between 2000 and 10,000 Å. The lines have been identified as transitions from two highly excited (11.1 eV) energy levels of Fe II (Johansson and Jordan, 1984). The appearance of these bright UV lines in Eta Carinae has been explained as line fluorescence generated through photoexcitation by H Lyα in a Bowen mechanism, in spite of a wavelength detuning of $\Delta\lambda \approx 2.1$–$2.4$ Å

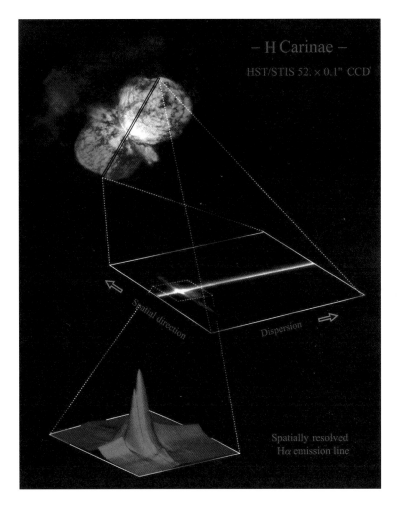

Fig. 10.1 General view of a two-dimensional (angle, wavelength) observation along the slit (upper image), which is passing through the central star of Eta Carinae and (bright strip on the middle image). The blobs are resolved from the star by the STIS camera on board of HST. The wavelength range is chosen to include the strong hydrogen Balmer α line (from Ishibashi with courtesy and permission). (See Plate 3)

(Johansson and Hamann, 1993). The blob spectrum contains a number of Fe II lines from highly excited levels in the near-IR, which are also explained by fluorescence generated by H Lyα. The small width of the bright UV lines of Fe II implies such a relatively low temperature of both the Fe$^+$ ions and the electrons, that it is not sufficient for collisional excitation to populate the highly excited levels of Fe II.

This fact and the extraordinary strength of the two features at $\lambda = 2507$ Å and $\lambda = 2509$ Å in the blob spectrum of η Car suggest a selective excitation mechanism

134 *Astrophysical lasers*

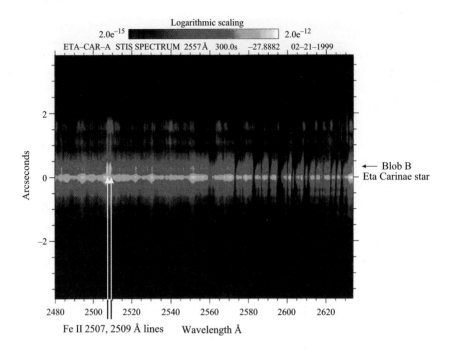

Fig. 10.2 Extract from the high resolution UV spectrum of the massive star eta Carinae recorded with the space telescope imaging spectrograph (STIS) onboard the Hubble Space Telescope. The vertical scale gives the distance (in arcseconds) from the central star (located at 0) to the Weigelt blobs and other gaseous condensations of matter expelled from the star. Of special interest is Weigelt blob B at about 0.2" "above" the star, showing a distinct emission line spectrum on a very weak continuous background. Two extraordinary features around 2507.5 and 2509.1 Å (vacuum rest wavelengths) in the blob spectrum required a specific excitation mechanism to be explained. (See Plate 4)

(Fig. 9.1). As a possible explanation of the extraordinary strength and anomalous intensity ratios of these lines and their satellites, natural lasing of Fe II in the UV range was suggested (Johansson et al., 1996). However, based on theoretical atomic data, lasing is in fact impossible because of the inappropriate decay times for the upper and lower states (opposite in sign to the requirement (6.1)). This puzzle stimulated our research of unusual radiative processes in the η Car environment, particularly in the Weigelt blobs.

10.1 Structure of H II/H I regions in the Weigelt blobs

The first spectral observation of η Car with HST led to crude estimations of a density between 10^7 and 10^{10} atoms/cm^3 for the blobs (Davidson, 2001b), and they were based on values of the optical density τ_0 of H Lyα. This can compensate for the wavelength detuning of H Lyα at 1215.671 Å relative to the wavelengths of two closely spaced spectral transitions in Fe II, namely, $a^4D_{7/2} \to 4p^4G_{9/2}$ (1218.213 Å) and $a^4D_{7/2} \to 5p^6F_{9/2}$

(1217.848 Å) (Johansson, 1978), which absorb the H Lyα radiation. The corresponding wavenumber detuning is $\Delta\nu \cong 160$ cm^{-1}. For the long-wavelength edge of the H Lyα emission line of neutral hydrogen to coincide with the comparatively narrow absorption lines of Fe II at around 1218 Å (their wavenumber spacing amounts to some 20 cm^{-1}), it is necessary that its optical density is about 1 for the detuning $\Delta\nu$ relative to the H Lyα line center:

$$\tau(\Delta\nu) = \sigma(\Delta\nu) N_{HI} D \simeq 1, \tag{10.1}$$

where N_{HI} is the density of the *neutral* hydrogen, and $\sigma(\Delta\nu)$ is the resonance scattering cross-section in the far ($\Delta\nu \gg \delta\nu_D$) Lorentz wing of the natural radiative broadening of H Lyα. The value of $\sigma(\Delta\nu)$ is determined by the known expression (Mihalas, 1978):

$$\sigma(\Delta\nu) = \sigma_0 \left(\frac{\delta\nu_D}{\delta\nu_{rad}}\right)\left(\frac{\delta\nu_{rad}/2}{\Delta\nu}\right)^2 = \sigma_0 \frac{\delta\nu_D \delta\nu_{rad}}{(2\Delta\nu)^2}, \tag{10.2}$$

where the first ratio is the Voigt factor, the second one is an effect of detuning, and $\delta\nu_{rad} = \gamma = A_{21}/2\pi c = 2.5 \cdot 10^{-3}$ cm^{-1} is the radiative width of H Lyα. From Eqs. (10.1)–(10.2) we get a simple estimation of the required optical density at the line center of H Lyα (2.62):

$$\tau_0 = \sigma_0 N_{HI} D = \frac{\sigma_0}{\sigma(\Delta\nu)} = \frac{(2\Delta\nu)^2}{\delta\nu_{rad}\delta\nu_D} \simeq 7 \cdot 10^6. \tag{10.3}$$

The cross-section $\sigma_0 \simeq \frac{1}{4}\lambda^2(\delta\nu_{rad}/\delta\nu_D) \simeq 1.4 \cdot 10^{-14}$ cm^2, if we insert $\delta\nu_D = 1.8 \cdot 10^{11}$ Hz $= 6$ cm^{-1} for $T_H \simeq 10^4$ K. It means that $N_{HI}D \simeq 5 \cdot 10^{20}$ cm^{-2} in blob B, which is emitting intense UV lines of Fe II. For the size of blob B, $D \simeq 10^{15}$ cm, the required density of neutral hydrogen $N_{HI} \simeq 5 \cdot 10^5$ cm^{-3}.

For a given value of N_{HI} it is possible to estimate the value of N_{HII} on the basis of photoionization/recombination steady-state balance (Johansson and Letokhov, 2001c). However, the homogeneous photoionization model for the blob is not valid. This conclusion follows from the ratio of the resonant radiative cross-section for Lyα at the center of line, σ_0(Lyα), and the average value of the photoionization cross-section above the ionization limit σ_{ph}(Ly$_c$) $\simeq 3 \cdot 10^{-18}$ cm^2 (or σ_i in Section 4.1). These parameters determine the optical density τ_{ph} for the photoionizing radiation (Lyman continuum) in the spectral range ($\nu_c, 2\nu_c$), which provide the main contribution to the photoionization rate W_{ph} (or W_i in Section 4.1):

$$\tau_{ph} \simeq \frac{\sigma_{ph}}{\sigma_0}\tau_0(Ly\alpha) \simeq 1.4 \cdot 10^3 \gg 1. \tag{10.4}$$

The radiation from η Car can thus photoionize only a small front part of blob B. This follows from a basic requirement (Eq. 10.3) for radiation transfer broadening of Lyα, which should excite the absorption line of Fe II. So, the border (Strömgren radius) between the H II zone (the active zone considering the generation of Lyα) and the passive H I zone lies inside blob B. The stellar radiation can homogeneously ionize the blob only at EUV frequencies $\nu \overset{\sim}{>} 10\nu_c$, where $\tau_{ph}(10\nu_c) \simeq 1$. However, the intensity of the radiation from η Car having such an extreme energy is negligible and cannot contribute to the formation of bright UV Fe II lines.

If we assume that the radiation of the central star η Car does not suffer from any substantial weakening while traveling the comparatively short distance $R_{\rm b} \simeq 3 \cdot 10^{16}$ cm to the nearest and brightest blob B, then the hydrogen, which is located in the front part of blob B facing η Car, should be in its ionized state H II. For this reason, the total hydrogen concentration $N_0 = N_{\rm HI} + N_{\rm HII}$ must be substantially higher than the estimate given above for H I based on the width of H Lyα. A simple estimate for the ratio between the H II and H I densities under steady-state conditions is determined by the ratio between the photoionization rate $W_{\rm ph}$ of the H I atom by radiation (at $h\nu > 13.6$ eV) from η Car and the recombination rate $W_{\rm rec}$ of the H II ion:

$$\frac{N_{\rm HII}}{N_{\rm HI}} = \frac{W_{\rm ph}}{W_{\rm rec}} \gg 1. \qquad (10.5)$$

The recombination rate of the H II ions depends on density $N_{\rm HII}$ because the electron density $n_{\rm e}$ in the electrically neutral nebular medium is equal to $N_{\rm HII}$:

$$W_{\rm rec}({\rm H\,II}) = \alpha N_{\rm HII}({\rm s}^{-1}), \qquad (10.6)$$

where $\alpha(T_{\rm e})$ is determined by the approximate expression (4.9) and $T_{\rm e}$ is the electron temperature of the blob. Thus, the density of H II is defined by an expression similar to (5.4) type:

$$N_{\rm HII} = \left(N_{\rm HI} \frac{W_{\rm ph}}{\alpha} \right)^{1/2}. \qquad (10.7)$$

The photoionization rate $W_{\rm ph}$ is governed by the flux of photons with $h\nu > 13.6$ eV from η Car, taking into account the dilution factor w, determined by (5.6), where $r_{\rm s}$ is the radius of the photosphere of η Car and $R_{\rm b}$ is the distance from the blob B to η Car. $W_{\rm ph}$ also depends on the photoionization cross-section $\sigma_{\rm ph}(\nu)$:

$$W_{\rm ph} = \Omega \int_{\nu_{\rm c}}^{\infty} \sigma_{\rm ph}(\nu) P(\nu) \, d\nu \simeq \sigma_{\rm ph} I_{\rm ph}, \qquad (10.8)$$

where $P(\nu)$ is the spectral brightness (photons/cm$^2 \cdot$s$\cdot$sr$\cdot$Hz) of the black-body radiation from the photosphere of η Car, $\sigma_{\rm ph}$ is an average effective photoionization cross-section introduced above, and $I_{\rm ph}$ is the integrated intensity (photons/cm$^2 \cdot$s) of the stellar radiation at the blob surface in that spectral range, which gives the main contribution to the photoionization rate. If one describes the radiation spectrum of the photosphere of η Car as that of a black body with a temperature of $T = 30 \cdot 10^3$ K, one can then approximately estimate the photoionization rate $W_{\rm ph}$ to be 0.02–2 s^{-1} for $R_{\rm b} = 3 \times (10^{16}$–$10^{15}$ cm). For other more remote blobs the value $W_{\rm ph}$ decreases in proportion to $(R_{\rm b}/R)^2$, where R is the distance of another blob from the central star in η Car.

The steady-state density of H atoms in blob B, required in accordance with Eq. (10.3), is sufficiently high and can completely absorb the ionizing radiation of

η Car. Indeed, the order of magnitude of the photoionization depth $\ell_{\rm ph}$ in blob B may be estimated from the following expression, similar to (5.13):

$$\ell_{\rm ph} \simeq \delta\ell_{\rm ph}\frac{W_{\rm ph}}{W_{\rm rec}} = \delta\ell_{\rm ph}\frac{N_{\rm HII}}{N_{\rm HI}}, \qquad (10.9)$$

where $\delta\ell_{\rm ph}$ is the thickness of the transition layer (H II/H I region) between the completely ionized H II region and the region of very weak ionization of H I (by the weak and extremely short-wavelength ($h\nu \gg 13.6$ eV) radiation of η Car), and defined in the same simple approximation as (Mihalas, 1978):

$$\ell_{\rm ph} \cong \frac{1}{\sigma_{\rm ph} N_0} = \frac{1}{\sigma_{\rm ph}(N_{\rm HI} + N_{\rm HII})}, \qquad (10.10)$$

where $\sigma_{\rm ph}$ is the photoionization cross section above the photoionization limit. For simplicity $\sigma_{\rm ph}$ is given by some average value of $\sigma_{\rm ph} \simeq 3 \cdot 10^{-18}$ cm^2.

We will first estimate the critical hydrogen density, $N_0^{\rm cr}$, at which the photoionization depth $\ell_{\rm ph}$ satisfies the condition

$$\ell_{\rm ph} \simeq D, \qquad (10.11)$$

which means that the boundary of the Strömgren sphere is exactly at the rearmost layer (opposite to η Car) of the blob B, so that the entire volume of the blob is subject to photoionization. We use Eqs. (10.7)–(10.9), considering that the density of H II in the H II/H I region suffering photoionization is equal to N_0, to obtain

$$N_0^{\rm cr} \simeq \left(\frac{W_{\rm ph}^2}{\sigma_{\rm ph} D \alpha}\right)^{1/2} = \left(\frac{I_{\rm ph}}{\alpha D}\right)^{1/2}. \qquad (10.12)$$

For the blob B we can adopt $I_{\rm ph} \simeq 3 \cdot 10^{16}$ ph/cm$^2\cdot$s, taking into account the dilution factor $w \simeq 10^{-6}$. Correspondingly, from Eq. (10.12) we get the value $N_0^{\rm cr} \simeq 10^7$ cm^{-3}. This means that only at $N_0 < N_0^{\rm cr}$ is the entire volume of the blob B uniformly photoionized, and at $N_0 > N_0^{\rm cr}$, it is only the outermost layer of the blob with a thickness of $\ell_{\rm ph}$ that is photoionized.

As an illustration, Fig. 10.3 presents the optical density τ_0, both in the photoionized region H II/H I and in the H I region, as a function of the total hydrogen density N_0 in the blob under consideration. At $N_0 < N_0^{\rm cr} = 10^7$ cm^{-3} the entire volume of the blob is uniformly photoionized, and the remaining small proportion of neutral hydrogen atoms provides for an optical density τ_0 whose value is lower than the necessary values given by Eq. (10.3). Higher τ_0 values can only be reached at $N_0 > N_0^{\rm cr}$, when only the outermost layer of the blob is photoionized. In the transition region where $N_0 \simeq N_0^{\rm cr} \simeq 10^7$ cm^{-3}, the photoionization limit passes inside the volume of the blob, i.e., in the distance range between $R_{\rm b}$ and $R_{\rm b} + D$ from η Car. In this ionized region, the optical density τ_0 reaches its steady-state value, as the reduction of the photoionization volume is being offset by the rise of the recombination rate $W_{\rm rec}$, which can be seen from the right-hand bottom branch of the curve $\tau_0(N_0)$ in Fig. 10.3. In

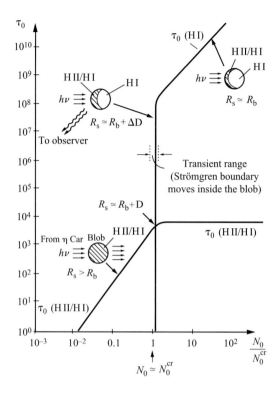

Fig. 10.3 Optical thickness, τ_0, of H I Lyα (for the center of the line) as a function of the total density, N_0, of neutral (H I) and ionized (H II) hydrogen in blob B. N_0^{cr} is the maximum (critical) density, for which the whole blob volume is ionized. At $N_0 > N_0^{cr}$ the blob volume is partially ionized, in the front layer of the blob (from Johansson and Letokhov, 2001c).

this H II/H I region, Lyα photons are created and they diffuse isotropically, i.e. into the practically non-ionized H I region as well, where the intensity of the Lyα radiation is naturally weakened. This H I region has a huge optical density τ_0 for the photons, which can reach the required values given by Eq. (10.3). The optical thickness of the transient layer H II/H I with physical width $\delta\ell_{ph}$ can be estimated as:

$$\tau_0^{tr} \simeq N_{HI}^{tr} \delta\ell_{ph} \sigma_0 \simeq \frac{1}{2} \frac{\sigma_0}{\sigma_{ph}} \simeq 2 \cdot 10^3, \tag{10.13}$$

where $N_{HI}^{tr} \simeq \frac{1}{2} N_0$ is the average concentration of H I in the transient layer, and $\delta\ell_{ph}$ is determined by Eq. (10.12). The high density of H I compensates for the small width of this layer and as a result the optical thickness of the transient layer τ_0^{tr} is almost equal to the optical thickness τ_0 of the whole active zone. Nevertheless, these values of τ_0 are much less than the required amount given in Eq. (10.3) for the Lyα excitation of Fe II. At the same time, the remaining, dissipating (passive) volume of the blob is also large enough to provide for $\tau_0 \simeq 10^7$ (the vertical branch of the curve in Fig. 10.3) and the appropriate broadening of the Lyα line as a result of the Doppler diffusion of the radiation. The Lyα radiation, diffusing from the active region into this passive volume, can excite the UV fluorescence in Fe II in a photoselective fashion, thanks to the compensation of the detuning between the wavelengths of Lyα and the absorption line of Fe II. The Fe II ions are always being formed in the passive region of the blob on account of photoionization by the radiation of η Car in the range 7.6 eV $< h\nu <$ 13.6 eV that passes through the H II/H I region without suffering practically any absorption through photoionization. Hamann et al. (1999) also noted that the Fe II lines do not form in H II regions, but rather in partially ionized zones behind the H II/H I recombination front.

The above crude qualitative analysis shows that intense UV Fe II lines can be observed only in those blobs whose hydrogen density $N_0 > N_0^{cr}$, which provides for $\tau_0 \gtrsim 10^7$ in the passive H I zone of the blob. These estimates have been modified to account for the effect of a stellar wind surrounding the Weigelt blobs (Johansson and Letokhov, 2004c).

10.2 High effective H Lyα temperature in the Weigelt blobs

According to the simple estimations in Section 10.1 the column density of neutral hydrogen $n_H = N_H D$, for WBs (particularly, WB "B") near η Car is estimated to be higher than $4 \cdot 10^{21}$ cm^{-2}, where N_H is the H I density and D is the size of the WB. This means that the WB is optically thick for the Lyman continuum ($\lambda <$ 912 Å), i.e.

$$\tau_c \simeq \sigma_{ph}(\nu c) N_H D \gg 1, \tag{10.14}$$

where $\sigma_{ph}(\nu_c)$ is the photoionization absorption cross section near the ionization limit. Eq. (7.14) leads us to conclude that two spatially separated but adjacent regions are formed in the WB: a hot H II region facing the central star and a cold H I region behind it. These two regions are separated by the Strömgren boundary that passes inside the WB. The cold H I region is responsible for the numerous narrow Fe II lines that can be divided into two categories: i) fluorescence lines from high-lying levels photoexcited by Lyα and ii) forbidden lines from low-lying, metastable levels.

The condition in Eq. (10.14) ensures the formation of a cold H I region where the Fe atoms are ionized by radiation with $\lambda >$ 912 Å, which passes through the H II region. This fact provides the basis for the physical model of the formation of the $\lambda =$ 2507 Å and $\lambda =$ 2509 Å lines (Johansson and Letokhov, 2001c; Klimov et al., 2002) in the WBs at distances $R_b \sim (10^2 - 10^3) r_s$ from the central star, where r_s is the radius of η Car. Another critical requirement is that the distance to the star R_b is small enough to ensure a high Lyα spectral temperature at the frequency of the Fe II absorption line. The Lyα temperature T_α in the WB region can be estimated by the evaluation of the radiation from η Car with $\lambda < \lambda_c$ absorbed in the WB (Johansson and Letokhov, 2004a).

Let us consider a spherical WB with a diameter $D \ll R_b$ (Fig. 10.4). Condition (7.14) is satisfied if the hydrogen density N_H exceeds a critical value (10.12). The integrated intensity of the hydrogen-photoionizing stellar radiation with $\lambda >$ 912 Å

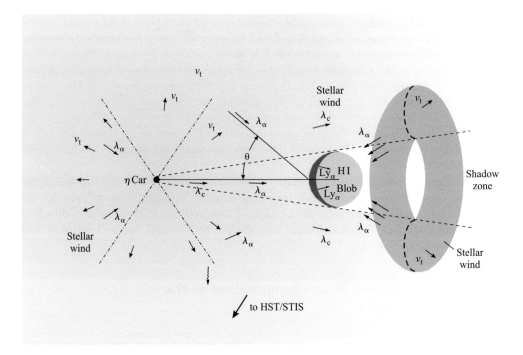

Fig. 10.4 The geometry of Eta Carinae and the Weigelt blob that blocks the Lyman continuum radiation so that the front part (the H II region) is completely ionized and the back part (the H I region) contains Fe$^+$ ions. The Fe$^+$ ions are selectively photoexcited by Lyα photons that diffuse either from the H II region of the blob or from the nearly torus-shaped region of the stellar wind of Eta Carinae behind the blob, which is also ionized by the Lyman continuum.

that reaches the WB surface is

$$I_{\rm ph} = \Omega \int_{\nu_c}^{\infty} P(\nu, T_{\rm s})\, d\nu \simeq \Omega \mathbf{P}(\nu, T_{\rm s})\, \Delta\nu_{\rm ph}, \tag{10.15}$$

where $P(\nu, T_{\rm s})$ is the spectral Planck intensity distribution (photons·cm^{-2}·s^{-1}·sr^{-1}·Hz^{-1}) for a star with a photospheric temperature $T_{\rm s}$. For an estimation we can use the effective frequency range $\Delta\nu_{\rm ph} \cong 7$ eV above the ionization limit, where $P(\nu)$ and $\sigma_{\rm ph}(\nu)$ are large enough. At $T_{\rm s} = 30 \cdot 10^3$ K, $\mathbf{P}(\nu_c, T_{\rm s})\Delta\nu_{\rm ph} \cong 10^{23}$ photons · cm^{-2}s^{-1}.

For $N > N_{\rm H}^{\rm cr}$, the neutral H I component of the WB absorbs almost all of the Lyman continuum radiation from the central star. The absorbed power is

$$P_{\rm abs}(\nu - \nu_c) \simeq \Omega_0 S_{\rm abs} \int_{\nu_c}^{\infty} P(\nu_c, T_{\rm s})\, d\nu$$

$$= \Omega \pi^2 D^2 P(\nu_c, T)\, \Delta\nu_{\rm ph}, \tag{10.16}$$

where $\Omega_0 = 4\pi\Omega$ is the solid angle subtended by the WB from the central star and $S_{\text{abs}} = (\pi/4)\,D^2$ is the area of the WB disk. The absorbed energy is reradiated in the hydrogen recombination spectrum, with the largest fraction ($\eta_1 = 0.7$) being emitted in the H Lyα resonance line. The maximum optical density $\tau_0(\text{Ly}\alpha)$ in the H II region at the center of the line is

$$\tau_0^m = \frac{\sigma_0}{\sigma_{\text{ph}}}(\nu_c) \simeq 5 \cdot 10^3, \qquad (10.17)$$

where $\sigma_0 = \sigma_{12}(\text{Ly}\alpha) = 1.4 \cdot 10^{-14}\,\text{cm}^{-2}$ is the resonant scattering cross section at the center of the line, $\sigma_{\text{ph}}(\nu_c) = 3 \cdot 10^{-18}\,\text{cm}^2$. Ly$\alpha$ radiation is diffusively confined in the H II region, but the diffusive confinement time is limited by the fact that the number of scatterings is limited. This limitation is attributable to the Doppler frequency redistribution during the scattering of Lyα photons, which ensures a relatively fast photon escape from the confinement region through the wings of the Doppler profile (Osterbrock, 1989). Since the optical density τ_0 is limited by $\tau_0^m = \sigma_0/\sigma_{\text{ph}}$ and the damping factor for Lyα is $a = \Delta\nu_{\text{rad}}/\Delta\nu_D$, the optical density in the Lorentz wings is $a\tau_0^m \simeq 1$. Therefore, Lyα photons escape from the H II region through diffusion with the Doppler width increasing by a factor of $\beta \simeq (\ln \tau_0^m)^{1/2} \simeq 3$ (see Section 2.3). Thus, the total power emitted by the WB surface in the Lyα line is

$$P_{\text{em}}(\text{Ly}\alpha) \simeq 4\pi S_{\text{em}} P(\nu_{\text{Ly}\alpha}, T_\alpha)(\Delta\nu_D \beta), \qquad (10.18)$$

where T_α is the Lyα effective spectral temperature and $S_{\text{em}} = \pi D^2$ is the area of the emitting surface of a spherical WB. In reality, S_{em} can differ slightly from πD^2 because of the peculiar shape of the emitting surface of the H II region (Fig. 10.4).

The mean Lyα spectral intensity at the WB surface is determined by the Planck distribution at frequency $\nu_\alpha(\text{Ly}\alpha)$ and by the spectral brightness temperature T_α, where the spectral broadening during the confinement is taken into account. In steady state, assuming that the absorption of the confined Lyα radiation in the H II region of the WB is negligible, we obtain

$$\eta P_{\text{abs}}(\nu - \nu_c) = P_{\text{em}}(\text{Ly}\alpha). \qquad (10.19)$$

Hence, the brightness temperature T_α for the Lyα radiation from the WB can be estimated from Eq. (10.19):

$$P(\nu_{\text{Ly}\alpha}, T_\alpha) = \frac{\eta_1}{4}\Omega\frac{\Delta\nu_{\text{ph}}}{\Delta\nu_D \beta}P(\nu_c, T_s). \qquad (10.20)$$

The dilution factor for the radiation from the central star is largely offset by the *spectral compression* of the absorbed Lyman-continuum energy into a relatively narrow recombination Lyα line. The compression factor f is

$$f = \frac{\Delta\nu_{\text{abs}}}{\Delta\nu_{\text{em}}} = \frac{\Delta\nu_{\text{ph}}}{\Delta\nu_D \beta} \simeq 10^3, \qquad (10.21)$$

where $\Delta\nu_{\text{abs}} \simeq \Delta\nu_{\text{ph}}$ is the effective spectral width of the photoionization absorption, $\Delta\nu_{\text{em}} = \beta\Delta\nu_D$ is the width of the Lyα spectrum, and $\Delta\nu_D = 6\,\text{cm}^{-1}$ is the

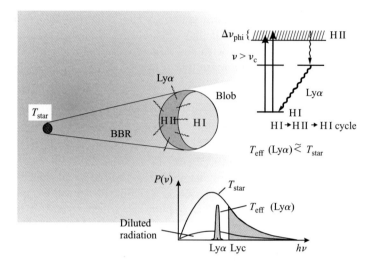

Fig. 10.5 Illustration of the "spectral compression" effect in the blob in the vicinity of a hot star which enhances significantly the effective temperature of hydrogen spectral lines due to photoionization-photorecombination.

Doppler width of the Lyα line. Fig. 10.5 illustrates the effect of spectral compression. The absorbed spectrally broad Lyman continuum with $\nu > \nu_c$ in course of the photoionization-photorecombination cycle is converted to a relatively narrow Lyα (or Lyβ) spectral line with effective temperature $T_{\text{eff}}(\text{Ly}\alpha) < T_s$. However, this temperature is comparable with the surface temperature of the star T_{eff}. For example, in the case of blob B η Car: $T_s \simeq 30{,}000$ K, $T_{\text{eff}}(\text{Ly}\alpha) = T_\alpha \simeq (10\text{--}15)10^3$ K. This effect is the key for optical pumping Fe II and O I astrophysical lasers (Chapters 11 and 12).

For the closest vicinity of the central star, it is convenient to introduce a spectral brightness conversion parameter γ_{br}:

$$\gamma_{\text{br}} = \eta_1 w \left(\frac{\lambda_{\text{Ly}\alpha}}{\lambda_c}\right)^2 \left(\frac{S_{\text{abs}}}{S_{\text{em}}}\right) f. \qquad (10.22)$$

This parameter includes the absorbed energy conversion ratio η_1, the spatial dilution factor w, the mode density enhancement factor $(\lambda_{\text{Ly}\alpha}/\lambda)^2$, the ratio of the absorption and emission areas, and the key spectral compression $f \gg 1$. The following equation relates the temperatures T_α and T_s (for $h\nu_{\text{Ly}\alpha} \gg kT_s, kT_\alpha$):

$$\frac{h\nu_{\text{Ly}\alpha}}{kT_\alpha} = -\ln \gamma_{\text{br}} + \frac{h\nu_c}{kT_s}. \qquad (10.23)$$

This relation is virtually equivalent to the expression for the source function in Lyα in terms of the source function in the Lyman continuum known for a nebula illuminated by a star (Sobolev, 1962, 1963).

Note that the intensity of the Lyα radiation inside the H II region of the WB is higher than on the surface of the WB by a factor of δ that accounts for the diffusive confinement of the radiation:

$$\delta = \beta^2 = \ln \tau_0^m \simeq 8\text{--}10. \tag{10.24}$$

According to the calculations by Auer (1968), the intensity of the Lyα radiation can be even a factor of 20 to 25 higher because of the confinement effect. Given this effect, the increase in the Lyα intensity described by Eq. (10.23) can be written in a more accurate form:

$$\frac{h\nu_{\text{Ly}\alpha}}{kT_\alpha} = -\ln(\gamma_{\text{br}}\delta) + \frac{h\nu_c}{kT_s}. \tag{10.25}$$

For the Weigelt blob B of η Car, $\gamma_{\text{br}}\delta \simeq 10^{-2}$, which corresponds to $T_\alpha \simeq (10\text{--}15) \cdot 10^3$ K inside the gas blobs. This value is comparable to or even higher than the electron temperature, implying a sharp change of the energy balance in the WB compared to a classical planetary nebula. Actually, the ratio μ of the Lyα radiation energy density inside the WB, $E(\text{Ly}\alpha)$, to the free energy density, E_{fr}, stored in charged particles (H II ions and electrons) becomes equal to

$$\mu = \frac{E(\text{Ly}\alpha)}{E_{\text{fr}}} = n_e \alpha_H \frac{D}{c} \delta \frac{\nu_{\text{Ly}\alpha}}{\nu_c}, \tag{10.26}$$

where n_e is the electron density (or the density N_H, N_{HII}). This ratio has a simple physical meaning: the ratio of the lifetime of a Lyα photon inside the WB, $\tau_{\text{ph}} \frac{D}{c} \delta$, due to the diffusive escape through the spectral line wings, to the ion recombination time, $1/(n_e \alpha_H)$. If we assume for the WB B of η Car that $n_e \simeq 2 \cdot 10^8$ cm^{-3} and $D = 10^{15}$ cm, then this ratio will be $\mu = 10$. The ratio of the density of $E(\text{Ly}\alpha)$ to the density of the charged-particle kinetic energy E_{kin} is even higher, because $h\nu_c \gg kT$. The density of the Lyα energy $E(\text{Ly}\alpha)$ is higher than the density of the Lyman continuum energy $E(\text{Ly}_c)$ at least by a factor of $\delta \simeq \ln \tau_0^m$ or even more (Auer, 1968). Therefore, the following equation is valid for an optically thick (for Ly$_c$) WB:

$$E(\text{Ly}\alpha) \gg E(\text{Ly}_c), \quad E_{\text{fr}} \gg E_{\text{kin}}. \tag{10.27}$$

This equation means that the photoprocesses in the WB are governed by Lyα and the Lyman continuum, Ly$_c$, radiation. Such Weigelt blobs may be called *radiation-rich* WBs, to distinguish them from standard *thermal* planetary nebulae (Aller, 1984).

11
Astrophysical lasers on the Fe II lines

Astrophysical lasers operating in optical Fe II lines in the range 0.9–2 μm in stellar ejecta of Eta Carinae (η car) were discovered (Johansson and Letokhov, 2002, 2003a) on the basis of observations with the Space Imaging Spectrograph (STIS) on board the Hubble Space Telescope (HST). This unique instrument provides high spectral resolution in a broad wavelength range (1150–10400 Å) and a near-diffraction-limited angular resolution (~0.1 arcsec) (Kimble et al., 1998). Beginning in 1998, the eruptive star η Car became one of the most remarkable targets for this instrument (Gull et al., 1999). It is an exceptionally interesting astrophysical object expelling enormous amounts of material into its surroundings. In the immediate vicinity of the central star at a distance of 100–1000 stellar radii ($r_s = 3 \cdot 10^{13}$ cm), compact gas condensations, called the Weigelt blobs B, C, and D, of exceptional brightness have been discovered (Weigelt and Ebersberger, 1986). More detailed observations (van Boekel et al., 2003) at near-IR wavelengths with high angular resolution confirmed a complex morphology of bright blobs and the existence of stellar wind in the vicinity of η Car on the scale of 5 milliarcsec (Fig. 5.9). The blobs and surrounding stellar wind are unique astrophysical plasmas having a higher density $N_H \gtrsim 10^8$ cm^{-3} than planetary nebulae ($N_H \simeq 10^4$ cm^{-3}) and located only a few light days away from the luminous central star. The sequence of basic radiative processes in the Weigelt blobs initiated by stellar black-body (Planck) radiation acting on H I and Fe I and followed by pumping due to an accidental resonance (PAR) between Fe II and Lyα is shown on Fig. 11.1. A modelling of the spectrum of the Weigelt blobs (Verner et al., 2002) has verified the importance of Lyα pumping to reproduce the fluorescent Fe II lines and continuum pumping to reproduce low-excitation Fe II as well as [Fe II] lines. As N_H is relatively high, the blob will almost completely absorb the Lyman-continuum radiation from the central star. The Strömgren border, separating the regions of completely ionized (H II region) and neutral hydrogen (H I region), is thus located inside the blob. The H II region is adjacent to the H I region. The intensity of the Lyα recombination line in the H II region of the blob is 10^3–10^4 times higher than that coming from the central star (Johansson and Letokhov, 2001e; Klimov et al., 2002). We get a unique situation in the H I region where a partially ionized mixture of many elements is exposed to an intense Lyα flux coming from the H II region. This is similar to a typical "laser" situation where the pumping flash lamp irradiates the nearby active laser medium.

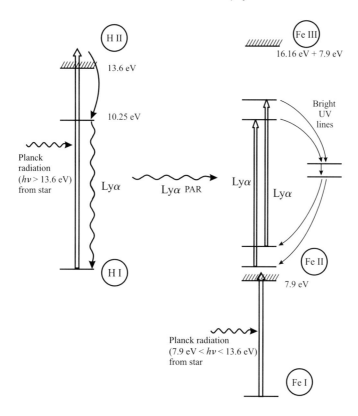

Fig. 11.1 Illustration of the main radiative processes in the Weigelt blobs, involving photoionization of H I and Fe I by stellar black-body radiation and subsequent resonant excitation of Fe II by intense H Lyα radiation.

Preliminary data on *laser effects in the Weigelt blobs* have been published in the form of short letters (Johansson and Letokhov, 2002, 2003a). The more detailed treatment of the laser schemes related to the energy level diagram of Fe II, a detailed analysis of the amplification in various schemes, and a discussion of the spectral width of laser lines were presented by Johansson and Letokhov (2004c). Below we give a description of the Fe II astrophysical laser based on that paper.

11.1 Radiative excitation and relaxation of Fe II levels

A large number of observed as well as predicted absorption lines of Fe II fall within a spectral width of 3 Å around H I Lyα. The transitions start from low metastable states as illustrated in Fig. 9.1 (Johansson and Jordan, 1984). There are about 15 emission lines/Å observed in the laboratory spectrum of Fe II in the wavelength region around Lyα. The number is probably even higher for absorption lines, some of which could be pumped by Lyα in astrophysical plasmas. Consequently, numerous Lyα pumped fluorescence lines of Fe II have been identified in spectra of various emission line sources, e.g. chromospheres of the Sun and cool stars (Jordan, 1988a, b; Harper et al., 2001), symbiotic stars (Hartman and Johansson, 2000), and the environment of AGNs (Netzer, 1998).

The Weigelt blobs outside η Car are rather special because of their high optical density for Lyα, $\tau_\alpha \approx 10^8$, and their short distance to the central star. The ratio of the Einstein coefficients for Lyα and the pumped absorption lines of Fe II is $\approx 10^2$, and since all Fe atoms in the HI region are ionized the abundance ratio $N_{Fe+}/N_H \approx 10^{-4}$. Since the optical density for the Fe II absorption lines is in the range $\tau_{Fe\,II} \approx 10$–100, the resonant intervals within the wide Lyα profile are fully absorbed. A significant density of absorption lines should result in a total absorption of Lyα in the HI region of the blob.

The photoselective excitation rate of state 4 is defined by

$$W_{exc}^{14} = A_{41} \frac{g_4}{g_1} \left[\exp\left(\frac{h\nu_{14}}{kT_\alpha}\right) - 1 \right]^{-1}, \qquad (11.1)$$

where the indices 1 and 4 correspond to the level notations in Figs. 11.2b, c. The Einstein coefficient for spontaneous decay is $A_{41} = 1.2 \cdot 10^7$ s^{-1} (Kurucz, 2002), and $T_\alpha = T_{eff}(HI\,Ly\alpha)$ is the brightness (or effective) temperature of Lyα inside blob B. According to the qualitative picture given above, the stellar Lyman continuum radiation is transformed into intense Lyα radiation. A more detailed analysis of this *"spectral compression" effect* is presented in (Johansson and Letokhov, 2004a), and it shows that T_α is *comparable with the effective temperature of the stellar photosphere* (Section 10.4). In the case of η Car, we adopt $kT_\alpha \approx 1.0$–1.5 eV $(T_\alpha \approx (12$–$18) \cdot 10^3$ K), which means a photoselective excitation rate $W_{exc}^{14} \approx 10^3$–$10^4$ s^{-1}. We must emphasize that, without compensation for the dilution factor by the "spectral compression" effect the value of W_{exc}^{14} will be about 1 s^{-1}, making it impossible to provide an inverted population in the $3 \to 2$ transition.

The numerous possible pathways of the radiative relaxation (decay) of the Lyα-excited Fe II levels can be grouped into two categories as illustrated in Fig. 11.2:

(1) spontaneous decay back to the low metastable states without any accumulation in higher states (Fig. 11.2a);

(2) spontaneous decay via long-lived higher states ("bottleneck"), which cause an inverted population with an associated stimulated emission (Figs. 11.2b, c).

The first type of pathway (Fig. 11.2a) is dominating for all photoexcited states, and the type illustrated in Fig. 11.2b occurs as a result of *accidental level mixing* (ALM), discussed in Section 11.2. The general case is a combination of (1) and (2), where the branching fractions from level 4 determine the population in 3. The branching fractions for the decay feeding the long-lived states are thus determined by the strength of the level mixing. In case (2), i.e. those Lyα pumped Fe II levels having large branching fractions to feed the *pseudo-metastable (PM) states*. It should be pointed out that the continuum pumping, shown by Verner et al. (2002) to increase the population of the low metastable states through cascading, cannot directly populate the PM states owing to parity considerations.

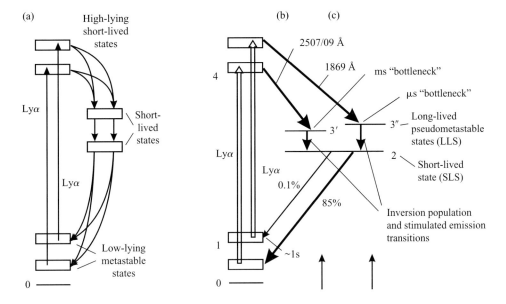

Fig. 11.2 Three types of radiative decay schemes of high-lying Fe II levels photoexcited by Lyα: (a) a fast radiative decay down to metastable states via short-lived levels without any "bottleneck" effect; (b) a radiative decay via high, pseudo-metastable states (the "bottleneck" effect) with formation of an inverted population; (c) a radiative decay with "bottleneck" effect, inversion of population and return to the initial state.

If level 2 in case (2) decays to the initial state, i.e. the level being pumped, we obtain a closed radiative cycle (Johansson and Letokhov, 2003a). This is determined by the possible decay channels of state 2 and their branching fractions, which may vary among the set of short-lived states (lifetime of a few ns) fed by the decay from the PM states. The distribution of the decay of state 2 is also important for permanent pumping of the low-lying metastable states, with lifetimes of the order of one second. Two examples of pathways containing radiative "bottlenecks" are shown in Figs. 11.2b, c. They represent closed radiative cycles, but with different efficiencies because the branching fractions differ considerably in those transitions that bring them back to the initial level. These two pathways of radiative decay also have different lifetimes of the "bottleneck" levels 3′ and 3″: a ms-lifetime for level 3′ and a μs-lifetime for 3″ (Figs. 11.2b, c). The branching fractions for the weakly closed cycle (0.1%) and for the strongly closed cycle (85%) are indicated in Fig. 11.2b, c.

The rate of collisions between excited Fe⁺ ions and H atoms, with a maximum density of 10^{10} cm^{-3}, is less than one per second. The presence of forbidden lines from the long-lived metastable states (lifetime of about 1 s) verifies that the relaxation of the excited Fe⁺ ions is purely radiative (Fig. 11.3). Most important are the "closed" radiative pathways including Lyα pumping, spontaneous decays and stimulated emission,

148 *Astrophysical lasers*

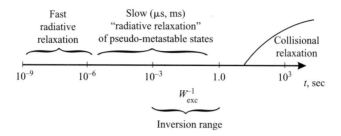

Fig. 11.3 Time scale of spontaneous radiative and collisional relaxation of highly excited Fe II states. An intermediate range including pseudo-metastable states is favorable for obtaining an inverted population in the $3 \to 2$ transition in Fig. 11.2.

since they provide a collision-free and pure radiative conversion of intense Lyα radiation into the intense and anomalous $\lambda\lambda 2507/09$ UV lines of Fe II (Johansson and Letokhov, 2003a).

11.2 Spectral peculiarities in Fe II

Two properties, one atomic and one astrophysical, make the spectrum of Fe II very dominant in the spectra of cosmic sources. A complex atomic structure with a large density of energy levels and a high cosmic iron abundance result in numerous lines in ultraviolet and optical spectra – absorption lines in stellar atmospheres and emission lines in nebular regions. Although Fe II has been thoroughly studied in the laboratory and more than 1000 energy levels have been found, ultraviolet stellar spectra still contain unidentified lines of Fe II.

The specific atomic astrophysics problem discussed in this Chapter involves two special features that are extremely rare in simple spectra but characteristic for complex spectra: *accidental level mixing* (ALM) and the presence of "*pseudo-metastable*" (PM) states. Level mixing is a well-known effect, which in its simplest form depends on deviations from a pure coupling scheme for the electrons, and a final state is described as a linear combination of, for example, pure LS states. Such mixing of levels may be well predicted by theory even for complex systems. Perturbation and level mixing as a result of configuration interaction are also in general taken care of by detailed calculations. However, two energy levels having the same parity and J-value may also mix if they accidentally fall very close to each other in energy. They can have completely different LS character and belong to different configurations. Such an ALM is very difficult to predict in quantum mechanical calculations.

The presence of PM states is the crucial feature in the radiative processes discussed in this paper, but before we define the concept "pseudo-metastable" we give a brief description of metastable states. A metastable state in an atom (ion) is a state that cannot decay to a lower state by electric dipole (E1) radiation, i.e. obeying the selection rules for parity and J-value. Hence, all states of even parity that are located below the lowest state of odd parity are by definition metastable, and they have radiative lifetimes of the order of 1 s. Metastable states are generally collisionally deexcited but in low-density astrophysical plasmas the probability for collisions is very low. Under such nebular conditions the metastable states decay radiatively in electric quadrupole (E2) or magnetic dipole (M1) transitions observed as forbidden lines. There are 62 metastable states in Fe II belonging to the three configurations $3d^64s$, $3d^7$ and $3d^54s^2$. One of the three low even-parity configurations containing the metastable states is the ground configuration $3d^64s$. In Fig. 3.7 we show a condensed energy level diagram for Fe II, arranged according to the "branching rule" for the build-up of the structure of complex spectra. Thus, by starting with the next higher ion, Fe III, we add a 4s-electron to each LS term (the parent term) of the ground configuration $3d^6$. The resulting LS terms in Fe II are represented by one box for each parent term. All 4s boxes are connected by a dotted line in Fig. 3.7. Since the 4s binding energy is independent of the parent term the pattern of connected 4s boxes mimics the pattern of the parent terms. The next higher configuration in the diagram is the odd-parity 4p configuration, represented by gray boxes. Thus, according to the definition all levels in the boxes located below the lowest gray 4p box are metastable. These are placed in a marked box in the diagram.

Owing to the extended structure of the parent terms the 4s configuration extends above 4p, i.e. the highest 4s boxes are located above the lowest 4p box (see Fig. 3.7). This is a special feature of the transition elements. In Fe II, six observed (and two unknown) LS terms of the $3d^64s$ configuration are located above the lowest LS terms of $3d^64p$, which means that a 4s electron can make a *downward* transition to a 4p state. The lifetime of these high 4s states is of the order of 1 ms, three orders of magnitude shorter than for pure, metastable states but six orders of magnitude longer than for ordinary, excited states. Because of their long lifetimes we call them PM states, and they are marked in a dashed box in Fig. 3.7. It looks as if the PM states keep a fraction of the metastability of their sister LS terms in the same configuration, even if LS-allowed decay channels become available. Without looking at the spectroscopic origin of these PM states, but just their LS assignment, it is not possible to distinguish them from other LS terms. However, the transition probability of their decay to 4p is six orders of magnitude smaller than for normal allowed transitions. It turns out that this is the crucial feature behind the explanation of the laser action in Fe II observed in Eta Carinae.

The two cases of four-level schemes in Fe II involved in the radiative processes are shown in Figs. 9.1a and 9.2b. The low 4s box (at 1 eV) in Fig. 9.15a is the a^4D term in the lowest 4s box in Fig. 3.7. It can be pumped by Lyα to several 5p levels around 11 eV, which is discussed in detail in Section 11.3. Owing to a near coincidence in energy, there is an ALM between some of the lowest 5p levels and some high 4p levels, as indicated in Fig. 3.7, and a mixing of states from two consecutive members (4p and 5p) in a Rydberg series is very rare. The ALM means that the high 4p levels (denoted 4p" in Fig. 9.1a), which belong to LS terms of the $(b^3F)4s$ and $(b^3F)4p$ subconfigurations, are also radiatively populated by Lyα. The fast decay from 5p is to 5s (in the near-IR) and back to 4s via 4p. The fast decay from 4p" is down to 4s" in Fig. 9.1a, i.e. along the vertical line down to $(b^3F)4s$ and $(b^3P)4s$ in Fig. 3.7. With all other strong 4s–4p transitions the corresponding lines fall around 2500 Å. These 4s" levels are the PM states discussed above, and they can decay in slow transitions to the low 4p states at 5 eV (40,000 cm^{-1}). The 4p levels have fast decays back to the metastable 4s levels, as shown in Fig. 9.1a.

The case outlined by the scheme in Fig. 9.1b starts from a higher metastable 4s level (marked 4s'), which is excited up to a 4s4p state by Lyα radiation. The natural decay for this pumped state is down to a 4s^2 state at about 7 eV. This level belongs to the 3d^{5}4s^2 configuration, which is not included in Fig. 3.7. It has about the same properties as the PM states discussed in the first case above, but the lifetime is about three orders of magnitude smaller, or about 1 µs. The 4s^2 state decays in a "two-electron jump" to 4p, i.e. to the same 4p levels as for the 4s" state in the first case. The 4p level has a branch to the metastable 4s' level and even in this case we can find a closed cycle for the radiative excitation and deexcitation.

11.3 Population inversion and amplification in Fe II transitions

There are a number of decay schemes in Fe II creating an inverted population due to a radiative "bottleneck". The main schemes are shown in Figs. 9.1a, b, where few selected energy levels and radiative transitions from the level- and line-rich spectrum of the complex system of Fe II illustrate a photo-selective excitation by Lyα and a subsequent cascade of radiative decay channels. A remarkable feature in the decay schemes is the existence of PM states. They decay relatively slowly (radiative lifetimes from 10 µs to 1 ms) in permitted transitions to *lower, rapidly decaying* states (lifetimes of a few ns). However, the time elapsing between successive collisions between excited Fe$^+$ ions and H atoms, the main component of the WB with a maximum density of $N_H = 10^{10}$ cm^{-3}, is one second or even longer. Therefore, the relaxation of the Lyα excited Fe$^+$ ions occurs without collisions in a purely radiative way.

The cascade of radiative decay of the Lyα excited states is characterized by a peculiar time hierarchy as illustrated in Fig. 11.3. On the time scale given in the figure one can distinguish between three relaxation regions: a fast spontaneous decay, a slow spontaneous decay, and a much slower collisional relaxation. Therefore, the spontaneous radiative decay from the Lyα pumped states results inevitably in an inverted population in the transition from the PM state (the 3 $\rightarrow$ 2 transition in Fig. 11.2), which acts as a "bottleneck" for radiative decay in the absence of collisions. If the inverted population density and the optical path length for this transition are large enough, the spontaneous radiation will be amplified owing to stimulated emission in the atomic ensemble. The stimulated emission enhances the intensity of the weak lines generated by the slow spontaneous radiative decay of the PM states, and it becomes comparable with the intensity of the fast, spontaneous transitions populating them.

A steady state population inversion is thus achieved in the 3 $\rightarrow$ 2 transition with a density of

$$\Delta N \approx N_3 - N_2 \approx N_3 \qquad (11.2)$$

since $N_2 \ll N_3$ owing to the much faster decay of level 2. The small influence of the statistical weights has been neglected as their ratio is close to one. Since the short-lived levels 4 and 2 can be regarded as empty relative to the long-lived levels 1 and 3 the radiative excitation W_{exc}^{14} produces the following steady-state distribution of the populations N_1 (in the initial level 1) and N_3 (in the long-lived excited level 3):

$$N_3 = N_1 W_{\text{exc}}^{14} \tau_3, \qquad (11.3)$$

where $A_{43} \gg W_{\text{exc}}^{14}$, i.e. coincides with (6.7). Since the rate of stimulated emission in the $3 \rightarrow 2$ transition, W_{32}, is not included in Eq. (11.3), all Fe II ions can be accumulated in level 3 if $W_{\text{exc}}^{14} \gg 1/\tau_3 (N_3 \gg N_1)$. In the lasing volume, however, W_{32} can be much higher than $1/\tau_3$, and a fast closed radiative cycle will therefore maintain the population of level 1 in such a way that

$$N_3 = N_1 \frac{W_{\text{exc}}^{14}}{W_{\text{exc}}^{14} + \tau_3^{-1}}, \qquad (11.4)$$

or even

$$N_3 = N_1, \qquad (11.5)$$

when the excitation rate is large ($W_{\text{exc}}^{14} \gg 1/\tau_3$).

The linear amplification coefficient for the $3 \rightarrow 2$ transition is defined by the standard expression

$$\alpha_{32} = \sigma_{32} \Delta N, \qquad (11.6)$$

where ΔN is the inverse population density defined by Eq. (11.2). The stimulated emission cross-section σ_{32} is given by (6.5) with degeneracy factor g_3/g_2. At the temperature of $T \approx 100\text{--}1000$ K in the relatively cold H I region, $\Delta\nu_D \approx (300\text{--}1000)$ MHz, i.e., $\sigma_{32} = (0.6\text{--}2) \cdot 10^{-13}$ cm^2. Thus, the amplification coefficient may be estimated by the expression

$$\alpha_{32} = \sigma_{32} (W_{\text{exc}}^{14} \tau_3) \beta N_{\text{Fe}}, \qquad (11.7)$$

where β is the fraction of the Fe$^+$ ions in state 1 relative to all Fe$^+$ ions (all irons atoms in the H I region of the blob are supposed to be ionized).

The fraction β is governed by the excitation rate of level 1 and its radiative lifetime τ_1. The excitation rates for collisional population of level 1 (i.e. by recombination of Fe III and electron collisions) are negligible in comparison with the radiative decay rate $1/\tau_1$ because of the low electron density ($n_e \cong 10^4\text{--}10^5$ cm^{-3}) in the H I region. The electrons are generated in the photoionization process of iron as well as other elements having an ionization potential of IP < 13.6 eV. However, the most important channel for populating the metastable state 1 is the radiative decay of the high Fe II states selectively excited by Lyα or other intense lines. These excitation channels can provide for an excitation rate $> 1//\tau_1 \approx 1$ s^{-1} and hence sustain the relative population of state 1 at a level of $\beta \approx 10^{-2}$. This would correspond to an approximately equal distribution of the Fe$^+$ ions among the 60 metastable states, including state 1. Leaving the calculation of a more accurate value of β for future modeling, we adopt here a qualitative estimate of $\beta \approx 10^{-2}$. An evidence for a substantial population in the metastable states is provided by the observation of strong, forbidden lines, e.g. from state 1, in the optical region of HST spectra of the blob (Zethson, 2001).

With these assumptions, we can use Eqs. (11.1) and (11.6) to estimate the amplification coefficient for the $3 \to 2$ transition to $\alpha_{32} \approx (3 \cdot 10^{-18} - 10^{-16}) N_{\text{Fe}} (\text{cm}^{-1})$, where $N_{\text{Fe}} \approx 10^{-4} \cdot N_{\text{H}}$. To create a cold H I region in the blob the hydrogen density N_{H} must exceed a critical value, the so-called critical density $N_{\text{cr}} \approx 10^8 \text{cm}^{-3}$ (Johansson and Letokhov, 2001c, e). Thus, for $\alpha_{32} \geq (3 \cdot 10^{-14} - 10^{-12})$ cm^{-1} and a blob diameter of $D \approx 10^{15}$ cm, which can be regarded as the size L of the amplifying region, $\alpha L \approx (30\text{--}100)$ at an effective blob temperature of Lyα of $T_\alpha \approx (12\text{--}18) \cdot 10^3$ K. The amplification coefficient α_{32} is, according to Eq. (6.5), sensitive to the spectral width of the amplification line. From this point of view the cold H I region is more suitable as a lasing medium than the hotter H II region, causing a higher Doppler width in the $3 \to 2$ transition.

Let us emphasize that the amplification coefficient α_{32} according to Eqs. (11.7) does not depend strongly on the Einstein coefficient A_{32}, since $A_{32} \sim 1/\tau_3$ as long as the branching fraction for the $3 \to 2$ transition is dominating in the decay of state 3. The (αL)-values obtained correspond to fairly high values of the linear (non-saturated) amplification $K = \exp(\alpha L)$. In the saturation regime, however, the intensity of the weak line λ_{32} (in photons/cm$^2 \cdot$ s) approaches that of the strong line λ_{43}. Thereafter, the amplification regime becomes saturated. This means that the rate of stimulated transition approaches the pumping rate of level 3 provided by the spontaneous $4 \to 3$ decay. Under such conditions the intensities of both these lines grow in proportion to the propagation length L.

11.4 Radiative cycling with stimulated emission

Many of the radiative decay schemes of Lyα pumped Fe II levels contain three spontaneous transitions before the electron ends up in a low-lying metastable state. In some of these schemes there is a high probability that the final state of the electron is the same as the initial state, i.e. there is a high branching fraction of the final transition connecting states 2 and 1. This is the case illustrated in Fig. 9.1a. However, the resonant photoexcitation of an atom (ion) and the subsequent spontaneous decay may lead to an accumulation of atoms in an intermediate long-lived state, in which the atom resides for a long time ("a dark state"). Such a bottleneck in the spontaneous decay chain limits the cycling rate of photoexcitation + fluorescence, i.e., limits the efficiency of the conversion of UV photons into less energetic fluorescence from the absorbing atom. This effect is illustrated in Fig. 11.4 in a simplified scheme of a four-level atom, which represents a number of real cases in multilevel atoms (ions). Let us consider the "cycling effect" from a more general point of view.

The bottleneck in the radiative decay chain, e.g., level 3 in Fig. 11.4a, is fed by photoexcitation in $1 \to 4$ and a fast spontaneous decay, $4 \to 3$. This leads to an accumulation of atoms in the pseudometastable state 3 having a lifetime $\tau_3 \gg \tau_4$. For a high photoexcitation rate $W_{14}^{\text{exc}} \gg A_{32}$, the intensities of the fluorescent lines $4 \to 3$ and $2 \to 1$ are limited by the intensity of the "weak" transition $3 \to 2$ owing to the transfer of atoms from the initial state 1 to the long-lived state 3. In a stellar atmosphere, for example, the density is high enough for collisional relaxation $3 \to 2$ to take place (Fig. 11.4b). It is, however, effective in a cyclic process (Thackeray, 1935) only if state

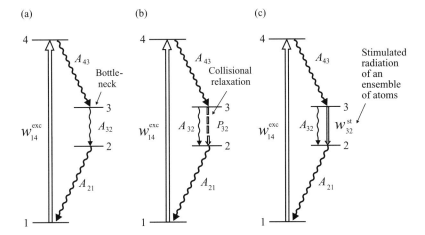

Fig. 11.4 A radiative cycle in a four-level atom having allowed radiative transitions $1 \to 4$, $4 \to 3$, $2 \to 1$ and a metastable (pseudo-metastable) state 3 with a slow spontaneous radiative decay $3 \to 2$: (*a*) an isolated atom without collisions (on the time scale A_{32}^{-1}); (*b*) an atom in a high-density environment, which provides a fast collisional relaxation in the transition $3 \to 2$; (*c*) a collision-free ensemble of atoms with an inverted population in the $3 \to 2$ transition and a significant amplification, which provides a fast stimulated radiative relaxation $3 \to 2$.

2 is the ground state of the atom. Otherwise, the collisions will open the closed cycle by relaxation from state 3 to other excited states. The accumulation of atoms in state 3 creates a population inversion in the transition $3 \to 2$ at an appropriate column density $\Delta N_{32} L$, where ΔN_{32} is the population density and L is the size of the ensemble of atoms. If $W_{14}^{exc} \gg A_{32}$, the rate of stimulated transitions $W_{32}^{st} \gg A_{32}$, and the channel of weak spontaneous decay will be filled up by fast stimulated decay (Fig. 11.4*c*). Thus, the rate for radiative cycling becomes enhanced as well as the energy transfer from the pumping UV line to the fluorescence lines. We showed (Johansson and Letokhov, 2003a) that such a radiative cycle with stimulated radiation operates in Fe II pumped by intense HLyα radiation in compact gas condensations. This mechanism explains the origin of two unusually bright fluorescence lines at 250.7 and 250.9 nm observed in spectra of high spectral and spatial resolution recorded by STIS onboard the Hubble Space Telescope (Kimble *et al.*, 1998).

11.4.1 Strong radiative cycle

Let us discuss the closed cycle containing Lyα absorption, spontaneous decay with emission of the bright UV $\lambda\lambda 2507/2509$ lines, stimulated emission at 1.68 and 1.74 µm, and a fast UV transition to the initial state. The diagram in Fig. 11.5 includes a few of the known energy levels of Fe II, which, together with the inserted transitions and atomic data, are relevant in the present study of a strong radiative cycle. The broad (a few hundred cm^{-1}) Lyα line almost coincides in wavelength with two Fe II transitions level $a^4 D_{7/2}$ (level 1 in Fig. 11.5) to the close levels $(^5D)5p\,^6F^0_{9/2}$ and $(b^3F)4p\,^4G^0_{9/2}$

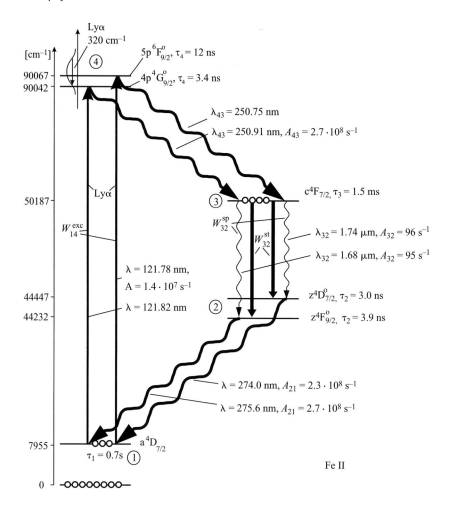

Fig. 11.5 A partial energy level diagram (with parameter values) of Fe II showing a closed loop with strong radiative cycling. The cycling includes a selective photoexcitation of the highest level 4 by intense Lyα radiation and a millisecond pseudo-metastable "bottleneck", causing population inversion. (Lifetimes and A-values are from Bergeson et al., 1996; Kurucz, 2002; Raassen, 2002).

(marked 4 in Fig. 11.5), the frequency difference (detuning) being $\Delta\nu = -160$ cm^{-1} and -185 cm^{-1}, respectively. The detuning is compensated by the Doppler shift or the broadening of Lyα (see Section 11.6). The line at 2509.1 Å to state 3 (c^{4}F$_{7/2}$) is the main decay channel from (b^{3}F)4p^{4}G$^0_{9/2}$ (state 4) with a branching fraction of about $\gamma \approx 0.9$. Owing to the small energy difference there is an ALM between the two energy levels in state 4, (^{5}D)5p^{6}F$^0_{9/2}$ and (b^{3}F)4p^{4}G$^0_{9/2}$, resulting in similar decay schemes (Johansson, 1978). As discussed above, level 3 is a PM state, with a lifetime of 1.5 ms, but in contrast to a pure, metastable state it can decay slowly by electric dipole radiation to the short-lived states 2 ($z^4D^0_{7/2}$ and $z^4F^0_{9/2}$). To get strong radiative cycling it is important that state 2 decays by returning a large fraction of the Fe$^+$ ions to the initial state 1, which is metastable with a radiative lifetime of about one second. Assuming the blob conditions discussed in the previous section 7 ($N \approx 10^9$–10^{10} cm^{-3}) the time scale for collisions is much longer than the time scale for radiative decay of state 3. Under these conditions the closed radiative cycle $1 \to 4 \to 3 \to 2 \to 1$ has a "bottleneck" in the $3 \to 2$ transition. The linear amplification coefficient for the $3 \to 2$ transitions at 1.74 μm and 1.68 μm (Fig. 11.5) is defined for $W^{14}_{\text{exc}} \ll \tau_3^{-1}$ (a rather moderate requirement) by Eqs. (11.4) and (11.7), rewritten as:

$$\alpha_{32} L = \left[\frac{\lambda_{32}^2}{2\pi} \frac{A_{32}}{2\pi \Delta\nu} \right] \tau_3 (\beta N_{\text{Fe}}). \tag{11.8}$$

From Eq. (11.8.) we get that the amplification coefficient α_{32} (cm^{-1}) for the scheme in Fig. 11.5 is largely determined by the factor ($A_{32}\tau_3$). Its maximum value is $A_{32}\tau_3 = 1$ in the ideal case, i.e. when there is only one radiative channel of spontaneous decay from state 3. In the real case (see Fig. 11.5) discussed here the factor $A_{32}\tau_3 \approx 0.15$. Rough estimates of $\alpha_{32} L \geq 10$, where $L \approx 10^{15}$ cm is the length of the amplifying medium, can be obtained for $T_\alpha > 10{,}000$ K and $N_{\text{H}} \simeq 10^8$ cm^{-3} or $T_\alpha > 8000$ K and $N_{\text{H}} \simeq 10^9$ cm^{-3}.

For an ensemble of atoms with sufficient density and size the channel $3 \to 2$ will open for a fast stimulated decay, and the duration of the whole radiative cycle is determined by the rate of the slowest excitation channel rather than by the W^{14}_{exc} rate of the slowest spontaneous decay A_{32}. For a high brightness temperature of Lyα in the range $T_\alpha \approx 15 \cdot 10^3$ K the excitation rate is $W^{14}_{\text{exc}} \approx 5 \cdot 10^3$ s^{-1}, giving a duration of the total radiative cycle in Fe II of $\tau_{\text{cycle}} \approx 2 \cdot 10^{-4}$ s. Hence, the Fe ions can undergo a maximum number of cycles ($\tau_1/\tau_{\text{cycle}} \approx 3.5 \cdot 10^3$) during the lifetime of the initial state τ_1.

A large number of radiative cycles containing a stimulated channel provide a large intensity enhancement of the $\lambda\lambda 2507/2509$ fluorescence lines ($4 \to 3$) due to the suppressed accumulation of Fe$^+$ ions in the PM state 3 and the corresponding depletion of the initial state 1. If the density N_{Fe} and the size L are not sufficiently large for the ensemble of atoms the amplification will be small and the stimulated channel will not operate. In such a case the radiative "bottleneck" limits the rate of the radiative cycling in Fe II to the time $\tau_1 = 1.5$ ms and the intensities of the two UV lines λ_{43} become normal.

High-resolution spectra of the blobs outside η Car, spatially resolved from the central star, show the anomalous brightness of the two $\lambda\lambda 2507/2509$ lines. Further observational evidence in the blob spectrum for the radiative cycle is the presence of one "bottleneck" line at 8674.7 Å (Zethson, 2001), which is an intercombination line with an extremely low transition probability $A_{32} \approx 13$ s^{-1} (Raassen, 2002). The "bottleneck" transitions around 1.7 μm indicated in Fig. 11.5 have been observed in ground-based spectra of η Car (Hamann et al., 1994). The $3 \to 2$ transition from c^{4}F$_{7/2}$ at $\lambda = 1.74$ μm and 1.68 μm appear relatively stronger than lines from the other fine structure levels of c$_4$F. Since the ground-based spectra of η Car contain integrated light from a larger region than the Weigelt blobs discussed here the absolute intensities cannot be compared with the intensities measured in the HST spectra.

Moreover, according to the observations made in (Gull et al., 2001), the UV lines of the 2→1 transition at 2740 and 2756 Å have an integrated intensity comparable with that of the bright lines of the 4→1 transition. The width of these spectral lines is great because of the substantial optical density in the 4→1 transition and the corresponding resonance transfer radiation broadening. Resonance radiation trapping increases the effective lifetime of level 2, but still it remains much shorter than the long lifetime of level 3 and does not prevent the formation of an inverted population in the 3→2 transition.

11.4.2 Weak radiative cycle

In some of the schemes of Lyα pumped Fe II levels there is a low probability that the final state of the electron is the same as the initial state, i.e. there is a low branching fraction of the final transition connecting states 2 and 1. Such a case is illustrated in Fig. 9.1b. A more detailed scheme of the microsecond "bottleneck" with relevant atomic data is given in Fig. 11.6. The primary spontaneous decay of state 2 is to the metastable state 1a at 1 eV, and a weaker branch populates the initial state (1b at about 3 eV) of the closed radiative cycle. In this case, state 1a is the same as state 1 in the millisecond "bottleneck" discussed in previous Section 11.4.1, which gave a strong radiative cycling. The low branching fraction of 0.1 % for the 2→1b transition at 5116 Å produces a closed cycle but a weak radiative cycling. Some feeding of the initial level, state 1b, is provided by the radiative cycle discussed in Section 11.4.1 which means that there is only need for an initial population in state 1a to start the pumping processes of both the ms- and μs-"bottleneck" cycles.

The Lyα radiation populates a high-lying state of Fe II from the metastable state a^{4}G$_{11/2}$. This excitation scheme is operative in various objects, verified by the observation of the spontaneous radiative decay in the primary cascade as a fluorescence line at 1869 Å (Johansson and Jordan, 1984) as well as the secondary cascade at 9997 Å seen in many objects as a prominent feature. Based on the extended laboratory analysis of Fe II (Johansson, 1978), the 9997 Å line was first identified (Johansson, 1977) in the near-IR spectrum of η Car (Thackeray, 1969). A plausible explanation of its appearance in the spectra of some astrophysical objects was ascribed to the excitation scheme shown in Fig. 11.6 (Johansson and Zethson, 1999).

According to theoretical calculations (Kurucz, 2002), the main radiative decay from state 4, 3d^5(a^{2}F)4s4p(^{3}P)^{4}G$^0_{11/2}$ (hereafter sp^{4}G), occurs to the fine-structure levels of b^{4}G with branching fractions of 0.83 to $J = 11/2$ and 0.04 to $J = 9/2$. The most natural representative of state 3 in Figs. 9.1b and 11.6 is thus b^{4}G$_{11/2}$ with a radiative lifetime of 11 μs. Hence, it has a relatively slow spontaneous decay, $A_{32} = 8 \cdot 10^4$ s^{-1}, to level 2 (z^{4}F$^0_{9/2}$), which has a much shorter lifetime ($\approx$3 ns).

Using HST/STIS, Gull et al. (2001) have recorded spectra of blob B in η Car at high angular and spectral resolution. The spectrum contains the intense 9997 Å line (see Fig. 11.7), which can be excited by Lyα radiation according to the excitation and decay scheme shown in Fig. 11.6. The 9997 Å line is thus a transition from a microsecond PM state, which belongs to a different configuration (3d^{5}4s^2) from the millisecond

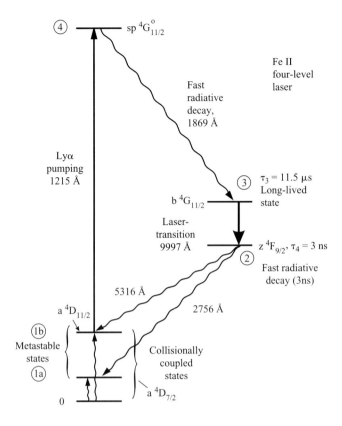

Fig. 11.6 Diagram showing the creation of population inversion in the 3 → 2 transition of Fe II because of a microsecond "bottleneck". Since only a small fraction of the Lyα pumped Fe II ions return to the initial state, 1b, the closed loop 1b → 4 → 3 → 2 → 1b generates a weak radiative cycling.

PM state in Section 11.4.1. Since the lower state of the 9997 Å transition is short-lived (a few ns) an inverted population is built up, and the gain is determined by the rate of photoexcitation provided by the Lyα radiation. The high intensity is naturally explained by stimulated emission of radiation, and the amplification coefficient of this transition is about one order of magnitude higher than in the case of the ms "bottleneck" because of the factor $(A_{32}\tau_3)$ in Eq. (11.8).

11.4.3 Combination of strong and weak cycles

The existence of a great number of metastable states and the high density of high-lying energy levels in Fe II cause several coincidences between absorption lines and Lyα and thereby the generation of inverted population. In this Section we give examples of cascade schemes from the same millisecond "bottleneck" but with both strong and weak radiative cycles.

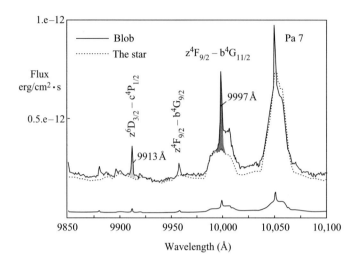

Fig. 11.7 Comparison of HST/STIS spectra of the central star (flux scale on the right-hand axis) and Weigelt blob B (flux scale on the left-hand axis) in η Car showing the intensity enhanced blob lines at 9913 and 9997 Å (see details in text).

Fig. 11.8 presents two transition schemes, which provide for the amplification of the extremely weak Fe II lines at 9391.5 Å, 9617.6 Å and 9913.0 Å and some other lines around 2 μm. The difference between the two groups is only the PM states, which have different J-values but belong to the same LS term. The two schemes represent a combination of the previous ones discussed in Sections 11.4.1 and 11.4.2 due to the fact that there are two sets of possible "bottleneck" $3 \rightarrow 2$ transitions. In one set, state 2 consists of sextet levels explaining the extremely low transition probability A_{32} in the intercombination lines at $\lambda\lambda 9391/9617$ in Fig. 11.8a and $\lambda 9913$ in Fig. 11.8b. Since the sextet levels decay mainly to the ground term in Fe II and only in a very slow decay to the initial level, the closed $1 \rightarrow 4 \rightarrow 3 \rightarrow 2 \rightarrow 1$ radiative cycle becomes very weak. However, as with the case described in Section 11.4.1 PM states can also decay to quartet levels in state 2, which combine strongly with the initial state. The corresponding PM transitions occur in the near-IR region around 2 μm.

The amplification coefficient for the 2 μm lines should be of the same order of magnitude as for the 1.7 μm lines in Fig. 11.5, whereas it is reduced for the lines in the range 9300–9900 Å owing to the smaller A value. The reduction of the Einstein coefficients A_{32} has, in principle, no effect on the amplification coefficient, provided that the upper level 3 decays radiatively only to one low laser level. However, this is not the case here, and the faster decay to the quartet levels in a time of $\tau_3 \approx A_{32}^{-1}$ reduces the amplification coefficient (11.6) for the transitions to the sextet levels by a factor of $A_{32}\tau_3 \approx 10^{-2}$. But, the large gain margin is sufficient to ensure $\alpha L \geq 10$, even for the weak lines, and all three lines mentioned above are observed and identified in the blob spectrum of η Car (Zethson, 2001). Two of them, the laser lines at 9617 Å and 9913 Å, are shown on Fig. 11.9 as intense features in the blob spectrum but not

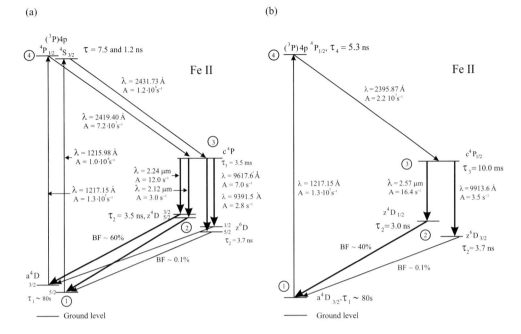

Fig. 11.8 Partial energy diagrams of Fe II showing two examples of combined strong and weak radiative cycling due to alternative decay routes of the two pseudo-metastable states $c^4P_{3/2}$ (a) and $c^4P_{1/2}$ (b). The "bottleneck" effect generates inverted population and lasing in the $3 \to 2$ transitions to sextet and quartet levels, which have quite different branching fractions (BF) to the initial state 1.

observed in the stellar spectrum. Moreover, the two lines at 9617.6 Å and 9391.5 Å should be very weak in spontaneous emission and have an intensity ratio of 2.5, determined by the branching ratio (Gull et al., 2001) as intense lines with equal intensities. This is naturally explained by their formation through a stimulated mechanism, which operates under saturated excitation conditions with a common pumping source and a common upper level. The "peculiar" intensity ratio observed for the $\lambda\lambda 9391/9617$ lines forms perhaps the strongest argument for lasing (Hillier, private communication). The laser lines around 2 μm in the strong radiative cycle of Figs. 11.8a, b appear, like the 1.7 μm lines in Fig. 11.5, as prominent features in the ground spectrum of η Car (Hamann et al., 1994). The absence of the laser lines in the stellar spectrum (see Fig. 11.9) is due to collisional relaxation of the PM states in the stellar atmosphere/wind implying a Boltzmann population distribution.

11.5 Effective temperature of Lyα from radiative cycle

The functioning of a radiative cycle with stimulated Fe II emission initiated by selective excitation by Lyα according to the scheme in Fig. 11.5 gives an alternative method of estimating the effective temperature of Lyα, independently of the estimation described

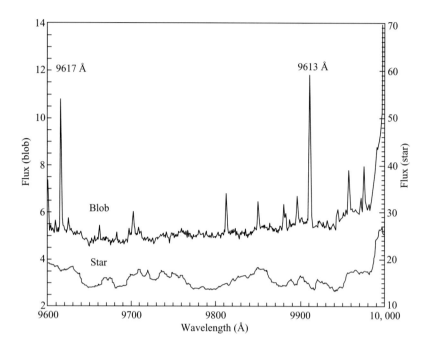

Fig. 11.9 Comparison of HST/STIS spectra of the central star (flux scale in right axis) and the Weigelt blob B (flux scale on left axis) in η Car with the two lasing lines $\lambda\lambda 9617, 9913$ generated in the level schemes in Fig. 11.8 and discussed in Section 11.4.3.

in Section 10.2 (Johansson and Letokhov, 2004a). The selective photoexcitation by Lyα radiation transfers Fe II from the low-lying metastable state 1 to the short-lived odd states 4 ($5p^6F^0_{9/2}$, $4p^4G^0_{9/2}$). These levels decay to state 3 (predominantly to $c^4F_{7/2}$ and, to a lesser extent, to $c^4F_{9/2}$), emitting intense $\lambda\lambda 2507/2509$ Å lines. For $T(Ly\alpha) = T_\alpha \gtrsim 12 \cdot 10^3$ K, the photoexcitation rate $W_{\text{exc}}(1 \to 4)$ exceeds the radiative decay rate for states 3:

$$W_{\text{exc}} = A_{41} \left[\exp\left(\frac{h\nu_{14}}{kT_\alpha}\right) - 1 \right]^{-1} \gtrsim \frac{1}{\tau_3} \simeq 10^3 \text{ s}^{-1}, \qquad (11.9)$$

where $A_{41} = 1.4 \cdot 10^7$ s^{-1} is the Einstein coefficient for the $4 \to 1$ transition. Thus, if condition (11.9) is satisfied, then Fe$^+$ ions will be accumulated in the pseudo-metastable $c^4F_{9/2,7/2}$ states. The lifetimes of these levels are of the order of a millisecond, because they are associated with the truly metastable Fe II levels, but they lie above the next, higher configuration of opposite parity in the complex atomic structure (state 2 in Fig. 11.5). This accumulation of ions in "pseudo-metastable" states is unusual and is an important ingredient of a model deriving the origin of the anomalous $\lambda\lambda 2507/2509$ A lines. It should be emphasized that at densities $N_H \ll 10^{13}$ cm^{-3}, collisions give no contribution to the relaxation of the long-lived Fe II states.

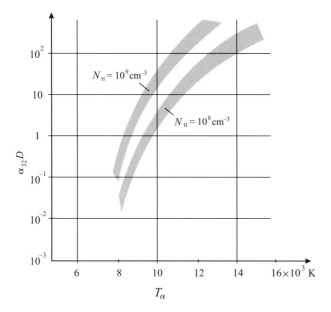

Fig. 11.10 The exponent $(\alpha_{32}D)$ in the amplification factor $K = \exp(\alpha_{32}D)$ versus the effective (spectral) radiation temperature T_α for two hydrogen densities N_H.

The accumulation of Fe II in the 'pseudo-metastable' state 3, $c^4F_{7/2}$, automatically leads to population inversion levels with respect to the low-lying short-lived states 2 ($z^4D^0_{7/2}$ and $z^4F_{9/2}$). For a sufficiently large WB and high Fe II density, the arising amplification increases the intensity of the spontaneous radiation due to *stimulated* transitions whose rate is much higher than the spontaneous decay rate. Note that the inverse population of $\Delta N_{32} = N_3 - N_2$ emerges always, irrespective of W_exc, because the lifetime of the lower level 2 is very short (3–4 ns). Satisfying condition (11.9), the inverse population is at a maximum, whereas for $W_\mathrm{exc} \ll 1/\tau_3$, the amplification factor α_{32} for the $3 \to 2$ is given by (11.7). The stimulated radiation cross section for the $3 \to 2$ transition is $\sigma_{32} \simeq (1.2$–$3.7) \cdot 10^{-16}$ cm^2, and $N_1 = f_0 N_\mathrm{Fe}$ is the population density of the initial level 1, where f_0 is the fraction of the Fe$^+$ ions in state 1. Fig. 11.10 shows a plot of the exponent $G = \sigma_{32}D$ in the amplification factor $K = \exp(\alpha_{32}D)$ for the $3 \to 2$ transition against the spectral temperature T_α, for two hydrogen densities N_H in the WB and for $D = 10^5$ and $f \simeq 10^{-2}$. The uncertainty in $(\alpha_{32}D)$ is attributable to the temperature uncertainty in the H I region, $T = 100$–1000 K, and the corresponding uncertainty in the Doppler width for the $3 \to 2$ transition, $\Delta \nu_\mathrm{D} \simeq (200 - 600)$ MHz. A decrease in N_3 for $W_\mathrm{exc} \ll 10^3$ s^{-1} at $T_\alpha \leq 12 \cdot 10^3$ K may well be offset by an increase in Fe II density to achieve the required amplification factor. For example, at $T_\alpha \simeq 10^4$ K, the excitation rate is $W_\mathrm{exc} = 10^2$ s^{-1}; to obtain a high amplification factor and, accordingly, the $1 \to 4 \to 3 \to 2 \to 1$ radiative cycle involving stimulated radiation, the population of the initial state N_1 should be increased by a factor of 10 by increasing the WB density and, accordingly, N_Fe by a factor of 10. If, however,

T_α is lower than 8000–9000 K, a large decrease in amplification takes place, which is difficult to offset by increasing N_1. Therefore, an independent estimate of $T_\alpha \gtrsim 10^4$ K follows from the experimental observation of the intense Fe II $\lambda\lambda 2507/2509$ Å lines that arise from cyclic transitions.

11.6 Sources of Lyα radiation pumping of Fe II in the Weigelt blobs

The blobs in η Car are embedded in the photospheric radiation and the wind from the central star. Let us speculate about possible sources for Lyα radiation capable of pumping Fe II within the blobs. Firstly, the Lyα radiation emitted by the photosphere of η Car can excite Fe$^+$ ions in blobs located at distances of $R_b \approx (300\text{--}1000) \cdot r_s$, where $r_s \approx 3 \cdot 10^{13}$ cm is the stellar radius. But, even at such a relatively close location of the blob a dilution factor of $w = (r_s/2R_b)^2 = 10^{-6}\text{--}10^{-5}$ substantially weakens the intensity of the Lyα radiation reaching the blob and makes it incapable of producing the bright 2507/2509 Å lines (Johansson and Letokhov, 2001e).

The stellar wind of η Car is an important characteristic of the star's surroundings, the mass loss rate of the star being substantial (Davidson and Humphreys, 1997). Spherical (Hillier et al., 2001) as well as aspherical (Smith et al., 2003) models of the stellar wind of η Car with a terminal velocity of $v_t \simeq 500\text{--}1000$ km · s^{-1} have been considered. In accordance with models by Lamers and Cassinelli (1999) of the stellar wind its velocity is almost close to the terminal value everywhere between η Car and the WB. The Lyman continuum radiation from the star, Ly$_c$, photoionizes the stellar wind and, after the subsequent recombination, a wide Lyα profile is produced. Another potential source of ionizing radiation is a binary companion, but in the discussion below about Lyα formation inside the blob it is not important to distinguish between the individual components in a binary system. It is possible to consider three cases of irradiation of the blob by radiation from the stellar wind depending on the size of its Strömgren radius in the stellar wind, $\tilde{R}_{\text{Str}}$ relative to the distance R_b, between the blob and the star as illustrated in Figs. 11.11a–c.

(a) $\tilde{R}_{\text{Str}} \gg R_b$ (Fig. 11.11a). In this case the Ly$_c$ radiation reaches the blob without significant attenuation and can photoionize the front part of it. If the hydrogen density in the blob exceeds the critical density $N_{\text{cr}} \simeq 10^8$ cm^{-3} (Johansson and Letokhov, 2001c, e; Klimov et al., 2002) the Strömgren boundary of the blob, defined intersects the blob. The front part of the blob produces an intense Lyα radiation with $T_\alpha > 10^4$ K. Thus, the dilution of the photospheric Lyα radiation falling on to the blob is to a great degree compensated by spectral compression of absorbed broadband radiation to a $10^3\text{--}10^4$ times more narrow Lyα profile (Johansson and Letokhov, 2004a). Under these conditions the blob is very different from a typical planetary nebula since the dilution factor $w = (r_s/2L)^2$ is *many orders of magnitude greater* than for a planetary nebula $(w \simeq 10^{-15}\text{--}10^{-12})$.

(b) $\tilde{R}_{\text{Str}} \ll R_b$ (Fig. 11.11b). In this case the Lyman continuum is blocked out for the blob. This means that the entire volume of space between η Car and the blob is an ionized H I region, with the Strömgren boundary for the stellar wind being located between the star and the blob. Since hydrogen is not fully ionized, the blob does not

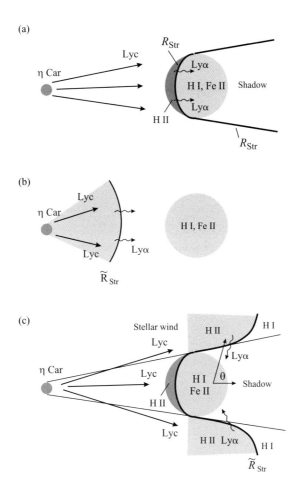

Fig. 11.11 A schematic model of a Weigelt blob, located at a distance R_b from the hot, luminous star η Car and subject to the stellar radiation and wind. (a) A local Strömgren boundary, R_{str}, between the H II and H I regions is located inside the blob. Stellar radiation in the Lyman continuum (Ly$_c$, $h\nu > h\nu_c = 13.6$ eV) photoionizes hydrogen only in the front part of the blob. Radiation in the range 7.6 eV $< h\nu <$ 13.6 eV passes through the H II region and photoionizes iron atoms, which have a density of $N_{Fe} \simeq 10^{-4} N_H$ in the H I region. Intense Lyα radiation from the H II region excites Fe II resonantly in the H I region (lasing volume). The Strömgren boundary for the stellar wind $\tilde{R}_{Str} \gg R_b$; (b) the stellar wind is optically thick for Ly$_c$ and the blob is a one-component H I medium. Only remote, blue-shifted Lyα radiation from the stellar wind can irradiate the H I region containing Fe II; (c) the Strömgren boundary in the wind lies behind the blob ($\tilde{R}_{Str} \approx R_b$), providing intense, red-shifted Lyα photons for efficient excitation of Fe II.

have two regions as in the previous case, where the H II and H I zones are adjacent to each other (Fig. 11.11a). A single-component blob (a H I region) is probably too cold to emit the bright UV Fe II lines. It will therefore make sense to assume that the optical density of the Lyman continuum $\tau(\lambda_c) \leq 1$ in the region between η Car and the blob.

(c) $\tilde{R}_{Str} \geq R_b + D$ (Fig. 11.11c). In this case the Strömgren boundary outside the blob is located slightly behind it. This can be considered as an optimal situation for pumping of Fe II in the blob for two reasons. First, the Lyα photons produced in this torus-like zone beyond the blob are red-shifted and $\Delta\lambda_\omega = \lambda_0 \left(v_t/c \right) \cos \Theta$, where $\Delta\lambda_\omega$ = the wavelength shift, λ_0 = the rest wavelength, v_t = the terminal stellar wind velocity and Θ is the viewing angle as indicated in Fig. 11.11c. The red-shifted and broadband Lyα radiation irradiates the H I region of the blob without any need for diffusion Doppler broadening. It can penetrate the whole H I region, since its optical density $\tau(\lambda_0 + \Delta\lambda)$ in the far wing of Lyα is not abnormally high, where $\Delta\lambda = 2.4$ Å is the shift of the Fe II absorption line relative to the center of Lyα.

Second, the dilution factor for the Lyman continuum irradiating the stellar wind (see geometry in Fig. 11.11c) will largely be compensated for by the conversion of Ly$_c$ into narrower Lyα radiation in the stellar wind in the vicinity of the blob. The optimal case is when the size of $\tilde{R}_{Str}$ coincides with the distance to the irradiated part of the H I region in the blob, which will require a significant density of the stellar wind near the blob. The critical hydrogen density in the stellar wind, $\tilde{N}_{cr}$, that can provide $\tilde{R}_{Str} \simeq R_b$ is connected with the critical density of the blob itself, N_{cr}, required to locate the Strömgren boundary inside the blob, by the approximate expression:

$$\tilde{N}_{cr}/N_{cr} \simeq (D/R_b)^{1/2}. \tag{11.10}$$

For the adopted values of the distance to the star $R_b \simeq (300–1000)$ $r_s \simeq 10^{16}$–$3 \cdot 10^{16}$ cm and the size of the blob, $D \simeq 10^{15}$ cm, $\tilde{N}_{cr}$ is some tenths of $N_{cr} \simeq 10^8$ cm^{-3}, i.e. a relatively high density of the stellar wind close to the blob. This very approximate estimate of $\tilde{N}_{cr}$ in the vicinity of the blob (0.15" from η Car) agrees with other estimates, based on observed and modeled spectra of the stellar wind (Hillier et al., 2001). The evaluation of Eq. (11.10) agrees with direct measurements of the size and shape of the stellar wind and the blobs in η Car (van Boekel et al., 2003). Of course, all these qualitative speculations should be the subject of detailed modeling of the stellar wind with the blob included in future considerations.

11.7 Andromeda spectral puzzle

The observation of laser effects in astrophysical conditions provides additional information about the physical conditions (density and kinetic temperature of the active zone, effective temperature of unobservable VUV resonance spectral lines) in gas condensations in the vicinity of hot stars. Moreover, the observation of unusual spectral lines and their interpretation in the context of stimulated emission phenomena makes it possible to detect extragalactic lasers and the gas condensations themselves (Johansson and Letokhov, 2003b). Let us consider this interesting possibility.

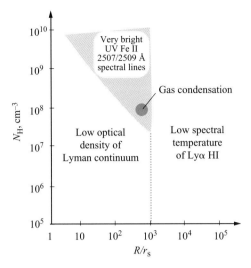

Fig. 11.12 A diagram of the hydrogen density versus the distance from the central star showing the parameter space in which the anomalously bright $\lambda\lambda 2507/09$ lines appear. Adequate parameter values for η Car are given, and the parameter range for the Weigelt blobs is indicated.

Davidson (2001a) has pointed out the amazing similarity between the UV spectra of AE And in M 31 and the core of η Car according to data obtained by Szeifert et al. (1996). He emphasized that this intensity anomaly was at the time known only for these two stars and required special attention. The appearance of the anomalously bright 2507/2509 Å UV spectral lines in the Weigelt blobs has been explained by four effects (Chapters 10, 11): (1) photoselective pumping of highly-excited levels of Fe II by H I Lyα VUV 1215 Å line; (2) high effective temperature ($10-15 \cdot 10^3$ K) of Lyα in the Weigelt blobs; (3) an inversion of population and stimulated emission on highly excited levels of Fe II (from medium-excited pseudo-metastable states of Fe II); (4) a strong radiative cycling in a closed four-level schemes (all levels excited) involving both spontaneous and stimulated transitions.

According to the discussion in Chapters 10 and 11 one can state that the anomalously bright $\lambda 2507/09$ Fe II lines are intrinsic characteristics of gas condensations (blobs), provided that the following two conditions are satisfied: (1) the hydrogen concentration in the blob is high enough for the Lyman continuum radiation to be blocked and a quasi-neutral H I region to be formed therein, and (2) the blob is located close enough to the central star for an H II region to form near the blob (either within its front part or in the stellar wind region near its rear part) and emit intense Lyα radiation (in the red wing of Lyα) capable of photoexciting Fe II in a selective way. These conditions are qualitatively illustrated in Fig. 11.12 by means of two main parameters – the distance R_b from the central star in terms of its photospheric radius r_s and the critical (minimal) hydrogen density N_H for the given distance R. The hydrogen density

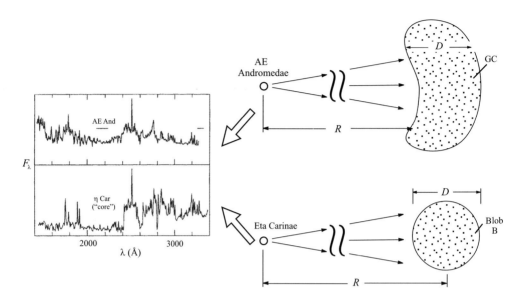

Fig. 11.13 Ultraviolet spectra of AE And in M 31 and a 0.5″ spatial region of η Car (adapted from Szeifert et al., 1996); right: geometry of the compact gas condensations illustrating the site of the anomalously bright $\lambda\lambda 2507/09$ lines.

N_H should exceed some critical value N_0^cr determined by Eq. (11.10). Equation (11.10) means that the local Strömgren boundary separating the H II and H I regions is inside the blob. For blob B in η Car, with $D \leq 10^{15}$ cm and $T_\mathrm{s} = 30 \cdot 10^3$ K, the critical density $N_0^\mathrm{cr} \geq 10^8$ cm^{-3}. The maximal distance R is determined by the requirement of a minimum effective temperature of Lyα, $T_\alpha \geq 10^4$K. The approximate value in Fig. 11.12 is based on observational data for the distance of blob B from η Car.

Davidson (2001a) has pointed out the amazing similarity in the UV spectra around 2500 Å between AE And in M 31 and η Car and its surroundings. However, the spectrum of AE And represents integrated light of the whole object in contrast to the spatially resolved HST spectra of η Car. To the left in Fig. 11.13 we show fragments of these spectra according to data from Davidson et al. (1995) and Szeifert et al. (1996). In both cases the $\lambda\lambda 2507/09$ Fe II feature is very bright. Davidson (2001a) emphasized that this intensity anomaly was at the time known only for these two stars and required special attention.

AE And is an extremely luminous blue star with a temperature probably as high as that of η Car ($T_\mathrm{s} \approx 30 \cdot 10^3$ K). The strength of the $\lambda\lambda 2507/09$ feature suggests that the immediate vicinity of AE And hosts gas condensations with a hydrogen column density n_H greater than the critical density, defined by condition (11.10).

Due to the huge distance, AE And and its surroundings are observed in integrated light, and it is not possible to associate the radiation observed with any specific region of the object. This means that the intensity of the $\lambda\lambda 2507/09$ lines from possible gas condensations should exceed that from AE And itself. The lines from the gas condensations would otherwise be indiscernible from the stellar black-body radiation in the same spectral interval.

In the HST/STIS spectra of η Car, the Weigelt blobs are spatially resolved from the central star itself, which allows us to directly study the conditions where the $\lambda\lambda 2507/09$ lines are formed. Fig. 11.13, however, shows for comparison a 0.5" spatial extract of the HST/FOS spectra of η Car recorded with the 0.3" circular aperture (Davidson et al., 1995) and the FOS spectrum of AE And (see Fig. 3 in Szeifert et al., 1996). The similarity of these spectra around 2500 Å may imply similarities in the plasmas where the $\lambda\lambda 2507/09$ lines are formed. For example, blob B of η Car intercepts the radiation of the star within a significant solid angle of $(D/R_\mathrm{b})^2 \approx 0.07$, where the distance to the star is $R_\mathrm{b} \approx 4 \cdot 10^{15}$ cm. Under these conditions intense $\lambda\lambda 2507/09$ lines can appear even against the stellar black-body radiation. In the case of AE And, the angular size of a potential GC seen by the star the star should not be smaller than for the η Car case.

Based on the model Johansson and Letokhov (2001c, e; 2003a) developed for the formation of the $\lambda\lambda 2507/09$ lines in the Weigelt blobs, and the similarity between the spectra of η Car and AE And in M 31 (Szeifert et al., 1996), one can consider the observation of very intense $\lambda\lambda 2507/09$ lines as a method for seeking such circumstellar gas condensations even without access to high spectral and spatial resolution. The anomalous brightness make the lines appear in spite of stellar background radiation. The spike seen in low resolution (Fig. 11.13) does not provide any information about the intensity ratio of the $\lambda\lambda 2507/09$ lines and their intensities relative to their satellite lines. These ratios will be different when the spectrum refers to a local region spatially resolved from the central star (as is the case of the blobs in η Car) and when the spectrum represents the whole extended object, i.e. radiation from the stellar photosphere and from the surrounding gas condensations.

This example demonstrates a new indirect way to detect a hidden spatial inhomogeneity (gas condensation) in the vicinity of an extragalactic blue star without having access to high spatial resolution for direct observation of such a spatial structure.

12
Astrophysical laser in the O I 8446 Å line

The next astrophysical laser, on line 8446 Å O I has been discovered by means of HST/STIS spectral data also by Johansson and Letokhov (2005c). However, the laser effect on this line was predicted many years ago by Letokhov (1972a) on the basis of the multiple astronomical observations of the anomalous intensity ratio of the O I lines.

12.1 Spectral anomalies in O I in stars

The anomalous intensity ratio between the O I triplets at 8446 and 7774 Å (Fig. 12.1) observed in spectra of various astrophysical sources has been thoroughly discussed in the astrophysical literature (Fig. 12.1). The two multiplets arise from the 3s-3p transition in the triplet and quintet systems of O I, respectively, and they have about the same excitation potential. However, the lines from the triplet system appear much stronger in emission than the quintet system lines in many astrophysical emission line spectra. This anomaly was first noticed by Hiltner (1947) in the spectra of χ Oph, and explained by Bowen (1947) as being due to selective photoexcitation of the triplet system by the wavelength coincidence of the bright H Lyβ 1025.72 Å and the 2p ^{3}P-3d ^{3}D transitions at 1025.76 Å. The subsequent decay of the 3d ^{3}D levels to 3p^{3}P creates overpopulation in the upper levels of the λ 8446 multiplet. The effect was studied and discussed for several Be stars (Slettebeck, 1951; Burbidge, 1952; Merill, 1956). In many of these objects, the 7774 Å line appears in absorption. In a study of near-IR spectra (8200–11200 Å) of 25 "peculiar emission-line objects" recorded at Haute Provence by Andrillat and Swings (1976) the O I λ 8446 line was present in all spectra, and anomalously strong in most of them. The phenomenon of the 8446 Å line has been studied for stellar atmospheres extensively (Strittmatter et al., 1977; Damineli, 2001). At the time when satellite spectra became available in the ultraviolet region, the decay of the lower levels of the λ 8446 lines, i.e. the 2p-3s triplet transition at 1302 Å, could be investigated. For example, Haisch et al. (1977) showed that the anomalously bright O I UV triplet in the spectrum of Arcturus could be explained by cascading from the Lyβ-pumped 3d levels. The O I λ1302 lines were later studied extensively in cool star chromospheres by Jordan and coworkers (see e.g. Jordan, 1988a), and detailed modeling of the lines in the Sun and stars was done by Carlsson and Judge (1993). Fluorescence due to Lyβ pumping was also shown to be active in the solar chromosphere (Skelton and Shine, 1982).

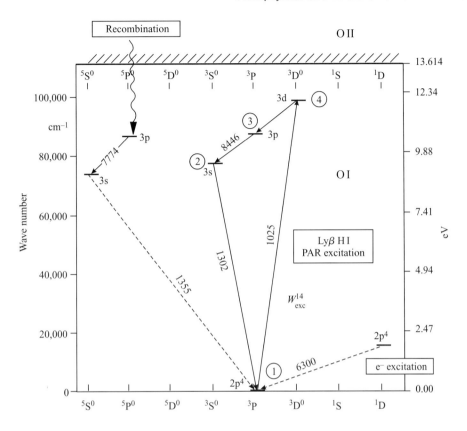

Fig. 12.1 Simplified energy level diagram of O I with relevant transitions inserted and excitation mechanisms indicated.

The O I 8446Å multiplet has also been observed in the Orion Nebula. By imaging the Orion Nebula in the O I 8446 Å line, Münch and Taylor (1974) found a unique filamentary structure not seen in the 6300 Å line of [O I]. They explained the high level of fluctuations as being due to the resonant excitation of 8446 Å by Lyβ. Grandi (1975) proposes direct starlight photoexcitation as the population mechanism. Grandi (1980) also examined spectra of 16 Seyfert galaxies, 13 of which showed broad O I λ 8446 emission. In all these cases, Grandi concludes that Lyβ pumping is the excitation mechanism, which requires large oxygen abundance, a source of Lyβ photons and a large depth in Hα. Kastner and Bhatia (1995) employed a detailed model of the oxygen atom under optically thin conditions to calculate fluorescent line intensities expected in the photoexcitation by accidental resonance (PAR) process in which Lyβ photoexcites O I. They applied their model to verify the operation of the PAR process behind the 8446 Å lines observed in spectra of classical novae.

A detailed analysis of the population of the O I levels under PAR excitation conditions has raised the question about an inverted population in the 3p^{3}P levels and, accordingly, the possibility of laser effect in the 8446 Å transition in astrophysical sources (Letokhov, 1972a; Lavrinovich and Letokhov, 1974). However, such an effect has earlier been very difficult to observe because of insufficient angular resolution and also because of the difficulty of distinguishing the PAR excitation mechanism from collisional excitation and recombination processes in stellar atmospheres.

A much more favorable situation exists for gas blobs at such an angular distance from a central star that observations of spectra from the central star and the gas blobs are spatially resolved. It has become possible mainly due two recent fundamental achievements in observational astronomy, both of which have been applied in observations of Eta Carinae. Firstly, the development of high-angular-resolution speckle interferometry made it possible to discover the Weigelt blobs located only some tenths of an arcsec from the central source (Weigelt and Ebersberger, 1986). Secondly, the Hubble Space Telescope (HST) with the High Resolution Spectrograph (HRS) and later the Space Telescope Imaging Spectrograph (STIS) (Kimble et al., 1998) made it possible to observe spectra of Eta Carinae at high spectral and angular resolution. The acquisition of such spectral data (Gull et al., 2001) made the interpretation of the photophysical processes occurring in the Weigelt blobs simpler and more certain. These observational circumstances made it possible to detect the inverted population and stimulated emission of the Fe II spectral lines in the wavelength range 0.8–2 μm (Johansson and Letokhov, 2002, 2003a, 2004c) (Chapter 11).

Johansson and Letokhov (2005c) explained, within the framework of a simple model and using HST/STIS data (Gull et al., 2001), the behaviour of the O I λ8446, O I λ6300 and O I λ7774 lines in spectra of the Weigelt blobs in Eta Carinae. They demonstrated that the population inversion and stimulated emission occurs in the 3p ^{3}P-3s ^{3}S transition (λ8446) generated by H Lyβ radiation in a PAR excitation mechanism. This is, to a certain degree, a natural extension of the concept of the Fe II laser operation in the Weigelt blobs.

12.2 Excitation mechanism of O I in the Weigelt blobs

The Weigelt blobs are characterized by a relatively high (10^7–10^8 cm^{-3}) hydrogen density (Davidson and Humphreys, 1997). The front part of the blobs absorbs the Lyman continuum radiation at λ < 912 Å ($I_{H I}$ = 13.595 eV) (Chapter 10). The ionization potential of O I is somewhat higher ($I_{O I}$ = 13.616 eV), so, in principle, oxygen should remain unionized in the H I region of the Weigelt blobs. This would be a unique situation wherein the O II spectrum should be absent and the PAR excitation of O I should be distinct. In other words, the anomalous behavior of the intensities of the O I λ8446 lines should be extreme. This is confirmed in spectra obtained with the HST/STIS instrument (Gull et al., 2001).

Fig. 12.2 illustrates qualitatively the geometry of the irradiation reaching one of the Weigelt blobs by the photospheric radiation from Eta Carinae. It is similar to the irradiation geometry of the Fe II astrophysical laser studied in detail by Johansson and Letokhov (2004c). The Lyman continuum radiation photoionizes the front part of the blob, but fails to penetrate deep inside it because of the high hydrogen density. In other words, the Strömgren boundary is located in the front part of the Weigelt blob (Fig. 12.2). As a result, oxygen atoms (in contrast to their iron counterparts) cannot

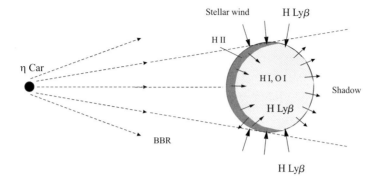

Fig. 12.2 Geometry of the irradiation of a Weigelt blob by radiation from the central source in Eta Carinae and by the H Lyβ radiation coming from the frontal H II region of the blob and from the stellar wind.

be photoionized by radiation at $\lambda > 912$ Å. Therefore, the H I region of the blobs is also an O I region with a temperature T_{bl} of a few thousands K (Verner et al., 2002). The role of other processes that can ionize oxygen (e.g. charge exchange with H II) is not significant in this particular case because of lack of H II inside the blob. This is a key point in the explanation of the mechanisms responsible for the appearance of anomalies in the O I spectrum.

Fig. 12.1 presents a simplified energy level diagram of O I, the relevant quantum transitions and three processes whereby low-lying forbidden transitions and allowed transitions between highly excited states can be excited. In the absence of O$^+$ ions, radiative recombination cannot populate the 3p ^{5}P term, and, accordingly, the O I 7774 Å quintet transition should not appear in the spectrum of the Weigelt blobs. This is confirmed by the HST/STIS data (Gull et al., 2001) in this spectral region as shown in Fig. 12.3b). On the other hand, the presence of Lyβ radiation from the H II region of the blob and from the stellar wind provides for resonant photoexcitation of 3d ^{3}D levels. By cascading, the 3p ^{3}P levels get populated, which is implied by the appearance of the $\langle$O I$\rangle$ 8446 Å fluorescence line, as shown by the HST/STIS spectral data (Gull et al., 2001) in Fig. 12.3a. Thus, a characteristic of the Weigelt blobs is an extreme ratio between the intensities of the O I spectral lines λ8446 and λ7774, much higher than that observed in stellar atmospheres. In many cases, the 7774 Å line is observed in absorption in stellar spectra owing to the long lifetime of the lower quintet level. It is quite natural that both O I excitation mechanisms – PAR excitation and recombination excitation in the presence of O$^+$ ions – are effective in the outer atmospheres of stars.

The HST/STIS Weigelt blob spectrum (Gull et al., 2001) presented in Fig. 12.3c shows the forbidden O I 6300 Å line. If collisionally excited, this is evidence of a substantial electron temperature ($T_e \approx$ a few kK) in the H I region of the Weigelt blobs (Verner et al., 2002). The energetic edge of the distribution of electrons with such a temperature is high enough to provide collisional excitation of the low-lying metastable levels in O I, which explains the appearance of the [O I] 6300 Å line. It

172 *Astrophysical lasers*

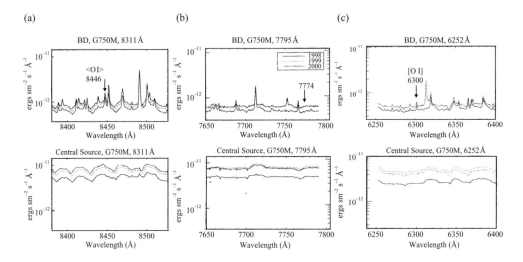

Fig. 12.3 Narrow wavelength intervals of the HST/STIS spectra of the Wegelt blobs B and D (top) and the central source (bottom) of Eta Carinae containing the O I line relevant to (*a*) laser action (8446), (*b*) a recombination line (7774), and (*c*) a forbidden line (6300) (spectra extracted from the paper by Gull *et al.*, 2001).

should also be pointed out that in the case of an optically thick 1302 Å line the trapping of resonance radiation can populate the $3s^3S$. With a branching fraction of the order of 10^{-5} the ^3S level can decay radiatively to $2p^{4\,1}D$ in the 1643 Å transition, thereby yielding a collision-free contribution to the population of the upper level of the forbidden 6300 Å line. Note, in this connection, that it would be interesting to analyze the kinetics of the intensity variation of this line during the course of the spectral events of Eta Carinae. Such an analysis would provide additional information about the major excitation process and possibly the temperature and electron density in the H I region of the Weigelt blobs.

12.3 Inverted population and amplification coefficient of the 8446 Å line

Fig. 12.4 presents a partial diagram of the energy levels in O I that participate in the PAR excitation by the H Lyβ radiation and formation of the intense ⟨O I⟩ 8446 Å fluorescence lines (Bowen, 1947). In this Chapter, we consider only the sharp resonant pumping by Lyβ of two O I lines, but for a broader Lyβ profile ($\Delta\lambda > 3$–4 Å), all 3d ^{3}D and 3p ^{3}P fine-structure levels will be populated.

The frequency detuning, $\Delta\nu$, between the centre of Lyβ at $\lambda = 1025.722$ Å and the O I absorption lines at 1025.762 and 1025.763 Å is only 0.04 Å. In the case of O I, there is no problem in compensating for the frequency detuning between the pumping radiation and the absorption lines compared to the case of the astrophysical laser

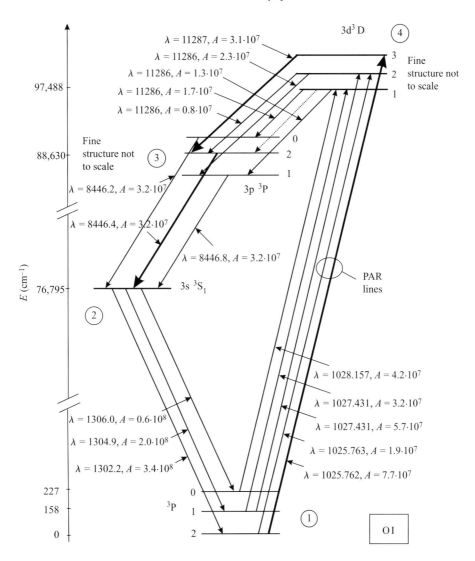

Fig. 12.4 Partial level diagram of O I including levels relevant to the inverted population between levels 3 and 2 due to resonant pumping by accidental resonance (PAR) by Lyβ. The A values are from the compilation by Wiese *et al.* (1996). (The fat lines indicate one of the pathways of pumping of fine-structure components.)

operating in Fe II. Two photoexcited levels (see 4 in Fig. 12.5), decay radiatively to two fine-structure levels, 3, by emitting radiation at wavelengths $\lambda_{43} = 11287$ and 11286 Å, which is outside the wide spectral interval observed with the HST/STIS facility. The next radiative decay channel is the $3 \to 2$ transition accompanied by the emission of radiation at $\lambda_{32} = 8446.4$ and 8446.8 Å. Most important is the fact that the lower

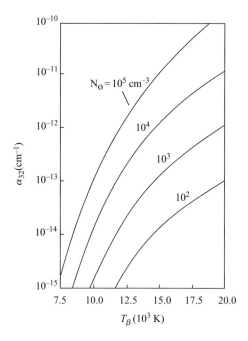

Fig. 12.5 The amplification coefficient α_{32} (cm^{-1}) of the $3 \rightarrow 2$ transition for the pathway indicated in (broad lines) as a function of the effective (spectral) temperature T_β of Lyβ radiation for various densities of O I.

levels of these transitions decay radiatively to the ground state within approximately 1 ns, i.e. 20 times faster than the decay of the upper levels, 3. For this reason, there always exists an inverted population in the allowed $3 \rightarrow 2$ transition, since the role of any collisional quenching on this timescale occurring in the Weigelt blobs is negligible. The possible role of resonance radiation trapping in the $\lambda 1302$ line is discussed below.

The inverted population in the $3 \rightarrow 2$ transition is defined by Eq. (11.2), where N_n is the density of oxygen atoms in level n. Once excited to level 4, the oxygen atoms can decay radiatively to the initial state 1 at the rate A_{41} or to level 3 at the rate A_{43}. Therefore, the steady-state population of level 3 is related to the excitation rate W_{exc}^{14} by the relation

$$N_3 = N_1 \frac{W_{\text{exc}}^{14}}{A_{41} + A_{43}} \frac{A_{43}}{A_{32}} \tag{12.1}$$

where N_1 is the density of oxygen atoms in the ground state, which is equal to the density N_O of O I in the Weigelt blobs ($N_2, N_3, N_4 \ll N_1 \approx N_O$), and W_{exc}^{14} is the rate of photoselective PAR excitation of O I by the Lyβ radiation. The rate W_{exc}^{14} is governed by the effective (spectral) temperature T_β of the Lyβ radiation according to Eq. (11.1).

The inverted population density of levels 3 and 2, ΔN_{32}, is defined by the expression

$$\frac{\Delta N_{32}}{N_O} \geq \frac{g_4}{g_1} \frac{A_{43}}{A_{32}} \frac{A_{41}}{A_{41}+A_{43}} \left[\exp\left(\frac{h\nu_{14}}{kT_\beta}\right) - 1\right]^{-1}, \quad (12.2)$$

which, considering that $h\nu_{14} \gg kT_\beta$, can be reduced to

$$\frac{\Delta N_{32}}{N_O} \approx \frac{g_4}{g_1} \frac{A_{43}}{A_{32}} \frac{A_{41}}{A_{41}+A_{43}} \exp\left(-\frac{h\nu_{14}}{kT_\beta}\right) \quad (12.3)$$

Most closely tuned to resonance with Lyβ are two of the five components in the $2p^4{}^3P$-$2p^3 3d^3D$ multiplet of O I, namely, those at $\lambda = 1025.762$ and 1025.763 Å (Fig. 12.4). In accordance with the schematic diagram of the cascading radiative decay of the excited levels, 4 (Fig. 12.4), population inversion occurs in two of the three fine-structure components of the $3 \rightarrow 2$ transition: at $\lambda = 8446.4$ and 8446.8 Å. Using the existing data Johansson and Letokhov (2005c) could not make a detailed analysis of these two pathways of PAR, yielding the two laser lines 8446.4 and 8446.8 Å and a comparison of anomalous intensity ratios expected from stimulated emission. This is not surprising, since, unfortunately, the spectral resolution of the HST/STIS data (Gull et al., 2001) is inadequate to discriminate between these two spectral lines and the unpumped line at 8446.2 Å. To estimate the magnitude of the ratio, let us consider the strongest PAR and cascade radiative decay pathway in Fig. 12.4 (solid lines):

$$1025.762 \text{ Å} \rightarrow 11287 \text{ Å} \rightarrow 8446.4 \text{ Å}$$
$$(1 \rightarrow 4) \quad (4 \rightarrow 3) \quad (3 \rightarrow 2) \quad (12.4)$$

For this pathway, we use the following Einstein coefficients in expression (12.3): $A_{41} = 7.7 \cdot 10^7$ s^{-1}, $A_{43} = 3.1 \cdot 10^7$ s^{-1} and $A_{32} = 3.2 \cdot 10^7$ s^{-1} (Wiese et al., 1996); and statistical weights: $g_1 = 5$ and $g_4 = 7$. In that case, eq. (12.3) assumes a simpler form,

$$\frac{\Delta N_{32}}{N_O} \approx 0.96 \exp\left(-\frac{h\nu_{14}}{kT_\beta}\right). \quad (12.5)$$

The amplification coefficient for the $3 \rightarrow 1$ transition, α_{32}, is defined by the standard expression (6.2), where σ_{32} is the cross-section of the $3 \rightarrow 2$ radiative transition, given by (6.5).

To make numerical estimates, we adopt the following parameters: (1) the density of oxygen atoms, $N_O \approx 10^{-3} N_H$ where N_H is the density of hydrogen atoms in the Weigelt blobs, $N_H \approx 10^7$–10^8 cm^{-3}, and (2) the Doppler width $\Delta\omega_D^{32} = 2\pi\Delta\nu_D^{32} = 2\pi \cdot 2.8 \cdot 10^9$ Hz for the H I region of the blob having a temperature of $T \sim 5 \cdot 10^3$ K. All these quantities should be considered approximate, but they are quite acceptable for estimation purposes. For example, we adopt the solar oxygen abundance $N_O \approx 10^{-3} N_H$ even if a lower value has been discussed for η Car. However, as shown in Fig. 12.5, amplification of the O I 8446 Å lines occurs in a wide range of oxygen densities, and an exact value for the O abundance in the Weigelt blob is not critical for the result of this analysis. For the cross-section σ_{32} we get the following value, typical of an allowed transition: $\sigma_{32} = 1.1 \cdot 10^{-12}$ cm². Fig. 12.5 shows the amplification coefficient α_{32} as a function of the effective (spectral) temperature T_β of the Lyβ radiation for various densities of oxygen N_O. The density $N_O = 10^5$ cm^{-3} corresponds to $N_H = 10^8$ cm^{-3} in the central part of the blob. The low density $N_O = 10^2$ cm^{-3} corresponds to the wings of the blob with the bell-shaped distribution of the density N_H. As one can see from Fig. 12.5, the gain per unit length, α_{32}, reaches 10^{-15} to 10^{-12} cm^{-1} at temperatures $T_\beta \approx (7.5-15) \cdot 10^3$. With the Weigelt blob diameter $D = 10^{15}$ cm, such a gain can give a linear amplification coefficient of $K = \exp(\alpha_{32} D) \geq \exp(1-10)$, which is quite a perceptible quantity.

12.4 Spatial features of the O I laser in the Weigelt blobs

The source of the Lyβ radiation is the H II region of the given blob (Fig. 12.2) as well as the stellar wind surrounding the blob. However, the Lyβ radiation is absorbed not only by the O I atoms, but also by H atoms in the H I region of the blob. To gain an insight into this situation, Fig. 12.6 presents the maximum-normalized absorption spectral lines of O I and H I (responsible for the PAR excitation) in the H I region of the blob at a temperature of 5000 K. Also shown in this figure are the Lyβ emission lines from the H II region of the blob at a temperature of 20,000 K and from the stellar wind. The widths of all these spectral lines are governed by the Doppler effect.

Let us estimate the absorption coefficient of the O I λ1025.77 line in the H I region of the Weigelt blobs. The absorption cross-section at the centre of this line is $\sigma_{14} \simeq 10^{-14}$ cm². With the O I density $N_O = 10^{-3} \times N_H = 10^5$ cm^{-3}, the absorption coefficient for the Lyβ radiation at the center of the O I 1025.77 Å line is $\kappa = 1/l^0_{\text{exc}} \sim 10^{-9}$ cm^{-1}. This means that level 4 is most effectively excited in the shell of the blob at a low density, N_O. When the detuning from the line center reaches

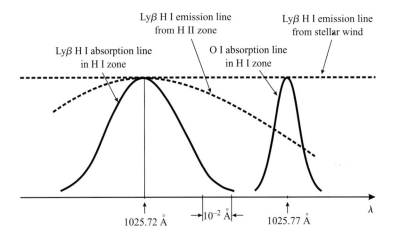

Fig. 12.6 Location and shape of the absorption and emission spectral lines of O I and H I in the narrow 1025 Å interval related to the PAR excitation of OI by Lyβ radiation.

as high a value as $\Delta\nu = 1.5\Delta\nu_D$, the absorption coefficient drops by a factor of 10^4, and the excitation depth of O I reaches $l^0_{\text{exc}} \simeq 10^{13}$ cm on account of excitation at the wings of the Doppler profile.

However, one should take into account the gradual decrease of the densities N_H and N_O on the periphery of the blob and the perceptible level of the amplification coefficient α_{32} at low N_O densities (Fig. 12.5). In the outer shell of the blob, the penetration of the Lyβ radiation, and hence the PAR excitation of O I by this radiation, is effective enough and, at the same time, the density of O I atoms is quite sufficient, for amplification to take place: $\alpha_{32} \approx 10^{-14}$ cm^{-1} at $T_\beta \approx 10{,}000$ K (Fig. 12.5). Fig. 12.7 qualitatively illustrates the radial distribution of the N_O density and Lyβ intensity in the blob that leads to the conclusion that the amplification α_{32} predominantly takes place exactly in the outer shell of the blob where $l^0_{\text{exc}} \simeq 10^{12}$–$10^{13}$ cm.

In the case of excitation of O I at a depth of $l^0_{\text{exc}} \ll D$, where D is the diameter of the blob, the maximum amplification length along the chord line of the amplifying shell is given by

$$L_{\text{amp}} \simeq 2\,(l_{\text{exc}} D)^{1/2}. \qquad (12.6)$$

For $D \approx 10^{15}$ cm and $l^\simeq_{\text{exc}} 10^{13}$ cm, the amplification length is $L_{\text{amp}} \simeq 2 \cdot 10^{14}$ cm, which is quite sufficient to obtain a substantial amplification at 8446 Å. At the effective (spectral) temperature of the Lyβ radiation within the Weigelt blobs, stimulated emission takes place inside a spherical shell with a ring-like stimulated emission zone

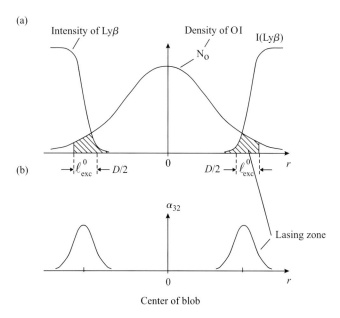

Fig. 12.7 The qualitative radial distribution of the O I density inside the blob (*a*), the corresponding intensity attenuation of Lyβ and the amplification zone (*b*) in a spherical shell.

for the observer. This effect can be observed with a ground-based telescope having a spatial resolution a few times better than D.

Secondly, let us estimate the effect of the presence of H I on the absorption of the radiation exciting O I. As can be seen from Fig. 12.6, the red Doppler wing of the Lyβ absorption line can have an effect on the absorption of radiation in resonance with the absorption line of O I because of the higher (by a factor of 10^3) density of H I. The absorption of the Lyβ radiation at the centre of the O I λ_{14} lines is lower than its absorption at the centre of the line by approximately a factor of 10^{-5}. The ratio between the absorption cross-sections at the centers of Lyβ and the O I λ_{14} lines is

$$\frac{\sigma_H(\text{Ly}\beta)}{\sigma_{OI}(\lambda_{14})} = \frac{A_{31}(\text{Ly}\beta)}{A_{41}(\text{O I})} \frac{\Delta\omega_D^O(\lambda_{14})}{\Delta\omega_D^H(\text{Ly}\beta)} = 0.2. \tag{12.7}$$

With $N_H/N_O \simeq 10^3$, the absorption coefficient of the Lyβ radiation by hydrogen, $x_H(\text{Ly}\beta)$, at the frequency of the O I absorption line centre is $x_H \simeq 10^{-2} x_{OI}$, and it is still lower in the red Doppler wing of the O I λ_{14} spectral line. Thus, the contribution of hydrogen to the absorption of radiation exciting O I is not more than 1 per cent of its absorption by O I itself, and has thus no effect on the excitation depth l_{exc} of the latter. As one can see from Fig. 12.6, the O I atoms can be excited to make the $1 \to 4$ transition by both the Lyβ radiation of high Doppler width that comes from the hot H II region of the blob and the Lyβ radiation of the stellar wind.

A very high optical depth of the pumping Lyβ radiation in the Weigelt blobs with hydrogen density $N_H \simeq 10^8$ cm^{-3} can affect the lifetime of the lower laser level 2 (3s $^3S^0$) and hence the population inversion due to trapping of radiation in the $1 \to 2$ transition (1302 Å line). In fact, Lyβ radiation can only penetrate into the outer shell of the blob, where the amplification of the 8446 Å line can take place, as shown in Fig. 12.7. The estimated hydrogen and oxygen densities in this shell region are $N_H \simeq 10^4$–10^5 cm^{-3} and $N_O \simeq 10$–10^2 cm^{-3}, respectively. These values correspond to an excitation length $l_{\text{exc}} \simeq 10^{12}$–$10^{13}$ cm. However, it is necessary to account for the possible influence of the optical depth in the O I λ1302 line, since a large optical depth and corresponding resonant trapping of radiation can enhance the effective lifetime of level 2, the lower level in the λ8446 laser lines (Holstein, 1947).

The radiative cross-section for the most intense fine-structure component of the resonance multiplet (the $1 \to 2$ transition in Fig. 12.4) is given by

$$\sigma_{21} = \frac{g_2}{g_1} \frac{\lambda_{12}^2}{2\pi} \frac{A_{21}}{\Delta\omega_D^{21}} = 4.6 \cdot 10^{-14} \text{ cm}^2, \tag{12.8}$$

where $g_2/g_1 = 3/5$, $A_{21} = 3.4 \cdot 10^8$ s^{-1}, and the Doppler width $\Delta\omega_D^{21} = 2\pi \cdot 1.9 \cdot 10^{10}$ Hz. Using these parameter values for the outer shell of the blob with local densities of $N_H \simeq 10^4$–10^5 cm^{-3} and $N_O \simeq 10$–10^2 cm^{-3} and with an excitation (pumping) length of $l_{\text{exc}} \simeq 10^{13}$–$10^{12}$ cm, the optical depth for the λ 1302 line is $\tau_{\text{od}}^{12} = \sigma_{12} l_{\text{exc}} N_O = 4.6$.

The enhancement of the effective lifetime, t_2^{eff}, of level 2 is determined by the following expression (Holstein, 1947; Molish and Oehry, 1998):

$$t_2^{\text{eff}} = \frac{\tau_{\text{od}}^{12} \log\left(\tau_{\text{od}}^{12}/2\right)}{1.05} t_2, \qquad (12.9)$$

where t_2 is the radiative lifetime of level 2.

The effective lifetime of level 2 thus becomes $t_2^{\text{eff}} \approx 2.6 t_2 \approx 4.4$ ns, i.e. $t_2^{\text{eff}} \ll t_3 = 31$ ns. Thus, the estimated influence of the resonant radiative trapping on the population inversion ΔN_{32} is about 10–20 per cent and cannot decrease the inverted population in the thin outer shell of the blob considered here (Fig. 12.7). Nevertheless, it is necessary to have in mind the sensitivity of the inverted population in the 8446 Å line for resonant radiative trapping in the 1302 Å line under different space plasma conditions.

12.5 Possible O I lasers in stellar atmospheres and Orion Nebulae

The O I transition at 8446 Å is ideal for lasing with optical pumping in a classical four-level scheme not only in the Weigelt blobs, but also in other conditions. In Sections 12.3 and 12.6 above, we have mentioned the anomaly in the behavior of the 8446 Å and the 7774 Å lines in spectra of Be stars (Merrill, 1956; Slettebak, 1951).

The appearance of the population inversion and the gain in the O I 3p ^{3}P-3s ^{3}S transition in atmospheres of Be stars was analyzed by Lavrinovich and Letokhov (1974). They considered a gas consisting of oxygen atoms at a density N_0 and temperature T_0 in the radiation field and with spectral density U and frequency $\omega \approx \omega_{41}$ in balance equations for the population in steady-state conditions. It is known that the pump threshold in four-level lasers is substantially determined by the population of level 2, which depends on the temperature of radiation acting at the $1 \rightarrow 2$ transition (in the case under study, the contribution of non-radiative transition is small). If the gas were in equilibrium with the radiation, the radiation temperature would be coincident with the gas temperature T_0. However, stellar regions show hotter radiation emitted by the photosphere at a temperature T_s, which noticeably exceeds T_0.

The density of O I atoms in the atmosphere located near the external boundary of the photosphere (in the inverting layer) is $N_0 \sim 10^5 - 10^6$ cm^{-3}, the gas temperature is $T_0 \sim 1.5 \cdot 10^4$ K, and the effective temperature of the Be star photosphere is $T_s \sim 3 \cdot 10^4$ K (Allen, 1973; Boyarchuk, 1957). According to calculations of Lavrinovich and Letokhov (1974), assuming that the radiation temperature in the $1 \rightarrow 2$ transition is $\sim T_s$, the required brightness temperature of the pump radiation at the threshold should be $4 \cdot 10^4$ K. However, to obtain the population inversion, for example, $\Delta N_{32} \sim 10^{-6} N_0$ this temperature should be greater ($\sim 4.1 \cdot 10^4$ K). The estimation of the possible gain α for the 8446 Å line of O I in the atmosphere of a Be star is quite favorable. For $A_{32} = 3 \cdot 10^7$ s^{-1} and $\Delta\omega_D \simeq 5 \cdot 10^{10}$ s^{-1}, the cross section for the $3 \rightarrow 2$ radiative transition is, according to (6.5), $\sigma_{32} \simeq 10^{-12}$ cm^2. For $\Delta N_{32} \simeq 1$ cm^{-3}, this corresponds to a gain at the line center $\alpha_{32} \simeq 10^{-12}$ cm^{-1}. Taking into account that the extension of the atmosphere of a Be star is about $10^{13}-10^{14}$ cm, we can conclude that an inverted population with a linear gain factor $\alpha L \simeq 10-100$ appears in the stellar atmosphere.

Note the possibility of photoexcitation of the photosphere by continuous radiation without a resonant coincidence, according to the scheme in Fig. 9.2d. To obtain the population inversion in this case, in some transition, the spectrum of continuous radiation should be in non-equilibrium. Mustel (1941) showed that the continuous spectrum of the photosphere can strongly differ from the Planck distribution. For example, in all symbiotic stars short-wavelength emission from a hot star irradiates the atmosphere of a cooler star. Under such conditions, laser action can also be observed in atoms and ions.

Another potential region for the appearance of population inversion and gain is found in, for example, planetary nebulae. In the Strömgren boundary region, the bright Lyβ line, generated in the H II region, irradiates the H I region, where oxygen is in the neutral state. The photoionization limit for the 8446 Å O I line ($I_{\mathrm{OI}} = 13.618$ eV) lies above the Lyman emission continuum limit (13.5985 eV). Under such conditions Lyβ should efficiently excite neutral oxygen in a comparatively thin Strömgren boundary region. Therefore, it is quite probable that a strange behavior of the 8446 Å O I line in the Orion Nebula observed by Münch and Taylor (1974) can be attributed to laser action. The authors of this paper discovered a unique intense structure in the intermediate H II/H I region only by observing the very intense 8446 Å line. At other lines of O I and at the Hα Balmer line (which should be present in the same regions as the intense Lyβ line), radiation has a spatially amorphous structure, without any filaments.

Münch and Taylor (1974) explained this unique fact by assuming the existence of solid particles in the intermediate region, which would in some way affect the morphology and the emission spectrum of O I, and they pointed out the necessity of a further study of this strange situation. High-resolution spectroscopic measurements with a Fabry–Perot interferometer (Münch and Taylor, 1974) showed that the width of the 8446 Å line was approximately 25% *narrower* than that of the 6300 Å line. It seems more natural to explain this fact by a gain in the 8446 Å transition, where a population inversion should exist. Of special attention is the remarkable fact that an (unidentified) point radiation source – "emitting object" (Gull et al., 1973) – is observed in the Orion Nebula, where the 8446 Å line is not observed(!). This suggests that the radiation pattern of the amplifying region can be anisotropic, with a maximum that is not directed to an observer. This possibility is worth studying in the future.

13
Narrowing of spectral lines in astrophysical lasers

In the absence of feedback, the amplifying medium of the astrophysical laser represents an amplifier of spontaneous emission (Section 6.5). Because of the large size of the astrophysical amplification region, the condition

$$\frac{a}{L} \gg \frac{\lambda}{a} \tag{13.1}$$

is always fulfilled, where a and L are the transverse and longitudinal sizes of the amplification region, so that the radiation divergence is determined by the geometry of the amplifying region. In the case of a nearly spherical geometry ($a \simeq L$), the radiation should be isotropic. This statement concerns, strictly speaking, a linear amplification regime.

In the saturated amplification regime, a small deviation from sphericity should be distinctly manifested, and we can assume, based on heuristic considerations, the existence of a peculiar instability of the spherical geometry of radiation from a saturated amplifier. The probability of the formation of spherical amplification regions even in boundary astrophysical regions, where the appearance of the population inversion is most probable, is very low. Therefore, it is most likely that the formation of anisotropic patterns of stimulated emission will depend on the geometry of the amplification regions, although it is unlikely that the directivity of radiation will be high. The latter remark is essential for the following qualitative explanation of the emission spectrum of a saturated amplifier when the amplifying spectral line is inhomogeneously broadened owing to the Doppler effect.

13.1 Role of geometry in the saturation regime on spectral narrowing

In the regime of linear amplification of spontaneous emission, the width of its spectral lines decreases because of a predominant amplification of emission at the line center, which is described by the following expression (Litvak, 1970; Casperson and Yariv, 1972; Allen and Peters, 1972a, b; Cook, 1977):

$$\Delta\nu = \frac{\Delta\nu_D}{(1+\alpha_{32}^0 L_{\text{lin}})^{1/2}}, \tag{13.2}$$

182 *Astrophysical lasers*

where α_{32}^0 is the linear coefficient per unit length (cm^{-1}) on linear amplification length L_{lin}. This occurs until the saturation of the inhomogeneously broadened gain line is accompanied by the "flattening" of its peak. This effect was considered for the astrophysical maser saturated amplifier (Litvak, 1970). That the narrowing of the spectral line ceases upon saturation and the spectrum broadens again up to the width of the gain line, is shown qualitatively in Fig. 13.1. This conclusion has been confirmed by many experimental data for astrophysical masers (Reid and Moran, 1981; Elitzur, 1992) with one exception. Barrett and Rogers (1966) observed an anomalously narrow emission line of the 18.5 cm OH maser, whose width corresponded to the Doppler width at the temperature $T \simeq 5$ K. This phenomenon is not explained so far. It was pointed out (Letokhov, 1996) that the conclusion of Litvak (1970) on limited narrowing is valid only for directional emission, when the saturation of the Doppler profile is accompanied by a flattening of its peak as shown in Fig. 13.1. In reality, the saturation of the Doppler profile under the action of isotropic radiation is not accompanied by its deformation but occurs uniformly (Fig. 13.2), while the narrowing of the isotropic radiation line is not followed by its broadening and should obey the law similar to (13.2):

$$\Delta \nu \simeq \Delta \nu_D \left(1 + \alpha_{32}^0 L_{\text{lin}} + \int_{L_{\text{lin}}}^{L} \alpha(r) dr \right)^{-1/2} \tag{13.3}$$

where L_{lin} is the length of linear amplification, and $L - L_{\text{lin}}$ of saturated amplification with an integrated value $\int_{L_{\text{lin}}}^{L} \alpha(r) dr$. The additional spectral narrowing of isotropic

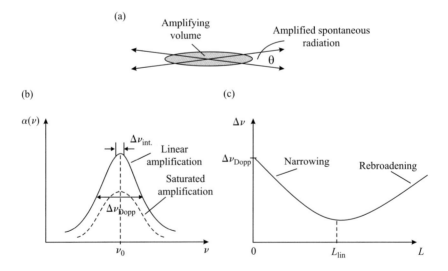

Fig. 13.1 (*a*) Elongated shape of the amplification volume, (*b*) change of the spectral profile of the inhomogeneously broadened amplification line under strong saturated conditions, and (*c*) evolution of the line width $\Delta \nu$ of the spontaneous radiation being amplified in such a way that the spectral line changes from an initial narrowing to a rebroadening at $L = L_{\text{lin}}$.

radiation in the saturated regime will stop the rebroadening of the spectrum and moreover will produce further narrowing of the spectral line (Fig. 13.2). This probably explains the anomalously narrow width of the emission line of the OH maser (Barrett and Rogers 1966). The fact that this case is quite rare confirms the conclusion that a nearly spherical geometry of the amplifying region occurs rarely because of the spherical nonstability of the saturation regime.

Thus, from this qualitative consideration we can, for simplicity, generalize a summary in the following three limiting geometrical cases: (a) elongated volume, (b) spherical volume, and (c) disk-like volume.

(a) *Elongated amplification volume* (Fig. 13.1a). In this case, the angular divergence Θ of the amplified radiation is governed by the angular spread of the lasing region:

$$\Theta \simeq \frac{a}{L}, \tag{13.4}$$

where a and L are the lateral and longitudinal sizes of the lasing volume, respectively. In rarefied gas condensations, the collisional broadening is very small. A more or less directed light beam interacts with the narrow spectral range $\Delta\nu_{\text{int}}$ determined by expression (6.12).

If $\Theta \ll \pi$, amplification gets saturated first at the center of the Doppler profile, and then the amplification saturation gradually spreads over the entire profile. As a result, the spectral profile of saturated amplification becomes flat (Fig. 13.1b), which gives rise to the reverse effect – a spectral rebroadening of the radiation being amplified under saturated conditions, the spectral rebroadening effect (Fig. 13.1c). This effect was analyzed by Litvak (1970) in the case of microwave space masers.

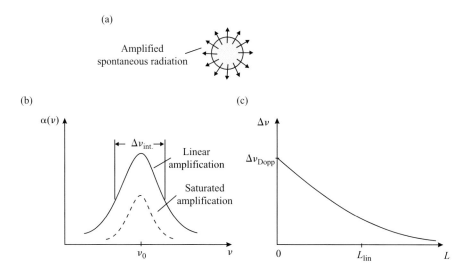

Fig. 13.2 (*a*) Spherical shape of the amplification volume, (*b*) absence of changes in the shape of the inhomogeneously broadened spectral line upon saturation by isotropic radiation, and (*c*) corresponding absence of the spectral rebroadening effect.

(b) *Spherical amplification volume* (Fig. 13.2a). With this geometry of the amplifying region, the amplified radiation is isotropic. The saturation of the Doppler profile under the effect of isotropic radiation occurs equally for the entire profile, i.e., without any change in the shape of the amplified line, as shown in Fig. 13.2b. In this case, the effect of rebroadening of the amplified radiation is absent (Fig. 13.2c), and the spectral width of the radiation is defined by (13.3). The spectral width $\Delta\nu$ for the isotropical case can be even less than the value defined by Eq. (6.67) and much less than $\Delta\nu_{\text{Dopp}}$.

(c) *Disk-like amplification volume.* This is an intermediate case between (a) and (b). In the direction along the disk plane, amplification saturation takes place under the effect of radiation with a geometrical angular divergence of $\Theta \sim a/b$, where a is the thickness of the amplifying disk and b is its diameter. When $\Theta \ll \pi$ amplification should be saturated first in the narrow central region of the Doppler profile after which the top of the profile should become flat, as in the case with an elongated volume (a). Thus, in the present case (c) the width of the laser line should be comparable with the Doppler width $\Delta\nu_D$. However, if saturation is attained in the direction perpendicular to the disk plane as well, the quasi-isotropic radiation will provide for saturation of the entire Doppler profile, as in the case with a spherical amplifying volume (b). One could then expect that the laser line rebroadening effect will be absent, so that the line will narrow in accordance with Eq. (13.3):

$$\Delta\nu \simeq \frac{\Delta\nu_D}{\sqrt{1+\alpha_{32}^0 a}} \tag{13.5}$$

For astrophysical lasers active in the wavelength range 0.9–2.0 μm in the Weigelt blobs of η Carinae (Chapters 11 and 12), any of the different types of geometry of the amplifying region mentioned above can be possible. What is important at the moment is that the laser lines observable in the range 0.9–2.0 μm can have a width, $\Delta\nu$, somewhere between $\Delta\nu_D$ down to $(0.1–0.3) \cdot \Delta\nu_D$. The magnitude of $\Delta\nu_D$ for the Fe II lines from the H I region of the blobs depends on the temperature T and can amount to $\Delta\nu_D \sim 300–1000$ MHz, since T is in the range 100–1000 K. Accordingly, the laser line width $\Delta\nu$ can lie between 30 and 1000 MHz. To measure such lines adequately requires a spectral resolution of $R \sim 10^7$, which is very difficult to achieve with the standard spectroscopic techniques.

Strictly speaking, it is necessary to solve the equation of radiation transport to find out the spectral bandwidth, intensity, and size of astrophysical lasers. This has been done by Litvak (1974b) for astrophysical OH and H_2O masers. He considered the equation of transfer radiation

$$\frac{d}{ds}I(\mathbf{r}, \mathbf{k}) = \alpha(\mathbf{r}, \mathbf{k})I(\mathbf{r}, \mathbf{k}) + G(\mathbf{r}, \mathbf{k}). \tag{13.6}$$

Here α is the local amplification tensor coefficient (cm^{-1}) for signals of wave vector $\mathbf{k}$ as influenced by the saturation effect of strong signals (in various directions) at position $\mathbf{r}$, and the function $G(\mathbf{r},\mathbf{k})$ is the source of radiation due to spontaneous emission and scattering. The derivative with respect to the ray path lengths, ds is justified only when wave properties are changing slowly over distances comparable to a wavelength. The tensor properties takes into account polarization properties.

The obtained results are quite interesting. Firstly, the actual size of hot laser (bright) spots are possibly 100–1000 times smaller than the size of the amplification zone (for sources near the compact H II regions). Secondly, appreciable saturation broadening restricts the spectral narrowing in accordance with Eq. (13.2). It is quite explainable by Eq. (6.14), that $\Delta\nu_{\text{hom}}$ is the full Doppler width and Doppler contour behaves as a "homogenized" spectral line in the isotropic radiation. According to Litvak (1974a, b), the partially saturated sphere shows the expected line narrowing in the hot spot and the line broadening in the surrounding saturation region. But the output bandwidth might be $1/\sqrt{2}$ times smaller than the Doppler width, while the bandwidth for filamentary masers stays on the Doppler bandwidth due to the rebroadening in the case of inhomogeneous Doppler broadening under the condition (6.15), as shown in Fig. 13.1.

The influence of the geometrical shape on the radiation properties (spectral width, apparent size of hot (laser) spot) of space masers was considered in later works (Bettwieser, 1981; Watson and Wyld, 2003) using extensive calculations for spherical and disk masers. In contrast to previous studies in which various approximations are made, the full equations were solved for frequency-dependent radiative transport that includes the thermal motion of the molecules and the high degree of saturation. The spectral line profiles for spheres and disks were found to rebroaden to the full Doppler width with increasing saturation in essentially the same way as the narrowing for a linear maser. The variation with frequency in the apparent angular sizes of masing spheres and thin disks was found to be negligible at frequencies within the spectral line at which the saturation is significant (Watson and Wyld, 2003). These results are quite striking because the isotropic spontaneous radiation and amplification should exclude the distortion of Doppler profile and rebroadening of amplifying isotropical radiation. Also the spectral line widths associated with some of the brightest astrophysical masers (the H_2O flares) are much less than the Doppler widths.

The solution of the transport equation (13.6) in the optical case will account for an effect of resonance scattering, which is essential in the laser case. The solution of one particular case without accounting for Doppler redistribution of frequency during the resonance scattering is presented in Chapter 14.

13.2 How to measure the narrow spectral line width of astrophysical lasers

The best proof of the appearance of the laser effect in certain spectral lines in astrophysical media would be the observation of the amplification-induced narrowing of the spectrum. Since the spectral line width may be considerably smaller than the Doppler width, such measurements can be carried out only with the aid of high-resolution spectral instruments (e.g., the Fabry–Perot etalon). Excellent results with such measurements were obtained by Münch and Taylor (1974). They showed that the width of the 8446 Å line from a source in the Orion Nebula is narrowed by approximately 25% compared to the 6300 Å line (see Section 12.4).

In the 1950s, R. Hanbury Brown and Q. Twiss proposed and developed a new method under the name *intensity interferometry* (Hanbury Brown and Twiss, 1957, 1958; Hanbury Brown, 1974), which is characterized by its very high spatial (angular) resolution. It produced accurate diameter measurements of 36 bright stars, using a pair of 6.5 m collectors and a 188 m baseline. The method was perhaps the first among a series of new techniques based on advanced technology that led to substantial progress in attaining levels of high and ultra-high resolution in astronomy (see review by Saha, 2002). For example, the speckle interferometry technique helped to discover the Weigelt blobs in the vicinity of Eta Carinae (Weigelt and Ebersberger, 1986). To master new wavelength regions and achieve high spectral resolution levels, the intensity interferometry method was modified to become heterodyne interferometry (Johnson et al., 1974). This technique uses a local monochromatic laser oscillator to produce beats between the light wave of the star of interest and the coherent laser wave of the local oscillator. The method can be considered intermediate between intensity interferometry and direct interferometry. Townes and co-workers made successful observations in the $10\,\mu$m infrared window of the atmosphere using a CO_2 laser as a local oscillator (Johnson et al., 1974; Townes et al., 1978). The baseline used consisted of a pair of auxiliary telescopes spaced a few meters apart at Kitt Peaks solar telescope. This technique provided the observation of CO_2 laser action in the mesospheres of Mars and Venus (Section 8.3).

The statistics of the fluctuations in the radiation emitted into a small solid angle by a noncoherent source coincides with the statistics of an equilibrium thermal radiation in a narrow spectral range (astrophysical amplifier or oscillator). For such light the correlation function for the intensity fluctuations

$$k(\tau) = \langle \Delta J(t+\tau) \Delta J(t) \rangle, \quad \Delta J(t) = J(t) - \langle J(t) \rangle \tag{13.7}$$

(the brackets indicate averaging over time) is connected with the spectral line width $\Delta \nu$ by the relation

$$k(\tau) = \langle J(t) \rangle^2 |\gamma(\tau)^2|; \quad \gamma(\tau) = \exp(-\tau \pi \Delta \nu), \tag{13.8}$$

here $\gamma(\tau)$ is the normalized function of mutual coherence (Mandel and Wolf, 1963). By measuring the intensity-intensity correlation function $k(\tau)$, we can obtain information about the spectral line width $\Delta \nu$. The spectrum constriction should manifest itself in the appearance of correlations with a characteristic time $T_{\rm cor} \approx 1/\pi \Delta \nu$. The variation of the baseline d of a pair of auxiliary telescopes enables the detecton of the disappearance of correlations, which corresponds to a diameter D of a radiation source:

$$D \simeq \lambda_0 L_0 / d \tag{13.9}$$

or to a large angular resolution $\varphi \simeq \lambda_0/d$ for baselines of a few and more meters. Figure 13.3 presents a schematic diagram of a Brown–Twiss–Townes (BTT) optical heterodyne interferometer that can be used to measure both the angular size and emission spectrum of the Fe II and O I laser line sources in the Weigelt blobs of Eta

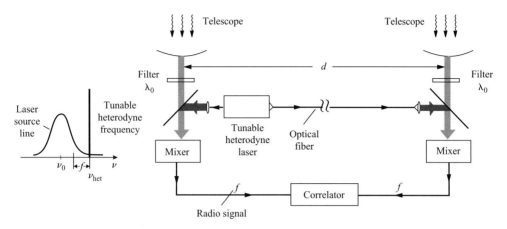

Fig. 13.3 Brown–Twiss–Townes optical laser heterodyne intensity correlation interferometer with two separated telescopes, local tunable laser diode heterodyne of frequency ν_{het}.

Carinae. A specific feature of such a correlation interferometer is the use of a 0.8–1 μm tunable monomode diode laser as a local oscillator and an optical fiber to transport this monomode laser radiation. This is much easier to accomplish than to transport the space laser radiation received by the telescopes. The distance d between the telescopes should meet the requirement in angular resolution that the radiation emitted by the Weigelt blob studied is separated from the photospheric radiation of Eta Carinae, i.e.,

$$d \simeq \lambda_0 \frac{L_0}{fD} \tag{13.10}$$

where $L_0 = 10^{22}$ cm is the distance to η Car, $D \sim 10^{15}$ cm is the diameter of the Weigelt blob, and f is the fraction of the blob region, in which the laser effect takes place to produce the radiation received by the ground-based telescopes. For estimation purposes, one can set $f = 1$, and the necessary distance between the telescopes will then be $d \sim 10^3$ cm. By choosing $d \sim 10^4$ cm, one can in principle analyze laser radiation coming from blob regions as small as one tenth of the size of the blob.

At the high-speed photomixer, the wavefront of the radiation being received should be matched with diffraction-limited accuracy to that of the local laser oscillator radiation over the entire area s of the photodetector in accordance with the antenna theorem for photomixing (Letokhov, 1965; Siegman, 1966). In that case, $s\Omega \simeq \lambda_0^2$, where Ω is the field of view or aperture. The heterodyne field of view will then be $\Theta \sim 1.2\lambda_0/a_0$, where a_0 is the diameter of the primary telescope mirror. Modern avalanche photodiodes (Prochazka et al., 2004) with a quantum efficiency $\eta \sim 1$ are most suitable to use as a high-speed photomixer. The photomixer signals of intermediate frequency $f = \nu_{\text{het}} - \nu$, where ν_{het} is the laser heterodyne frequency and ν is the frequency of the desired spectral line frequency of the astrophysical laser radiation being received, should be processed with a correlator. The measured correlation functions of the signal intensity fluctuations as a function of frequency (Fig. 13.4a) should provide information about the Fourier transformation of the spectral profile

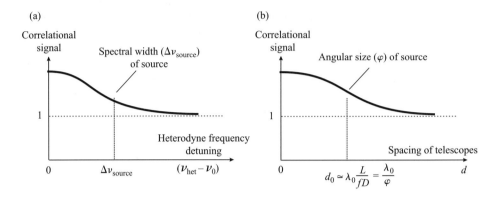

Fig. 13.4 Expected dependence of the correlation signal intensity as a function of (a) the heterodyne frequency detuning and (b) the spacing of telescopes d.

of the radiation being received. The spectral resolution R will be determined by the reception bandwidth B (Hz) of the photomixer radio signals:

$$R = \frac{c}{\lambda_0 B}. \tag{13.11}$$

At $B \sim 10^6$ Hz the spectral resolution $R \sim 3 \cdot 10^8$, which is sufficient to measure the emission spectrum of an astrophysical laser having a spectral width that is tens of times narrower than the Doppler width. The dependence of the correlation signal on the distance d between the telescopes should give information on the angular size φ ($\simeq \sqrt{\Omega}$) of the blob region, wherein the astrophysical laser of interest is active at the wavelength λ_0 under study (Fig. 13.4b). However, the key problem for the experiment proposed is the signal-to-noise ratio S/N that poses the requirement for the primary mirror diameter a_0 of the telescopes, the reception bandwidth B, and the observation time τ.

With heterodyne detection, one can control the intensity of the local laser oscillator and thus raise the signal level above the noise level of the electronic circuits used. As a result, one can reach the quantum noise limit, at which the signal-to-noise level is given by (Abbas et al., 1976; Rothermel et al., 1983)

$$\frac{S}{N} = \frac{\eta P}{h\nu B}, \tag{13.12}$$

where P is the power of the source, $B = \Delta\nu$ is the bandwidth of a single frequency resolution element of the mixing signal, and η is the quantum efficiency of the photodetector-mixer. If the signal is integrated over time τ, the S/N ratio is enhanced by a factor of $(B\tau)^{1/2}$, and becomes

$$\frac{S}{N} = \frac{\eta P}{h\nu} \left(\frac{\tau}{B}\right)^{1/2}. \tag{13.13}$$

The S/N ratio drops with the increasing bandwidth $B = \Delta\nu$. To measure the profile of a spectral line having a width of tens and hundreds of MHz, i.e., much wider than B, it is not necessary to scan the entire profile with a spectral resolution of B. It is natural to suppose that the spectral profile of the APL radiation is bell-shaped. Therefore, it is sufficient to measure the signals within very narrow bands B, a few Hz wide, at several points on the profile and integrate over a sufficiently long time at each sampling point. In that case, one can achieve a sufficient increase in the S/N ratio without any loss of essential information. To estimate the S/N ratio in the case of astrophysical laser lines from the Weigelt blobs, one can use the HST/STIS data for the integrated intensity of one of these lines (Gull et al., 2001). For example, the intensity of the 9997 Å line from the Weigelt blobs BD amounts to $2 \cdot 10^{-12}$ erg/cm$^2 \cdot$ s $\cdot$ Å. With the primary telescope mirror area $A \simeq a_0^2$ of the order of 1 m^2, one can expect a power of $2 \cdot 10^{-8}$ erg/s within the limits of the Doppler profile $\Delta\nu \ll 1$ Å. From this we get a lower limit estimate of the S/N ratio:

$$\frac{S}{N} \geq 1.2 \cdot 10^4 \eta \left(\frac{\tau}{B}\right)^{1/2}. \qquad (13.14)$$

For $\eta \sim 1$, $\tau \sim 10^2$ s, and $B = 1$ kHz, we get $S/N \sim 4 \cdot 10^3$.

The estimate based on Eq. (13.14) should be considered as very approximate, as it can be either greater or smaller by at least an order of magnitude for the following reasons. On one hand, the size of the laser volume in the Weigelt blob can be smaller than that of the whole blob whose emission is received by the HST/STIS. Therefore, the integrated intensity of the 9997 Å line can be lower than the value used for the estimation. On the other hand, the spectral width of the astrophysical laser radiation can be an order of magnitude smaller than 1 Å according to data by Gull et al. (2001) and even smaller than the Doppler width. It can enhance the S/N by a factor of (1 Å/$\Delta\lambda_{\text{Laser}}$). Furthermore, the divergence of the APL radiation in the case of an elongated volume for the amplifying medium can be much smaller than 4π. In any case, the estimates above point to the feasibility of the proposed experiment with the power of the radiation received varying over wide limits because of the existing variable parameter $(\tau/B)^{1/2}$. Let us underline that the proposed experiment can already be performed using existing pairs of ground-based telescopes in the southern hemisphere.

14
Possibility of scattering feedback in astrophysical masers/lasers

In this Chapter we consider the possibility that a maser/laser amplifier in space is converted to a generator (oscillator) by means of a positive energy feedback provided by a small scattering of electromagnetic radiation back to the amplifying medium.

Masers are based upon placing the amplifying medium in a resonance cavity (Gordon et al., 1954; Basov and Prokhorov, 1954), and lasers, upon placing this medium in an open Fabry–Perot cavity (Schawlow and Townes, 1958; Prokhorov, 1958). In both these cases, the oscillation of an electromagnetic field takes place in a limited number of resonant modes, and so the generated electromagnetic field is spatially coherent. Such a feedback may be called coherent and resonant (Fig. 6.5). This type of feedback fulfills two functions: (1) it returns some of the electromagnetic energy emitted by the active medium and (2) it forms a stable electromagnetic field configuration in the cavity because of the constructive interference between the incident and reflected waves.

However, other types of positive feedback are also possible, where there occurs a return of electromagnetic energy into the amplifying medium without any stable phase retention, i.e., without formation of a small number of resonant modes. This was demonstrated by Ambartzumian et al. (1970) using ruby crystals with a high amplification and with scattering surfaces. Escape of radiation from such a laser system because of scattering is the predominant loss mechanism for all modes. As a result, instead of individual high-quality resonances there appear a large number of low-quality resonances, which overlap and form a continuous flat spectrum of modes. The absence of resonant feedback means that the spectrum of the radiation being generated tends to be continuous, i.e., it does not contain discrete components at selected resonant frequencies (see review by Ambartzumian et al., 1970). The only resonant element left in the laser is the amplification line of the active medium. Therefore, after reaching a threshold the oscillation spectrum narrows continuously towards the center of the amplification line. It was found that the process of spectral narrowing is much slower than in an ordinary laser with resonant feedback, and that this limits the width of the spectrum in the pulsed mode operation. In the steady-state or continuous-wave (CW) mode the spectrum width is determined by fluctuations.

The statistical properties of the radiation in the case of nonresonant feedback are somewhat unusual for a laser. It has been shown that it possesses the statistical properties of the radiation of an extremely bright "black body" in a narrow range of

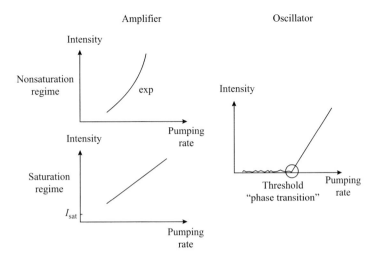

Fig. 14.1 Radiation intensity of a maser amplifier (left) and a maser oscillator (right) as a function of the pumping rate.

the spectrum. The radiation of such a laser has no spatial coherence and is not stable in phase. Since the only resonant element in a laser with nonresonant feedback is the amplification line of the active medium, the mean frequency of the radiation generated does not depend on the dimensions of the laser but is determined only by the frequency of the center of the resonant amplification line. If this frequency is sufficiently stable, the radiation of this kind of laser has a stable frequency.

Before going into the details of various scattering feedbacks it is worth mentioning other possible differences of amplification and oscillation regimes (Letokhov, 1996). First, unlike an amplifier, the oscillator has a sharp threshold. Fig 14.1 presents a qualitative relationship between the emission intensity of a maser amplifier and the pumping rate under saturated, and unsaturated conditions. If the pumping rate is localized near the threshold, the small variations of pumping rate will lead to significant variations of the output intensity of the maser/laser. Second, an oscillator is a dynamic system with corresponding dynamic properties due to positive feedback. In particular, the rapid change of pumping rate can lead to a transient variation of the output intensity of the maser/laser (Letokhov, 1967b). Fig. 14.2 illustrates qualitatively the dynamic response of a maser amplifier to a fast increase of the pumping rate. A sudden and very fast increase of the luminosity of a central star can induce a transient oscillation of the pumping rate of a nearby maser/laser. This phenomenon may occur in the vicinity of supernovae.

14.1 Noncoherent scattering feedback

The laser with nonresonant scattering feedback may be more properly referred to as a laser with *noncoherent feedback*. Such a laser can be built around an amplifying

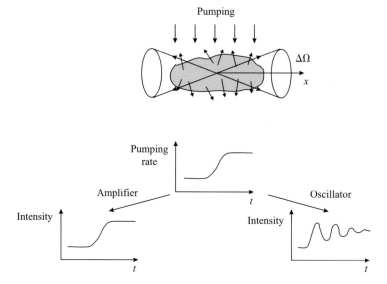

Fig. 14.2 Dynamical response of a maser amplifier (left) and maser oscillator (right) to a fast increase of the pumping rate.

medium with *nonresonant scattering particles* inside it, i.e., on the basis of a scattering medium with amplification (Letokhov, 1967a), shown in Fig. 14.3a. There is a variety of nonresonant scattering mechanisms – microparticles, Rayleigh scattering, etc. Let us consider briefly the specific features of such a laser with nonresonant, noncoherent feedback.

Take for the sake of simplicity an ensemble of identical dielectric particles with an average number per unit volume, N_0, and a complex permittivity $\epsilon = \epsilon_0 + i\,\epsilon''_\omega$, where $\epsilon''_\omega > 0$ in the vicinity of the frequency ω_0. Let Q_s be the scattering cross-section, and Q_ω the cross-section for amplification of light at a frequency ω of the particle. Here we consider the case when each particle scatters and amplifies. Naturally, the results obtained can easily be extended to the case of different amplifying and scattering particles, and also to the case of scattering particles immersed in a uniform amplifying medium.

Consider the case when the average dimension of the region occupied by the cloud is R, the mean-free path of a photon owing to scattering is $\Lambda_s = 1/Q_s N_0$, and the wavelength of the emission λ satisfies the inequalities

$$R \gg \Lambda_s \gg \lambda, \tag{14.1}$$

whereas the mean distance between the scattering particles $N_0^{-1/3}$ is much greater than λ:

$$N_0^{-1/3} \gg \lambda. \tag{14.2}$$

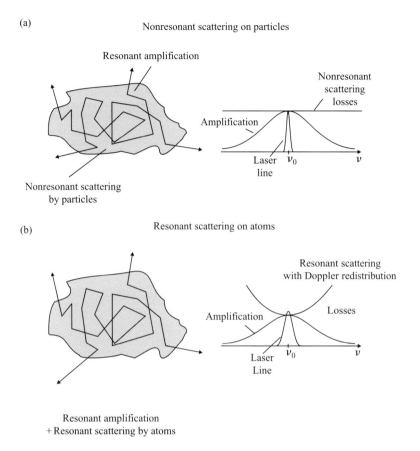

Fig. 14.3 Noncoherent scattering feedback by means of nonresonant scattering on particles (a), or resonant scattering on lasing atoms (molecules) (b).

Then the change in flux density $J_\omega(\mathbf{r},t)$ of photons of frequency ω at the point $\mathbf{r}$ can be described as follows in the diffusion approximation:

$$\frac{1}{c}\frac{\partial J_\omega(\mathbf{r},t)}{\partial t} = D\Delta J_\omega(\mathbf{r},t) + Q_\omega(\mathbf{r},t)N_0 J_\omega(\mathbf{r},t), \qquad (14.3)$$

where D is the diffusion coefficient, Δ is the Laplace operator, and c is the mean velocity of light in the region occupied by the particles. The frequency dependence of the photon density is connected with the resonant nature of the amplification. The cross-section $Q_\omega(\mathbf{r},t)$ does not remain constant, since the imaginary part of the permittivity $\epsilon''_\omega(\mathbf{r},t)$ depends on the photon flux density owing to the saturation effect.

The process of diffusion multiplication of photons described by Eq. (14.3) is in many ways reminiscent of the diffusion of neutrons in a homogeneous nuclear reactor. In the solution, we can therefore make use of certain results known for the diffusion of neutrons (Weinberg and Wigner, 1958). At the same time, in our case there are a number of essential differences connected with the resonant and non-stationary nature of the amplification cross-section $Q_\omega(\mathbf{r}, t)$.

The diffusion coefficient in an absorbing medium with anisotropic scattering takes the form (Weinberg and Wigner, 1958):

$$D = \frac{\chi_s}{3\Sigma\left(\Sigma - \bar{\mu}\chi_s\right)}, \tag{14.4}$$

where $\chi_s = Q_s N_0$; $\alpha_a = Q_s N_0$ are the macroscopic coefficients per unit length for scattering and amplification; $\Sigma = \chi_s + \alpha_a$; $\bar{\mu}$ is the mean cosine of the scattering angle. In the case we are discussing $\chi_s \gg \alpha_a$; and therefore:

$$D \approx \frac{1}{3\chi_s(1-\bar{\mu})} = \frac{\Lambda_s}{3(1-\bar{\mu})}. \tag{14.5}$$

For axisymmetrical scattering (Rayleigh type) $\bar{\mu} = 0$. In the case of preferential forward scattering ($\bar{\mu} > 0$) the diffusion coefficient increases.

The imaginary part of the particles' permittivity $\epsilon''_\omega(\mathbf{r}, t)$ in the case of homogeneous line broadening can be represented in the form

$$\epsilon''_\omega(\mathbf{r}, t) = a(\omega)\,\epsilon''(\mathbf{r}, t), \tag{14.6}$$

where $a(\omega)$ is the form-factor of the normalized absorption line shape (2.58). The quantity $\epsilon''(\mathbf{r}, t)$ is determined by the pumping power and depends on the photon intensity. The equation for $\epsilon''(\mathbf{r}, t)$ can be written in the form:

$$\frac{\partial\,\epsilon''(\mathbf{r}, t)}{\partial t} + \frac{1}{T_1}\,\epsilon''(\mathbf{r}, t) = -2\sigma_a\,\epsilon''(\mathbf{r}, t) \int a(\omega) J_\omega(\mathbf{r}, t)d\omega + \frac{1}{T_1}\bar{\epsilon}''(\mathbf{r}), \tag{14.7}$$

where $\sigma_a = \sigma(\omega_0)$ is the cross-section for radiative transitions of the ions in the dielectric particles responsible for resonance amplification; T_1 is the longitudinal relaxation time describing the spontaneous decay of amplification; $\bar{\epsilon}''(\mathbf{r})$ is a term proportional to the pumping power. In the steady-state case, in the absence of generation, $\bar{\epsilon}''_\omega(\mathbf{r}, t) = \bar{\epsilon}''(\mathbf{r})$.

The system of Eqs. (14.6) and (14.7) completely describes the stimulated emission of a cloud of scattering and amplifying particles. Boundary conditions must be set in order to solve it. The photon density is very low at the outer boundary of the medium. However, it cannot be zero, since photons diffuse from the cloud and can pass through the boundary surface. We can introduce the distance d, at which the photon flux extrapolated linearly outwards from the boundary of the medium would be zero. This distance, which is called the linear extrapolation distance, is defined by the expression (Weinberg and Wigner, 1958):

$$d = \frac{2}{3}\Sigma^{-1} \simeq \frac{2}{3}\Lambda_s. \tag{14.8}$$

The right-hand side of Eq. (14.3) describes the attenuation of the emission due to diffusion "spreading" and amplification. There is obviously a threshold at which the emission losses are compensated for by the amplification. At the threshold the amplification saturation can be neglected, i.e. $\bar{\epsilon}''_\omega(\mathbf{r},t) = \bar{\epsilon}''(\mathbf{r})$. Let us consider the case of uniform pumping $\epsilon''(r) = \epsilon''_0$, when the complex permittivity is the same for all scattering particles. Then the solution of the problem is reduced to the solution of Eq. (14.3) with a cross-section $Q_\omega(\mathbf{r},t) \equiv Q_\omega$ that is constant in ensemble and time. The general solution of Eq. (14.3) takes the form:

$$J_\omega(\mathbf{r},t) = \sum a_n \Psi_n(r) \exp\left[-\left(DB_n^2 - Q_\omega N_0\right)ct\right], \qquad (14.9)$$

where $\Psi_n(\mathbf{r})$ and B_n are the eigenfunctions and eigenvalues of the equation

$$\Delta \Psi_n(\mathbf{r}) + B_n^2 \Psi_n(\mathbf{r}) = 0 \qquad (14.10)$$

with the boundary condition $\Psi_n = 0$ at a distance d from the boundary, a_n are arbitrary constants determined by the initial distribution $J(\mathbf{r},t)$ when $t = 0$.

The threshold condition follows at once from (14.9):

$$DB^2 - N_0 Q_0 = 0 \qquad (14.11)$$

where B is the lowest eigenvalue B_n (usually $B = B_1$), $Q_0 = Q_{\omega_0}$, ω_0 is the frequency of maximum amplification.

If the region occupied by the ensemble of scattering particles takes the form of a sphere of radius R, then (Weinberg and Wigner, 1958):

$$\Delta \Psi_n(r) = \frac{1}{r} \sin \frac{n\pi r}{R}, \quad B_n = \frac{\pi n}{R}, \quad B = \frac{\pi}{R}. \qquad (14.12)$$

Thus, the threshold radius R_{thr} of such a random laser is determined by the expression

$$R_{\text{thr}} = \pi \left[3(1-\bar{\mu})\alpha_0 \chi_{\text{sc}}\right]^{-1/2} \qquad (14.13)$$

The cross-sections Q_s and Q_0 are determined by the geometry and the value of the complex permittivity $\epsilon = \epsilon_0 + i\epsilon''(\omega_0)$ of the scattering particles. For simplicity let us consider the case of spherical particles with a radius $d \gg 1/k = \lambda/2\pi$. For approximate estimates in the region $(\epsilon_0 - 1) \approx 1$, it is possible to use: $Q_s \approx S$; $Q_0 \approx 2\eta d\alpha_0 S$; $\bar{\mu} \approx 0$, where η is the mean transmission factor of the boundaries of a dielectric sphere. Then the expression for the critical size of the active scattering media becomes:

$$R_{\text{thr}} \approx g^3 \sqrt{\frac{32d}{3\eta\alpha_0}} \qquad (14.14)$$

where $g = N_0^{-1/3}/2d$ is the ratio of the average distances between particles to their diameter.

The emission spectrum of the laser with a nonresonant scattering feedback also has interesting features due to scattering. Let us, as above, consider them in the diffusion approximation. Note that, with the steady-state value for the integral intensity $J(r,t)$ the spectral density $J_\omega(r,t)$, may be non-stationary. The non-stationary nature of the

196 *Astrophysical lasers*

spectrum consists of a continuous narrowing of the emission spectrum after the start of generation. This is easy to check by examining the equation for the spectral density (14.3) for a stationary value of the amplification cross-section Q_0 and stationary spatial distribution of the photon density. We will neglect the distortion of Q_0 due to saturation, so $\Delta J_\omega = -B^2 J_\omega$. Equation (14.3) is then reduced to the following:

$$\frac{1}{c}\frac{\partial J_\omega}{\partial t} = -DB^2 J_\omega + a(\omega) Q_0 N_0 J_\omega. \tag{14.15}$$

Since $a(\omega_0) Q_0 N_0 = DB^2$,

$$J_\omega(t) \sim \exp\{-[a(\omega) - a(\omega_0)] Q_0 N_0 ct\}. \tag{14.16}$$

By expanding the function of the line form $a(\omega)$ near the center of the line ω_0 we find the law for spectrum narrowing in the steady-state generation:

$$\Delta\omega(t) = \frac{\Delta\omega_0}{\sqrt{(Q_0 N_0 ct/\ln 2)}}, \tag{14.17}$$

where $\Delta\omega_0$ is the full width of the amplification line at half-maximum.

Spectral narrowing proceeds, as usual, up to a certain fluctuation limit. In practice the most important part is played by the unavoidable fluctuations (Brownian motion) of the scattering particles, which leads to a random shift in the photon frequency due to the Doppler effect from the scattering particles. If at each scatter the frequency of a photon changes by a mean amount $\delta\omega \ll \omega_0$, then the change in the spectrum can be found by examining only the frequency diffusion of the photons, i.e., by examining the spatial and frequency diffusion separately. In this case we can use the frequency diffusion equation,

$$\frac{1}{c}\frac{\partial J_\omega}{\partial t} = D_\omega \frac{\partial^2 J_\omega}{\partial \omega^2} + Q_\omega N_0 J_\omega, \tag{14.18}$$

where D_ω is the frequency diffusion coefficient defined in the mode used by the expression

$$D_\omega = \frac{(\delta\omega)^2}{2\Lambda_s}. \tag{14.19}$$

The stationary solution of Eq. (14.18) takes the form

$$J_\omega = J_0 \exp\left[-\frac{(\omega-\omega_0)^2}{\Delta\omega_0}\sqrt{\frac{Q_0 N_0}{D_\omega}}\right] \tag{14.20}$$

or, by substituting the value D_ω and Λ_s, we obtain

$$J_\omega = J_0 \exp\left[-\frac{(\omega-\omega_0)^2}{\Delta\omega_0 \delta\omega}\sqrt{2\frac{Q_0}{D_s}}\right]. \tag{14.21}$$

The expression for the stationary width of the spectrum at half-maximum is of the form:

$$\Delta\omega_{st} = \left[(4\ln 2)\,\Delta\omega_0\,\delta\omega\sqrt{\frac{Q_s}{D_0}}\right]^{1/2}. \qquad (14.22)$$

The steady-state spectral width $\Delta\omega_{st}$ is the geometrical mean of the amplification line width $\Delta\omega_0$ (Doppler width $\Delta\omega_D$) and the frequency jump $\delta\omega$ during the frequency diffusion.

Laser action in scattering and amplifying media (random lasers) has been observed in the laboratory for laser crystal powder (Markushev et al., 1986) and in a laser dye solution with TiO_2 microparticles (Lawandy et al., 1994). They confirmed the main properties of a random media laser (threshold, narrowing of spectral lines etc.).

The diffusion model of a random laser does not account for random interference of photons, for instance the weak localization of photons in backscattering from amplifying media (Zyuzin, 1994). These effects are becoming more essential for small volume random lasers comparable with λ^3, when the number of spatial modes is not very high. In a random microlaser the scattering noncoherent feedback converts to scattering coherent feedback (Cao et al., 2003). However, this limiting case is very far from astrophysical conditions.

14.2 Resonant scattering feedback

A scattering feedback may have a *resonant* character as a result of the resonant scattering by the amplifying atoms (molecules) (Fig. 14.3b). The maximum resonant scattering cross-section, hence the minimum photon mean-free path, and accordingly the minimum scatter losses, coincides with the maximum amplification. For that reason, the lasing frequency coincides with the amplification line center.

To determine the resonance scattering coefficient per unit length one must know the cross-section for resonance scattering of radiation by a single particle. The cross-section for resonance scattering of radiation of frequency ω into 4π steradians is related to the cross-section of the radiative transition by (Letokhov, 1972b):

$$\sigma^{sc}_{nm}(\omega) = \sigma^{abs,em}_{nm}(\omega)\left(\frac{A_{nm}}{2\Gamma}\right), \qquad (14.23)$$

where the cross-section for a resonant radiative transition between two levels at the frequency ω is given by the expression (Heitler, 1954):

$$\sigma^{abs,em}_{mn,nm}(\omega) = \sigma_0\frac{\Gamma^2}{\Gamma^2 + (\omega-\omega_0)^2} = a(\omega)\,\sigma_0, \qquad (14.24)$$

$$\sigma_0 = \frac{\lambda^2}{2\pi}\frac{A_{nm}}{\Gamma}, \qquad (14.25)$$

where A_{nm} denotes the probability of a radiative transition between the two levels considered, and Γ is the homogeneous half linewidth, determined by the full levels falling inside a spectral interval of width 2Γ, corresponding to homogeneous broadening. In a line an inhomogeneous broadening $\Delta\omega_D$, because of the Doppler effect, say, σ_0 is determined by Eq. (6.3), and the Lorentz contour should be substituted by a Doppler contour with the width $\Delta\omega_D$. The resonance scattering cross-section is always smaller than the resonance emission or resonance absorption cross-section.

The coefficient for resonance scattering per unit length into a unit solid angle by particles in either the lower n or the upper m levels is equal to

$$\chi_{\rm sc}(\omega) = \sigma_{\rm sc}(\omega)(N_m + N_n). \tag{14.26}$$

The coefficient for resonance scattering into the solid angle $\Delta\Omega$ is related to the resonance amplification coefficient $\alpha_{mn} = \sigma_{nm}^{\rm em} N_m - \sigma_{mn}^{\rm abs} N_n$ by the expression

$$\chi_{nm}(\omega) = q_0 \left(\frac{\Delta\Omega}{4\pi}\right) \alpha_{mn}(\omega), \tag{14.27}$$

where the factor q_0 is defined as

$$q_0 = \frac{A_{nm}(N_m + N_n)}{\Gamma(N_m - N_n)} = \frac{A_{nm}}{\Gamma}\frac{N_{n,m}^0}{\Delta N_{nm}}, \tag{14.28}$$

where the effect of Doppler broadening is incorporated in the amplification coefficient $\alpha_{mn}(\omega)$, and $N_{n,m}^0 = N_m + N_n$. The coefficient χ_0 for resonance scattering per unit length into all directions will be

$$\chi_0 = q_0 \alpha(\omega). \tag{14.29}$$

The relationship between the resonance amplification and resonance scattering coefficients is defined by the product of the factors $(A_{nm}/\Gamma) \ll 1$ or $A_{nm}/\Delta\omega_D$ and $\left(\frac{N_{m,n}^0}{\Delta N_{mn}}\right) \gg 1$.

The expressions for threshold (14.13) and steady-state spectral width (14.22) can be modified for the case of scattering on lasing/masing particles using Eq. (14.27). For example, the threshold radius is

$$R_{\rm thr} = \pi(3\alpha_0\chi_{\rm sc})^{-1/2} = \pi/\alpha\sqrt{3q}. \tag{14.30}$$

At $\alpha_0 \gg \chi_{\rm sc}$ the lasing threshold with a noncoherent resonant scattering feedback is only reached with a high amplification per scattering length $l_{\rm sc} = 1/\chi_{\rm sc}$. It is precisely such a case that can be realized in space masers (Section 14.3). Resonant scattering is very weak in the long-wavelength region (for $\Gamma \gg A_{nm}$), but becomes quite substantial in the visible and UV regions (space lasers), where A_{nm} and Γ may be commensurable of the values.

The expression (14.22) for the steady-state width of the spectrum should account for a large value of the frequency jump $\delta\omega$ due to Doppler shift on scattering atomic particles: $\delta\omega \simeq \Delta\omega_D$. As a result the $\Delta\omega_{\rm st}$ is

$$\Delta\omega_{\rm st} \simeq \Delta\omega_D q^{1/4}, \tag{14.31}$$

i.e. the steady-state spectral width $\Delta\omega_{st}$ is comparable with Doppler width. It means that the Doppler scattering rebroadening compensates for the spectral narrowing due to resonant amplification. In other words the spectral output of a space laser amplifier can be much narrower than a space laser oscillator with resonant scattering on amplifying atomic particles except cases with $q \ll 1$.

14.3 Space masers with scattering feedback

For the sake of generality let us now consider the possibility of scattering feedback in space masers, eventhough the subject of this book is optical space lasers. The universally accepted theoretical approach to the space maser models is the traveling-wave amplifier model (Litvak, 1970; Elitzur, 1992). This approach explains the main emission characteristics of many observed maser sources. But thanks to the development of radio interferometric techniques and the creation of very-long-base radio interferometer systems (VLBI), powerful H_2O maser outbursts have lately been observed experimentally (see Clegg and Nedoluha, 1993). For example, interstellar masers have long been known to be variable on timescales from several years to a few days, but recent observations (Clegg and Cordes, 1991), have suggested that OH maser sources may show variability on timescales of a few *minutes*. The microwave emission of these powerful space masers exhibit new properties that make it greatly different from the emission of the earlier H_2O masers (Rowland and Cohen, 1986). Recent observations demonstrate the effect of very fast flares of H_2O masers. Attempts made to explain these flares within the framework of the standard space maser model in the form of a traveling-wave amplifier have run into serious difficulties. Thus, the effort of interpreting the compact OH and H_2O radio emission sources is still important.

A generator mechanism has long been suggested for the production of intense narrow-band radiation in population-inverted space maser media (Letokhov, 1966, 1972). Feedback in this case can be formed on the account of nonresonant scattering of radio emission by electrons or microscopic dust particles (Letokhov, 1966), or as a result of resonance scattering by the active molecules themselves (Letokhov, 1972b). Scattering in the maser model with a noncoherent feedback can help to explain unusual properties of compact space masers (Truitt and Strelnitski, 2000).

Resonance scattering over the characteristic amplification distance $1/\alpha(\omega)$ will return into the solid angle $\Delta\Omega$ according to Eq. (14.28). The required threshold amplification $K_0 = \exp[\alpha(\omega) L]$, on passage through the cloud, equals

$$K_0 \cong \frac{\alpha(\omega)}{\chi_{sc}(\omega)} = \frac{4\pi}{q\Delta\Omega}, \qquad (14.32)$$

where the factor q is defined by Eq. (14.28).

The relation can readily be derived for a simple configuration of an elongated amplifying cloud, in which one may neglect the angular dependence and consider the propagation of radiation beams of flux, P_+ and P_-, in two opposite directions along

200 Astrophysical lasers

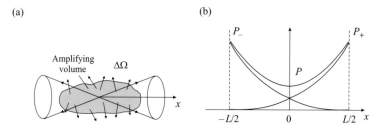

Fig. 14.4 Amplifying cloud with weak backscattering: (top) angular distribution of radiation; (bottom) intensity distribution along cloud P_+ and P_- being propagated in two opposite directions (linear regime) and total intensity $P = P_+ + P_-$.

the x axis (Fig. 14.4), with allowance for backscattering:

$$\frac{1}{c}\frac{\partial P_+}{\partial t} + \frac{\partial P_+}{\partial x} = (\alpha_0 - \chi_0)P_+ + \chi_{bs}P_-$$
$$\frac{1}{c}\frac{\partial P_-}{\partial t} - \frac{\partial P_-}{\partial x} = (\alpha_0 - \chi_0)P_- + \chi_{bs}P_+. \qquad (14.33)$$

Here, χ_{bs} is the coefficient for resonance backscattering per unit length into the solid angle $\Delta\Omega$ of the cloud, and χ_0 is the coefficient for resonance scattering per unit length in all directions. By solving the above system of equations, we can find a condition for exponential growth of power with time (a threshold condition):

$$K_0 = e^{(\alpha_0 - \chi_0)L} > \frac{\alpha_0 - \chi_0}{\chi_{bs}} \qquad (14.34)$$

or

$$K_0 > \frac{1-q}{q}\frac{4\pi}{\Delta\Omega}, \qquad (14.35)$$

where L is the length of the cloud and $\alpha_0 - \chi_0$ is the effective amplification per unit length corrected for losses by resonance scattering. If the medium contains additional nonresonant losses γ per unit length, the amplification α_0 in condition (14.34) should be replaced by an effective amplification $\alpha_0 - \gamma$.

One can obtain the value of the threshold amplification for other geometries of the amplifying cloud by considering the scattering of radiation in all directions in the presence of amplification, as has been shown in (Letokhov, 1967a). But, it is not possible to apply these results directly, because in the space maser $\alpha_0 \gg \chi_0$; i.e., we have weakly scattering active media. The generation threshold for the case of a weakly scattering medium may be written in the more simple form

$$q\left(\frac{\Delta\Omega}{4\pi}\right) \cdot \exp(\alpha L) > 1$$

$\uparrow$ $\quad\quad\quad\quad\quad\uparrow$

Probability   Amplification
of resonant   per pass
scattering   (14.36)

(1) *OH maser*. The transition probability for the strongest Λ-doublet lines of OH is $\Gamma_{12} = 8 \cdot 10^{-11}\text{sec}^{-1}$. The homogeneous width 2Γ of the transition depends, according to Eq. (14.36), on the combined contribution of the pumping rate (longitudinal relaxation) and the broadening effects of collision with atoms (transverse relaxation). The contribution of the pumping rate is quite different for different pumping mechanisms. In the model of pumping by ultraviolet radiation the contribution to the quantity Γ is 10^{-6}–10^{-8} sec, and the relative inversion reaches $\Delta N/N \approx 0.1$–0.01, where $N = N_n + N_m$ (Litvak et al., 1966). For pumping by infrared radiation this contribution to the homogeneous width, according to Litvak's (1969) estimates, would be 10^{-4} and 10^{-2} sec^{-1} for far- and near-IR radiation, respectively. In the case of collisions with H atoms the contribution to the homogeneous width depends on the density of atoms. If we adopt the quite high value 10^5 cm^{-3} for the density of hydrogen atoms in OH clouds, the contribution from collisions would yield a broadening of about 10^{-4} sec^{-1}. This contribution to Γ is decisive for the mechanism of pumping by ultraviolet radiation, and is also important for pumping by far-IR radiation. In these cases the quantity $\Gamma \approx 10^{-4}$ sec^{-1}, and with $\Delta N/N \simeq 0.1$ a cloud, having a geometrical solid angle of, say, $\Delta\Omega = 0.2$ sr, would require an amplification upon passage of $K_0 \gtrsim 10^8$, in order to be self-excited. Although many of the values ($\Delta N/N$, the density of H atoms, the divergence of radiation) adopted for this estimate are admittedly rather arbitrary, it seems entirely possible that the threshold amplification upon passage through the maser-generator model with feedback as a result of resonance scattering could explain the observations with less amplification than is required in the maser-amplifier model ($K_0 \approx 10^{11}$–10^{12} (Cohen et al., 1968)). If the homogeneous width of the transition has a higher value (for example, in the case of pumping by near-IR radiation $\Gamma \approx 10^{-2}$ sec^{-1} (Litvak, 1969)), the necessary amplification would increase ($K_0 > 10^{10}$ if we adopt the same estimates as before for $\Delta\Omega$ and $\Delta N/N$), and the proposed model will no longer offer an appreciable advantage over the model of amplification of a traveling wave.

(2) *H_2O maser*. In the case of galactic H_2O clouds (Knowles et al., 1969) the situation is more definite. Amplification will take place for the transition $6_{16} \to 5_{23}$ between high rotational sublevels of the ground vibration state (the energy of excitation is $\Delta E = 447$ cm^{-1}) (Cheung et al., 1969). The homogeneous width of the transition will be relatively large here, and will be determined by the spontaneous decay of the 6_{16} and 5_{23} states to the lower rotational states $5_{05}, 5_{14}, 4_{12}, 4_{32}$: $\Gamma_{10} + \Gamma_{20} \simeq \Gamma_{6_{16}-5_{05}} + \Gamma_{5_{73}-4_{14}} = 0.77 + 0.42 \simeq 1.2$ sec^{-1}. Since the radiative transition

$6_{16} \to 5_{23}$ has a probability $\Gamma_{12} = 2 \cdot 10^{-9}\,\mathrm{s}^{-1}\mathrm{g}^{-1}$, the factor q is given by

$$q = \frac{\Gamma_{12}}{\Gamma_{10} + \Gamma_{20}} \frac{N}{\Delta N} \simeq 10^{-8}, \qquad (14.37)$$

provided that $N_0/\Delta N$ is of the order of several units, and accordingly the threshold amplification will be $K_0 > 10^{10}$. Such an amplification is of the order of, or higher than, the amplification required for passage through the maser-amplifier model (Knowles et al., 1969).

A choice between the maser-amplifier model and the model of a maser with feedback can be made for galactic maser regions on the basis of special supplementary observations. For example, it would be important to measure the characteristic periods T_{var} of intensity variation when measuring the size L of compact emission regions by means of radio interferometry with a very long baseline (Cohen et al., 1968).

In principle the most important distinction between the maser amplifier and maser generator models is the difference between their emission spectrum widths. For the maser amplifier, the narrowing of the emission spectrum is due to the predominant amplification at the center of the amplification line, $\alpha(\omega)$. As a result, the emission line width $\Delta \omega$ is smaller than the Doppler width $\Delta \omega_{\mathrm{D}}$ by a factor of $1/\sqrt{\alpha_0 L}$ according to Eq. (13.2). The observed four- to five-fold narrowing of the emission spectrum agrees well with the estimate from Eq. (13.2). Under amplification saturation conditions, the radiation intensity $I \propto \alpha_0 L$ and thus $\Delta \omega / \Delta \omega_{\mathrm{D}} \propto I^{-1/2}$, which also agree with the experimental data of Rowland and Cohen (1986).

The calculation for a generator model with a strong resonance scattering yielded the following expression for the emission line width under steady-state conditions due to spontaneous emission (Lavrinovich and Letokhov, 1974; Letokhov, 1996):

$$\Delta \nu / \Delta \nu_{\mathrm{D}} = \left(\frac{N_{\mathrm{m}}}{N_{\mathrm{m}} - N_{\mathrm{n}}} \frac{1}{\langle n_0 \rangle} \right)^{1/2}, \qquad (14.38)$$

where $\langle n_0 \rangle$ is the emission power in terms of the number of photons per degree of freedom of a multiple-mode maser generator field. However, the expression in Eq. (14.38) can be used for an astrophysical maser with weak scattering feedback.

In the case of resonant scattering on the lasing atomic particles themselves the Doppler rebroadening will compensate for the one-pass amplification narrowing of the spectral width. As a result, the spectral line of laser emission of the total amplifying volume will be governed by Eq. (13.3) for a one-pass saturated amplifier. However, in the potentially possible case of nonresonant scattering on microparticles (dust) the contribution of Doppler broadening of scattered light is negligible. In this hypothetical case it is quite possible to obtain very narrow spectral lines in the generator regime.

The emission statistics of a generator with a noncoherent feedback were studied theoretically by Letokhov (1967b). The distribution function of the number of quanta in the k-th radiation mode, n_k, was found to be given by the expression

$$P(n_k) = \frac{\langle n_k \rangle^{n_k}}{(1 + \langle n_k \rangle)^{1+n_k}}, \qquad (14.39)$$

i.e., the distribution was found to be of the Bose–Einstein type. At $\langle n_k \rangle \gg 1$, Eq. (14.39) yields a Gaussian distribution. It was precisely this type of distribution that was experimentally observed in the case of the OH maser (Evans et al., 1972). However, the statistical distribution (14.39) is the same for amplified spontaneous emission. In other words, the statistical properties can not be used for the discrimination between amplifier and generator in noncoherent feedback regimes of space masers/lasers.

14.4 Space lasers with resonance scattering feedback

The concept of a maser with a noncoherent resonance scattering feedback can be extended to optical quantum transitions in atoms, for example, in stellar atmospheres (Lavrinovich and Letokhov, 1974). The attractiveness of the generator model lies in the fact that a high gain per pass is not required when the scattering is sufficiently effective, since above the threshold, when the gain per pass exceeds the loss per pass, the intensity grows in the course of many passes. The authors calculated the so-called threshold or the critical radius r_0, at which the spherical volume is on the threshold of self-excitation. For the case when $\alpha(\omega) \ll \chi(\omega)$ the solution of the problem is known (Letokhov, 1967a); however, depending on the values of the quantities determining $\alpha(\omega)$ and $\chi_{sc}(\omega)$, the relation between the latter quantities can be arbitrary, and it was necessary to consider the problem in its general form. The linear approximation, neglecting saturation effects, can be used in determining the threshold.

The equation satisfied by the radiation intensity J_ω at the frequency ω in the case of low particle densities N_0 and coherent resonant light scattering (scattering without a change in frequency) with a spherical indicatrix has the form (Sobolev, 1960):

$$\cos\theta \frac{\partial J_\omega}{\partial r} - \frac{\sin\theta}{r}\frac{\partial J_\omega}{\partial \theta} + \frac{1}{c}\frac{\partial J_\omega}{\partial t} = [\alpha(\omega) - \chi_{sc}(\omega)] J_\omega + \frac{\chi_{sc}(\omega)}{4\pi}\int J_\omega d\Omega, \quad (14.40)$$

where $d\Omega = \sin\theta d\theta d\varphi$ and c is the velocity of light.

Equation (14.40) admits the separation of variables. Denoting the separation constant by S, and representing the solution of the equation in the form $J_\omega = J_{0\omega}(t) J_{1\omega}(r, \theta, \varphi)$, we obtain for $J_{0\omega}(t)$ the equation

$$\partial J_{0\omega}/\partial t = cS J_{0\omega}, \quad (14.41)$$

whose solution is: $J_{0\omega}(t) = J_{0\omega}(0)\exp(cSt)$. For $S > 0$ this solution corresponds to the appearance of temporal instability – the exponential growth of the intensity in time or of the generation. The threshold regime corresponds to the value $S = 0$. We then have the following equation for $J_{1\omega}$:

$$\cos\theta \frac{\partial J_{1\omega}}{\partial r} - \frac{\sin\theta}{r}\frac{\partial J_{1\omega}}{\partial \theta} = (\alpha_\omega - \chi_{sc}) J_{1\omega} + \frac{\chi_{sc}}{4\pi}\int J_{1\omega} d\Omega. \quad (14.42)$$

The solution of Eq. (14.42) is a complex problem, and one of the most serious difficulties is the necessity for us to take into account the dependence of $J_{1\omega}$ on the angle θ. Averaging of $J_{1\omega}$ over the directions (i.e., over θ) will lead to a solution for the critical radius R_{cr}. The method of averaging adopted here is called Eddington's approximation. Carrying through a procedure similar to the one used by Sobolev (1963) it is possible to replace Eq. (14.42) with the following approximative equation:

$$\frac{d^2(rJ)}{dr^2} + 3\alpha(\chi_{\text{sc}} - \alpha)(rJ) = 0, \qquad (14.43)$$

where

$$J = \frac{1}{2}\int_0^\pi J_{1\omega} \sin\theta\, d\theta \qquad (14.44)$$

It is assumed that no radiation is incident on the medium from outside. Then $J_{1\omega}(R_{\text{cr}}, \theta) = 0$ for $\pi/2 < \theta \leq \pi$, and the approximate boundary condition can be written in the form

$$J(R_{\text{cr}}) = 2H(R_{\text{cr}}), H = \frac{1}{2}\int_{\frac{\pi}{2}}^{\pi} J_{1\omega} \sin\theta \cos\theta\, d\theta. \qquad (14.45)$$

Furthermore, it is necessary to impose on J the condition $J \to \text{const}$ as $r \to 0$, which corresponds to the requirement of finiteness of the solution at $r = 0$.

The solution to Eq. (14.43) has, under the indicated boundary conditions, the form

$$J = \frac{\text{const}}{r}\sin br, \quad b^2 = 3\alpha(\chi_{\text{sc}} - \alpha), \quad 0 \leq r \leq R_{\text{cr}}. \qquad (14.46)$$

The condition (14.46) leads to the following relation between R_{cr} and b:

$$\text{tg } bR_{\text{cr}} = 2\alpha bR_{\text{cr}}/(2\alpha - b^2 R_{\text{cr}}). \qquad (14.47)$$

This relation is essentially a *threshold condition*. In principle, all the solutions q_n, for which $\alpha > \alpha_{\text{thr}}(R_{\text{cr}})$ (where $\alpha_{\text{thr}}(R_{\text{cr}})$ is the threshold value) are possible. Physically, however, it is clear that the solution possessing the least threshold value $\alpha_{\text{thr}}(R_{\text{cr}})$ does not allow the other possible solutions (at least before the onset of saturation). Fig. 14.5 shows a qualitative plot of the behavior of R_{cr} as a function of α at fixed χ_{sc} for the solution possessing the least threshold. In the limiting case when $\alpha \ll \chi_{\text{sc}}$ (the region I) the result coincides with the form of the solution (14.14) in the diffusion approximation (Letokhov, 1967a).

Let us now estimate the critical dimension R_{cr} for the O I laser, using the estimates for α_0 obtained in Chapter 12. The parameter q in Eq. (14.29) equals 1 according to (14.28), because $A_{\text{nm}}/2\Gamma \simeq 1$ for pure radiative transversal phase relaxation and $\frac{\Delta N_{32}}{N_3+N_2} \simeq 1$ ($N_2 \ll N_3, \Delta N_{32} \simeq N_3$). It means that $\chi_{\text{sc}}^{(\text{cr})} \simeq \alpha(\omega)$, i.e., the resonant amplification and resonant scattering lengths are equal. According to results plotted in Fig. 14.5 the critical radius $R_{\text{cr}} \simeq 1/\alpha \simeq 1/\chi_{\text{sc}}$.

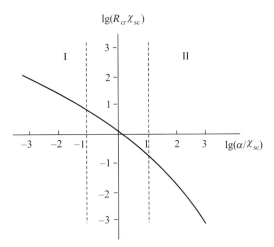

Fig. 14.5 The dependence of the critical (threshold) radius R_{cr} in units of the scattering length $1/\chi_{sc}$ on the ratio α/χ_{sc}. In region I the ratio $\alpha/\chi_{sc} \ll 1$ and $R_{cr}\chi_s \simeq \pi \left(3\alpha_0/\chi_{sc}\right)^{-1/2}$, whereas in region II the ration $\alpha/\chi_{sc} \gg 1$ and $R_{cr}\chi_s \simeq \alpha/\chi_{sc}$.

It is a rather surprising situation, when the amplification length $l_{amp} = 1/\alpha$ equals the scattering length $l_{sc} = 1/\chi_{sc}$, i.e. a single event of stimulated emission in the amplifying transition corresponds to a single event of resonant scattering of a photon on the atom in the upper state of the transition. Naturally, the critical radius R_{cr} is of about the same order as l_{amp} and l_{sc} in such a case. This situation does not exist in laboratory lasers, because $2\Gamma \gg A_{nm}$ for all known lasers. Actually, the time of phase relaxation (dephasing or decoherence) of the wave function of the quantum state of the lasing particle in laboratory laser media, $1/2\Gamma$, is much shorter than the radiative decay time of the quantum state, A_{nm}^{-1}. For example, the ratio $A_{nm}/2\Gamma \simeq 10^9(!)$ for Cr^{3+} ion in a ruby laser. Thus, the resonant scattering is not observable in laboratory lasers. Otherwise, it would be a huge hindrance for laser action. An astrophysical laser (not maser!) medium is a diluted astrophysical plasma, in which the radiative decay is a very fast relaxation (phase and population of quantum state) process, and the resonant scattering should be an essential feature of a space laser.

Diffusion light confinement inside a volume with the critical radius R_{cr} can lead to an independent action of many separated laser volumes in the saturation regime, i.e. the forming of *laser clusters* for a large size of the amplifying volume ($L \gg R_{cr}$). This effect would be similar to a spatial-temporal non-stability of nonlinear (saturated) active media having the characteristic time variation on the scale $\tau_{var} \simeq R_{cr}/c$ (a few minutes). High spatial and temporal resolution observations of the O I 8446 Å line (see Chapter 12) can perhaps detect this effect in the lasing Weigelt blobs of Eta Carinae.

The transition of an amplification regime to an oscillation regime for the case when $l_{sc} \simeq 1/\alpha$ cannot be accompanied by a narrowing of the spectrum. The Doppler rebroadening of resonantly scattered light on moving atomic particles will eliminate the spectral narrowing because of resonance amplification. However, the spectral narrowing of the amplification length is still retained in the case of weak scattering ($l_{sc} \gg 1/\alpha$). Moreover, the narrowing effect can be very large (up to 10^4) in the case of nonresonant scattering on microparticles (dust) owing to multipass propagation of light because of the absence of significant Doppler effect on scattering microparticles.

15
Nonlinear optical effects in astrophysical conditions

Studies of the possibility of nonlinear phenomena occuring in an astrophysical medium have a long history. The Soviet physicist Sergei Vavilov mentioned this possibility many years ago (before the laser era) in the case of the solar atmosphere with its high intensity of light radiation (Vavilov, 1960). The isotropic optical properties of an astrophysical plasma permit the existence of noncoherent nonlinear effects only when they do not require phase-matching conditions of the light waves. There are a number of such effects: absorption, excitation, and ionization in a two-photon process, and Raman scattering. The linear Raman scattering effect (Fig. 15.1a) was discovered in the nebula of a symbiotic star by Schmid (1989). This interesting phenomenon may deserve a more detailed study for the understanding of unidentified stellar emission lines. Two-photon absorption by vibrationally excited H_2 molecules was considered by Sorokin and Glownia (1995, 2000) as a possible explanation of the long-standing mystery concerning the origin of the optically diffuse interstellar absorption bands (DIBs) (Whittet, 1992). While intriguing, this suggestion overlooks several critical spectroscopic and astrophysical problems (Snow, 1995). Johansson and Letokhov (2001a, b, c) considered the resonance-enhanced two-photon ionization (RETPI) of atoms and ions in radiation-rich gas condensations. The presence of this effect was supported by the observation of an anomalous and variable intensity of the Si III intercombination line at 1892 Å in the Weigelt blobs of Eta Carinae (η car) (Johansson et al., 2006). We will discuss this effect below in more details.

15.1 Photoionization processes in radiation-rich astrophysical plasma

Photoionization processes have been known to play an important part in the physics of planetary and gaseous nebulae (Aller, 1956, 1984; Pottasch, 1984; Osterbrock, 1989), as they are responsible for the inflow of radiative energy from the central star (stars). The short-wavelength radiation provides the photoionization absorption of hydrogen and helium, whose atoms repeatedly take part in this process. The H and He ions recombine with electrons, thus providing for the permanent conversion of the absorbed stellar radiation into kinetic energy of the photo electrons produced, and radiative energy of the recombination transitions, especially the high-intensity EUV lines in the Lyman series of H I and He II and the resonance lines of He I. This line radiation suffers from

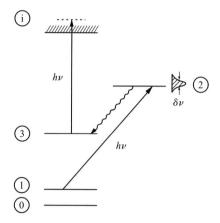

Fig. 15.1 Possible subsequent "bound-free" photoionization transition via metastable 3 state.

resonance scattering due to diffusion trapping in optically dense nebulae, which results in Doppler diffusive broadening of the spectral lines and an increase of their intensity inside the nebula. Both these effects are important for the radiative cooling of the nebula due to the escape of trapped photons through the wings of the spectral lines (the Zanstra effect) (Zanstra, 1949), and the escape of optically thin fluorescence lines of other elements produced in a Bowen mechanism (Bowen, 1935). The latter is caused by an accidental wavelength coincidence between an H or He line and an absorption line (1 → 2) of these elements (see Chapter 9).

The efficiency of the resonant excitation by the Bowen mechanism is especially illustrated by two UV Fe II lines around 2508 Å (Fig. 10.2) observed with the Hubble Space Telescope in high spectral and spatial resolution of the Weigelt blobs in η Car (Johansson et al., 1996). The abnormally high brightness of these lines is associated with the Bowen excitation of Fe II from the low-lying metastable state $a^4D_{7/2}$ (state 1 in Fig. 15.1) by H Lyα (Johansson and Hamann, 1993). To explain the high brightness of these lines and the abnormal intensity ratios between them and their satellite lines, Johansson and Letokhov (2001e) considered the possibility of a subsequent photoionization of Fe II from the long-lived states $c^4F_{7/2,9/2}$ (state 3 in Fig. 15.1), in which Fe II gets accumulated as a result of the fast UV 2 → 3 decay. It is important to note that the rate of the photoionization process involving diluted radiation from the central star is much slower than the photoionization rate associated with the intense radiation from the trapped H Lyα line, whose wavelength at 1215 Å is shorter than the threshold wavelength required to photoionize Fe II from the state $c^4F_{7/2}$ ($\lambda = 1244$ Å). This statement is based on a high effective temperature of Lyα in the Weigelt blobs due to effect of "spectral compression" of absorbed Lyman continuum radiation (Section 10.2). The rate of the subsequent photoionization of state 3 of Fe II is lower than or comparable with the rate of the stimulated transition (lasing) and radiative cycling of excited Fe II ions (Section 11.4).

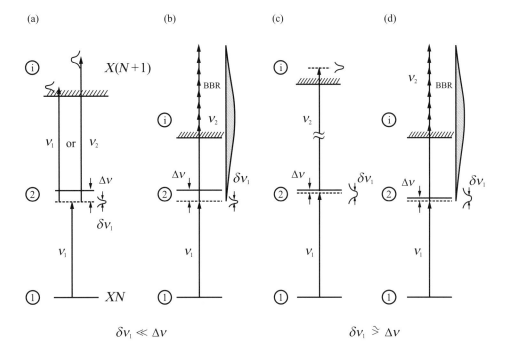

Fig. 15.2 Various schemes of resonance-enhanced two-photon ionization (RETPI) by two monochromatic spectral lines with various detuning $\Delta\nu$ of exciting spectral lines ν_1: $\delta\nu_1 \ll \Delta\nu$ in (a) and (b); $\delta\nu_1 \geq \Delta\nu$ in (c) and (d), and a small binding energy of the excited state in (b) and (d).

The slow recombination rate of the photoions produced in rarefied nebulae results in an accumulation at a comparatively low photoionization rate. Johansson and Letokhov (2001a) have considered the possibility that ions (for example Si II, Si III) can be created by resonance-enhanced two-photon ionization (RETPI) in a low-density ($N_H < 10^{10}$ cm^{-3}) astrophysical plasma, in which the H Lyα line is supposed to have a sufficiently high intensity. One example of such a plasma is gas condensations (blobs) near a hot star, and the best-known case is the Weigelt blobs near η Car (see Chapter 10). The photoionization-recombination cycle in these blobs leads to the formation of intense H I, He I, and He II resonance lines with effective spectral temperatures of the order of $(10-20) \cdot 10^3$ K close to the photospheric temperature of the star (Johansson and Letokhov, 2004b). This unique situation differs radically from the conditions in a typical planetary nebula located orders of magnitude further from the central star, where the main role in exciting and ionizing atoms is played by the electron energy, while the photon density is negligible (Pottasch, 1984; Osterbrock, 1989). The first indications of the RETPI effect in an astrophysical plasma have been found in spectra of the Weigelt blobs (Johansson et al., 2006).

Another example of a radiation-rich region is the environment of an active galactic nucleus (AGN), although the origin of the high-intensity radiation in these objects has a different nature (Osterbrock, 1989; Robson, 1986). In the broad-line regions of AGNs, the ionization parameter (the ratio of the density of the ionizing photons and the density of free electrons) can reach 10^{-2}–10^{-3}. High-intensity lines of H, He, and He II arise during the photoionization conversion of the bright, short-wavelength radiation of the AGN into recombination lines of these species. In such radiation-rich regions, photoionization processes can play a more dominant role than ionization by electron collisions (see, for example, Filippenko, 1985).

A high intensity of VUV resonance lines, e.g. H I, He I, and He II, in radiation-rich regions provides the possibility of a series of RETPI processes $X \to X^+ \to X^{2+}$ etc., which has also been considered by Johansson and Letokhov (2001c). In subsequent papers, this possibility was examined for the elements C, N, O (Johansson and Letokhov, 2001b, d) and the rare gases Ne, Ar (Johansson and Letokhov, 2004b), with the involvement of intense H Lyα, β, γ as well as He I and He II lines. These results are reviewed in the following Sections 15.2–15.4.

15.2 Rate of resonance-enhanced two-photon ionization in bichromatic radiation

Let an isolated atomic particle XN (a neutral atom or an ion in a given ionization state N) reside in an isotropic radiation field with a spectral intensity of $P_1(\nu)$ [in photons/cm$^2 \cdot$ s Hz]. Also let the spectral width of the radiation field $\delta\nu_1$ exceed the Doppler width $\Delta\nu_D$ of the allowed transition $1 \to 2$, and the shift $\Delta\nu$ (hereafter called "detuning") of the central frequency ν_1 of the radiation field from the allowed transition ν_{12} exceed the spectral width $\delta\nu_1$ (Figs. 15.2a, b). In that case, the one-quantum resonant excitation to level 2 is impossible because the energy defect $h\Delta\nu$ cannot be transferred to a third partner in a rarefied nebular medium. But resonance scattering of the radiation is possible here, and its probability is determined by the Lorentzian wing of the natural (radiative) broadening $\gamma = A_{21}$ of the transition $1 \to 2$, whose amplitude is proportional to $(\frac{\gamma}{2\pi\Delta\nu})^2$ (Weisskopf, 1933). The direct excitation of level 2 is only possible as a result of the two-quantum excitation of a virtual level with an energy of $2h\nu_1$ and the subsequent spontaneous emission of a photon with a frequency of $h(2\nu_1 - \nu_{12})$. According to Makarov (1983), the probability of this process is proportional to $(\frac{\gamma}{2\pi\Delta\nu})^4$, and it is substantially lower than the resonance scattering probability.

However, the atomic particle XN can virtually be in the excited state 2 with a probability of W_2. It is not difficult to calculate W_2 and express it in terms of the Einstein coefficient A_{21}. For simplicity we assume that the radiation field with the amplitude E is linearly polarized and acts in a coherent fashion at the frequency ν_1 detuned by $\Delta\nu$ from the frequency ν_{12}. Actually, the quantity W_2 is independent of the polarization of the radiation. In the case when the frequency detuning $\Delta\nu \gg \delta\nu_1$, the oscillations of the wave function amplitudes are faster than the oscillations of the field amplitude. Under such a condition the interaction of the radiation field with a two-level quantum system can be treated in a coherent way. According to perturbation theory we have

$$W_2 = \frac{1}{g_1} \sum_{M_1} \left(\frac{d^{(M_1 M_2)} E}{2\hbar \cdot 2\pi \Delta\nu} \right)^2, \quad (15.1)$$

where the summation is extended over all magnetic sub-levels of state 1, and $\Delta\nu$ is expressed in Hz. The quantity $d^{(M_1 M_2)}$ is the dipole moment of the transition between sublevels M_1 and M_2.

Next we want to express the sum of the squared dipole moments on the right-hand side of formula (15.1) in terms of A_{21}. We can use the well-known formula relating the matrix element of the dipole moment operator for a pair of non-degenerate sublevels to the rate $\gamma_{21}^{(M_1 M_2)}$ of spontaneous transitions between these sublevels (see, for example, Sobel'man, 1979):

$$\gamma_{21}^{(M_1 M_2)} = \frac{4\omega_{21}^3}{3\hbar c^3} \left(d_{21}^{(M_1 M_2)} \right)^2, \quad (15.2)$$

where $\omega_{21} = 2\pi\nu_{21}$. The total decay rate of any magnetic sublevel M_2 is

$$A_{21} = \sum_{M_1} \gamma_{21}^{(M_2 M_1)}. \quad (15.3)$$

We use another well-known formula for the matrix elements of vectors (see, for example, Landau and Lifshitz, 1958) and obtain

$$\sum_{M_1} \left(d_{21}^{(M_2 M_1)} \right)^2 = \frac{\hbar c^3}{4\omega_{21}^3} g_2 A_{21}. \quad (15.4)$$

We express E^2 on the right-hand side of formula (15.1) in terms of the radiation intensity I_1 (in photons/cm$^2 \cdot$ s) at the frequency $\omega_1 = 2\pi\nu_1$:

$$E^2 = \frac{8\pi}{c} \hbar\nu_1 I_1. \quad (15.5)$$

Finally, by substituting formulae (15.4) and (15.5) into Eq. (15.1), we get

$$W_2 = \frac{1}{32\pi^3} \frac{g_2}{g_1} \frac{\lambda_{21}^3}{\lambda_1} \frac{A_{21}}{(\Delta\nu)^2} I_1 \approx \frac{1}{32\pi^3} \frac{g_2}{g_1} \frac{\lambda_{21}^2 A_{21}}{(\Delta\nu)^2} I_1, \quad (15.6)$$

where g_i is the statistical weight of the ith level, and the lower level 1 is the ground state, and $\lambda_1 \approx \lambda_{21}$. Note that there is no need here to account for the Doppler broadening $\Delta\nu_D$ of the transition $1 \to 2$. The shift $\Delta\nu$ in the transition frequency of all atomic particles is much greater than $\Delta\nu_D$, and they are practically equally excited to the virtual level, no matter what velocity they have.

The ionization energy of the atomic particle in the virtually excited state is (IP $-$ $h\nu_1$), where IP is the ionization potential of XN. It is lower than either the photon energy $h\nu_2$ of the same radiation field or the photon energy of a second intense field with an average frequency ν_2, for which $h\nu_2 >$ (IP $- h\nu_1$). In that case, there exists a certain probability that the atomic particle will make a two-quantum transition through the virtual level to the ionization continuum at a rate $W_{\text{ph}}^{(2)}\,(\text{s}^{-1})$ given by:

$$W_{\text{ph}}^{(2)} = W_2 W_{\text{ph}}^{2i} \tag{15.7}$$

where $W_{\text{ph}}^{(2)}$ is the photoionization rate of the excited level 2 expressed in terms of the overlapping integral of the spectral radiation intensity $P_2(\nu)$ and the spectral dependence of the photoionization cross-section of the excited level, $\sigma_{2i}(\nu)$:

$$W_{\text{ph}}^{2i} = \int_{\text{IP}-h\nu}^{\infty} \sigma_{2i}(\nu) P_2(\nu)\, d\nu. \tag{15.8}$$

The radiation in the first-step excitation can also serve as a radiation in the second step, provided that $(\text{IP} - h\nu_1) < h\nu_1$. In the case of relatively monochromatic intense trapped lines inside the nebula, the spectrum width $\delta\nu_2$ is much smaller than the spectral resonances for photoionization from the excited state (Fig. 15.2a), so that the photoionization rate from this state can be written as

$$W_{\text{ph}}^{2i} \approx \sigma_{21}(\nu_2)\, I_2, \tag{15.9}$$

where I_2 is the intensity of the narrow-band radiation at the frequency ν_2 (in photons/cm$^2 \cdot$s). Based on these considerations, one obtains from formulae (15.6–15.9) an expression for the rate of the resonance-enhanced two-photon ionization of an atomic particle in a two-frequency isotropic radiation field (subject to the condition $(|\Delta\nu \gg \delta\nu_1|)$:

$$W_{\text{ph}}^{(2)} = \frac{\lambda_{21}^2}{32\pi^3} \frac{g_2}{g_1} \frac{A_{21}}{(\Delta\nu)^2} \sigma_{2i} I_1 I_2, \tag{15.10}$$

where $\Delta\nu$ is expressed in Hz. In the case of intense non-monochromatic photoionizing radiation (for example, diluted, but spectrally wide, black-body radiation from the central star as shown in Fig. 15.2b) it is necessary to use expression (15.8) for an estimation of W_{ph}^{2i}.

The intensity of a spectral line I_l can be compared with the intensity of the black-body radiation at the corresponding wavelength λ, and can be described in terms of the effective temperature T_{eff}:

$$P(\lambda, T) = \frac{8\pi}{\lambda^2} \frac{\delta\nu}{e^{h\nu/kT_{\text{eff}}} - 1}\ [\text{photons/cm}^2 \cdot \text{s}] \tag{15.11}$$

where $\delta\nu$ is the spectral line width [in Hz]. According to estimations (Johansson and Letokhov, 2004a) the effective temperature T_{eff} of Lyα inside the radiation-rich Weigelt blobs in η Car $T_{\text{eff}} \simeq (10-20) \cdot 10^3$ K. For example, for H Lyα ($\lambda = 1215$ Å) at $T_{\text{eff}} = 15 \cdot 10^3$ K and $\delta\nu \approx 25$ cm^{-1}, the intensity $P \approx 0.6 \cdot 10^{20}$ photons/cm$^2 \cdot$s, which corresponds to 10^2 W/cm^2 of CW (continuous wave) EUV radiation, a figure as yet unattainable in spite of the status of present day advanced laser technology. Such high radiation intensity at a comparatively small occupation number of the photon quantum state field mode, $\langle n \rangle \approx 5 \cdot 10^{-4}$, is explained by the large number of free space field modes within the limits of a solid angle of 4π steradians and the large spectral width $\delta\nu$.

In terms of the effective temperature T_{eff} of the equivalent Planck radiation (15.11), the RETPI rate (15.10) may be represented in the form

$$W_{1i} = \frac{2}{\pi} \frac{g_2}{g_1} \frac{\delta\nu_1 \delta\nu_2}{(\Delta\nu)^2} \frac{\sigma_{2i}}{\lambda_2^2} A_{21} \left[\exp\left(\frac{h\nu_1}{kT^1_{\text{eff}}}\right) - 1\right]^{-1} \left[\exp\left(\frac{h\nu_2}{kT^2_{\text{eff}}}\right) - 1\right]^{-1}. \tag{15.12}$$

Since $\delta\nu_1$ in the case of H Lyα and other, short-wavelength intense lines (H I, He I, He II) is much greater than kT^1_{eff} and kT^2_{eff}, Eq. (15.12) may be represented in the more simple form

$$W_{1i} \approx \frac{2}{\pi} \frac{g_2}{g_1} \frac{\delta\nu_1 \delta\nu_2}{(\Delta\nu)^2} \frac{\sigma_{2i}}{\lambda_2^2} A_{21} \exp\left(-\frac{h\nu_1}{kT^1_{\text{eff}}} - \frac{h\nu_2}{kT^2_{\text{eff}}}\right) \tag{15.13}$$

As an example, let us consider a sequence of ions having $\sigma_{2i} \approx 10^{-17}$ cm^2 and $A_{21} \approx 10^9$ s^{-1}. For H Lyα ($\lambda_{12} = 1215$ Å) with a spectral width $\delta\nu \approx 30$ cm^{-1} and a detuning of $\Delta\nu \approx 1000$ cm^{-1}, the rate $W^{(2)}_{\text{ph}}$ reaches values of 10^{-8}–10^{-7} s^{-1} at a temperature $T_{\text{eff}} \approx (13-15)\cdot 10^3$ K. This value of $W^{(2)}_{\text{ph}}$ is very small for laboratory laser experiments (Letokhov, 1987), but it is quite substantial under nebular conditions, for which the recombination rate W_{rec} is very low. At $W^{(2)}_{\text{ph}} > W_{\text{rec}}$ there will occur an accumulation of photoions in the next higher ionization state.

To understand the efficiency of the RETPI process, let us compare its rate W_{1i} with the rate of ionization by electrons. In both cases, there exists the exponential factor $\exp\left(-\frac{E}{kT}\right)$, where $E = h(\nu_1 + \nu_2) > $ IP, IP is the ionization potential, and T is the corresponding radiation or electron temperature. Thus, the main difference between collisional ionization by electrons and collisionless RETPI by photons lies in the difference between the pre-exponential factor denoted by A_e and A_{ph}, respectively. According to (15.13), in the case of RETPI

$$A_{\text{ph}} \approx \frac{2}{\pi} \frac{g_2}{g_1} \frac{\delta\nu_1 \delta\nu_2}{(\Delta\nu)^2} \frac{\sigma_{2i}}{\lambda_2^2} A_{21}, \tag{15.14}$$

and for the case considered above, $A_{\text{ph}} \approx (1-10)$s^{-1}. In the case of electronic excitation, the corresponding non-exponential factor is $A_e \approx \langle n_e v_e \sigma_e \rangle$, where v_e is the velocity of electrons with the energy $E > $ IP, n_e is the electron concentration, σ_e is the electronic ionization cross-section. For $n_e \approx 10^5$ cm^{-3}, $A_e \approx 10^{-2}$–$10^{-3} \ll A_{\text{ph}}$. Therefore, the rate of formation of ions by RETPI is higher than that through collisions with energetic electrons, and it is quite comparable with the recombination rate of the photoions produced. For this reason, the concentration of photoions (for example, $X(N+1)$) is much higher than that provided by collisions with the electrons having kinetic energy $E_{\text{kin}} > \text{IP}$, where IP is the ionization potential of XN. Accordingly, the recombination emission lines of XN become brighter. In addition, the concentration of the ions produced (for example, $X(N+1)$) becomes sufficient for a subsequent RETPI and as a result of that the appearance of bright emission lines from $X(N+1)$.

In some cases the frequency detuning $\Delta\nu$ is only a few tens of a cm^{-1}, i.e., it can be smaller than the spectral line width $\delta\nu_1$ (Fig. 15.2c). In such an exact accidental resonance, as is the case in the Bowen mechanism, the excited level is actually populated. The maximum excitation probability W_2 would be limited by the effective temperature T_{eff}^1 of the exciting radiation, provided that the radiation is in equilibrium with the resonant atoms:

$$W_2 = \left[\exp\left(\frac{h\nu_1}{kT_{\text{eff}}^1}\right) - 1\right]^{-1}. \tag{15.15}$$

In this case of exact resonance the biphotonic ionization rate increases and reaches, according to (15.7) and (15.15), its maximum

$$W_{\text{ph}}^{(2)}(1-i) = \sigma_{2i}(\nu_2) I_2 \left[\exp\left(\frac{h\nu_1}{kT_{\text{eff}}^1}\right) - 1\right]^{-1} \tag{15.16}$$

or

$$W_{\text{ph}}^{(2)}(1-i) = 8\pi \frac{\sigma_{2i}(\nu_2)}{\lambda_{12}^2} \delta\nu_2 \left[\exp\left(\frac{h\nu_1}{kT_{\text{eff}}^1}\right) - 1\right]^{-1}$$

$$\times \left[\exp\left(\frac{h\nu_2}{kT_{\text{eff}}^2}\right) - 1\right]^{-1} \tag{15.17}$$

As will be demonstrated below, in some cases the photon energy in the photoionization process might be very small (in the visible region) as shown in Figs 15.2b, d, and the diluted black-body radiation of the central star can contribute to the rate of step-wise photoionization. The rate of the RETPI process, which involves black-body radiation in the second step, depends upon the condition of the nebula, and in particular upon the distance from the central star.

Expression (15.10) for the probability of RETPI of an atomic particle per unit time can be written in the form

$$W_{1i}^{(2)} = \beta^{(2)} I_1 I_2, \tag{15.18}$$

where the coefficient $\beta^{(2)}$ (in cm$^4 \cdot$s) is

$$\beta^{(2)} \simeq \frac{\beta_{12}^2}{32\pi^3} \frac{g_1}{g_2} \frac{A_{21}}{(\Delta\nu)^2} \sigma_{2i}. \tag{15.19}$$

Expression (15.18) for $W_{1i}^{(2)}$ is presented for the case when the line width $\delta\nu_1$ at frequency ν_1 satisfies the condition $\delta\nu_1 \ll \Delta\nu$ and the photoionization cross-section σ_{2i} varies only slightly within the width of the line at ν_2 as is shown in Fig. 15.2a.

The cross-sections for two-photon absorption (in cm^2) at the frequencies ν_1 and ν_2 are given by the expressions

$$\sigma_1^{(2)} = \beta^{(2)} I_2 \qquad (15.20)$$

and

$$\sigma_2^{(2)} = \beta^{(2)} I_1. \qquad (15.21)$$

Accordingly, the coefficients for two-photon absorption per unit length are given by the expressions

$$\chi_1^{(2)} = \sigma_1^{(2)} N_0 = \beta^{(2)} I_2 N_0 \qquad (15.22)$$

and

$$\chi_2^{(2)} = \sigma_2^{(2)} N_0 = \beta^{(2)} I_1 N_0, \qquad (15.23)$$

where N_0 is the density of two-photon-absorbing particles (in cm^{-3}). The numbers of absorbed photons n_1 and n_2 per unit time per unit volume at frequencies ν_1 and ν_2 are the same, and are given by the expression

$$n_1 = n_2 = \chi_1^{(2)} I_1 = \chi_2^{(2)} I_2 = \beta^{(2)} I_1 I_2 N_0 = W_{1i}^{(2)} N_0. \qquad (15.24)$$

For example, if the probability $W_{1i}^{(2)} \simeq 10^{-5}$–$10^{-7}$ s^{-1} and the density of two-photon-absorbing particles is $N_0 \simeq 10^3$ cm^{-3}, the coefficients for the two-photon absorption are much smaller than 10^{-20} cm^{-1}.

15.3 RETPI of Si II in the Weigelt blobs of Eta Carinae

The RETPI process can compete with ionization by collisions between atoms (ions) and free electrons in a radiation-rich astrophysical plasma, in which the radiation energy density is comparable with or even higher than the energy density of free electrons. Such an astrophysical plasma can, for example, be represented by a gas cloud ejected from a hot star. The best-known case is the Weigelt blobs (WBs) near Eta Carinae. The emission line spectra of the WBs have been resolved from the radiation of the central star by means of the STIS two-dimensional spectrograph aboard the Hubble Space Telescope (Gull et al., 2001). The STIS observations show that the WBs emit a number of extremely bright spectral lines which need the unusual explanation, for example, of the RETPI process.

Simple estimates show that the effective spectral temperature of Lyα inside the blobs $T_\alpha^{\text{eff}} \simeq (10-15) \cdot 10^3$ K (Johansson and Letokhov, 2004a). This value is comparable to or even higher than the electron temperature, which implies a sharp change of the energy balance in the WBs in comparison with classical planetary nebulae. This change is explained by the fact that the dilution factor of the radiation reaching the blobs from the central star is compensated for by the effect of "spectral compression".

Thus, the Lyman-continuum radiation, which is absorbed in the photoionization of hydrogen, is emitted in the relatively narrow lines H Lyα, β, γ as a result of radiative recombination. The WBs may therefore be regarded as *radiation-rich* nebulosities to distinguish them from *thermal* planetary nebulae (Aller, 1984).

The WBs represent naturally a suitable astrophysical plasma in which the RETPI effect may occur. To verify this fact, one can, in principle, investigate the ionization equilibrium among various elements in the blobs aiming at finding some indications of the RETPI process. Such indications could be abundance depletion or enhancement of some particular ion compared to predictions from an ionization equilibrium governed by electronic ionization. Another possibility is to investigate spectral anomalies that are apparent for some ions. This approach is especially valuable, since RETPI is a purely radiative process depending on the intensity of the H Lyα, β, γ radiation. At the same time, there is the well-known "spectral event" in η Car, a periodic attenuation of the intensity of some lines occurring every 5.5 years and having a duration of about three months (Damineli, 1996). For many of these lines it has been shown that their intensities depend on H Lyα, β, γ radiation. Thus, data recorded during these three months allow for a discrimination of the excitation mechanism of the spectral lines in the WBs (Hartman et al., 2005). In general, the radiatively excited lines should vanish during the spectral event, whereas the recombination lines should remain, as the recombination time in the H I zone of the WBs exceeds three months.

Following this strategy of finding proofs of the existence of the RETPI effect, Johansson et al. (2006) proposed an explanation of the spectral anomaly of the Si III] intercombination line in spectra of the Weigelt blobs. The emission lines of various elements in the WBs were summarized by Zethson (2001). In particular, he noted that, in 1999, the Si III] $3s^2$ $^1S_0$-3s 3p $^3P_1$ intercombination line at 1892 Å was the third strongest emission feature in the satellite UV region of the observed spectrum (only surpassed by the 2507 and 2509 Å Fe II fluorescence lines). However, there is no sign of the Si III] line in the data recorded during the spectral event in 1998. This is perhaps the most striking example of the influence that the spectral event imposes on the WB spectrum.

The same effect was observed during the spectral event of 2003. Figures 15.3a and b show the drop of the intensity of the Si III] $\lambda 1892$ line according to the HST/STIS CCD-data of the Weigelt blobs taken during the event in June 2003, as part of the HST Treasury Project of η Car. The slit of the spectrograph was centered on Weigelt blob D, which is the main contributor to the observed line emission. For some position angles of the slit, the adjacent blobs B and C are not fully excluded from the field of view. Thus, B and C also contribute to the observed flux, but this contribution does not change the general behavior of the curves in Figs 15.3a and 15.3b. The rate of the intensity decrease for the Si III] line coincides approximately with that of 12 Fe II spectral lines excited by Lyα radiation (Hartman et al., 2005). This fact suggests that the excitation of the Si III] $\lambda 1892$ line should also be associated with Lyα radiation.

The ionization potential of Si I is 8.12 eV, i.e. it is lower than the ionization potential of H I. Therefore, ionization of Si I can be achieved by means of stellar Planck radiation in the spectra range 912 Å $< \lambda <$ 1527 Å that penetrates inside the

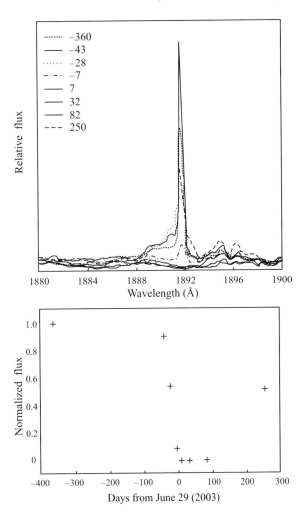

Fig. 15.3 Time variation of the intensity of the anomalous intercombination Si III] line at 1892 Å observed in the Weigelt blobs of Eta Carinae during the spectral event 2003. Upper: spectral profiles of the line; lower: the intensity change during a period of two years around the event. The data are part of the HST Treasury Project of Eta Carinae.

WB. The RETPI of Si II to form Si III can occur as a result of two-photon absorption of H Lyα radiation (Johansson and Letokhov, 2001a). But in that case, the excess of energy provided by the two Lyα photons after ionizing Si II to Si III, $2h\nu(\text{Ly}\alpha)-\text{IP}$ (Si II), amounts to 4.06 eV. This is insufficient to populate the $3s3p^3P_1$ state of Si III whose excitation energy is 6.553 eV. However, the RETPI of Si II under the effect of a combination of two different Lyman lines, Lyα and Lyγ, provides for the excitation of a state in the Si II continuum at an energy of $E = 6.608$ eV. This is $E = 0.055$ eV

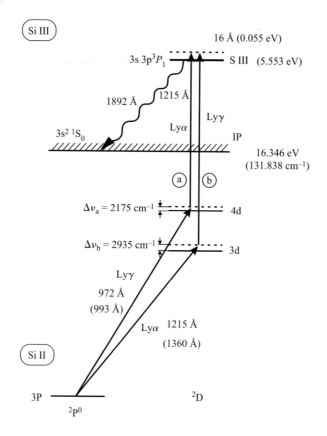

Fig. 15.4 Two possible pathways of resonance-enhanced two-photon ionization (RETPI) of Si II by H Lyα and H Lyγ radiation providing excitation of the 1892 Å Si III] intercombination line.

higher than the excitation energy of the triplet state. Based on this fact, two possible pathways of the RETPI of Si II are illustrated in Fig. 15.4.

The coupling between the excited state in the Si II continuum and the $3s3p\,^3P_1$ triplet state of Si III may prove quite enough for its excitation, followed by a radiative transition to the ground state in Si III, i.e. emission of the Si III] λ1892 intercombination line. This is the only allowed radiative decay channel of the $3s3p\,^3P_1$ state, and because of a relatively strong LS coupling the transition probability of this LS-forbidden line is $A = 1.67 \cdot 10^4$ s^{-1} (Kwong et al., 1983).

Considering that the Lyα and Lyγ spectral H lines are generated in the H II zone of the stellar wind as well, the energy difference in the excitation of the triplet state is even smaller than $E = 0.055$ eV (≈ 16 Å). With the terminal velocity v_{term} of the stellar wind from η Car being as high as $+625$ km/s (Hillier et al., 2001; Smith et al., 2003) the two photons, Lyα and Lyγ, irradiating the WB from the opposite side, reduce the difference to 10 Å, which increases the coupling between the continuum states of Si II and the triplet state of Si III.

Energetically, the RETPI process proposed could also populate the $J = 0$ and $J = 2$ levels of the 3s 3p ^{3}P term, yielding an energy difference $E = 0.022\,\text{eV}$ (≈ 6 Å) for the $J = 2$ level. However, the radiative decay of this level involves a forbidden transition at 1882.7 Å whose gA-value is six orders of magnitude smaller than the value for the observed $\lambda 1892$ intercombination line. The forbidden $\lambda 1882$ line is not observed, which could partly be due to collisional ion-electron coupling between the $J = 2$ and $J = 1$ levels.

The rate $W_{1i}(\text{s}^{-1})$ of the RETPI of Si II under the effect of the Lyα + Lyγ two-frequency radiation for each of the pathways, (a) and (b), of Fig. 15.4 is defined by the expressions, which follow from (15.13):

$$W_{1i}^a \simeq \frac{2}{\pi} \cdot \frac{\delta\nu_\alpha \delta\nu_\gamma}{(\Delta\nu_a)^2} \cdot \frac{\sigma_{2i}^a}{\lambda_\alpha^2} A_{21}^a \exp\left(-\frac{h\nu_\gamma}{kT_\gamma^{\text{eff}}}\right) \exp\left(-\frac{h\nu_\alpha}{kT_\alpha^{\text{eff}}}\right) \quad (15.25)$$

and

$$W_{1i}^b \simeq \frac{2}{\pi} \cdot \frac{\delta\nu_\alpha \delta\nu_\gamma}{(\Delta\nu_b)^2} \cdot \frac{\sigma_{2i}^b}{\lambda_\gamma^2} A_{21}^b \exp\left(-\frac{h\nu_\alpha}{kT_\alpha^{\text{eff}}}\right) \exp\left(-\frac{h\nu_\gamma}{kT_\gamma^{\text{eff}}}\right) \quad (15.26)$$

where $\delta\nu_\alpha$ and $\delta\nu_\gamma$ are the spectral widths of the Lyα and Lyγ lines, $\Delta\nu_a = 1800$ cm^{-1} and $\Delta\nu_b = 2920$ cm^{-1} are the frequency detunings of the Lyα and Lyγ lines relative to the 4d ^{2}D$_{3/2}$ and 3d ^{2}D$_{3/2}$ intermediate quasi-resonant levels, respectively. λ_α and λ_γ are the wavelengths of the Lyα and Lyγ lines, σ_{2i}^a and σ_{2i}^b the cross-sections for photoionization from the 4d ^{2}D and 3d ^{2}D, A_{21}^a and A_{21}^b the Einstein coefficients for the 4d ^{2}D $\to$ 3p ^{2}P and 3d ^{2}D $\to$ 3p ^{2}P radiative transitions, and T_α^{eff} and T_γ^{eff} are the effective (spectral) temperatures of the Lyα and Lyγ radiation, respectively. In the scheme in Fig. 15.4 we have only included parameter values for the ground state transitions, 3p ^{2}P$_{1/2} \to$ 4d ^{2}D$_{3/2}$ and 3p ^{2}P$_{1/2} \to$ 3d ^{2}D$_{3/2}$, but there are also contributions from the $3/2 \to 5/2$ and $3/2 \to 3/2$ fine structure transitions.

Let us make a very approximate estimate of the rates W_{1i}^a and W_{1i}^b of the RETPI of Si II under the effect of the Lyα and Lyγ radiation with the spectral widths $\delta\nu_\alpha \simeq \delta\nu_\gamma \simeq 500$ cm^{-1}, which corresponds to the widths of these lines in the H II region of the stellar wind in the vicinity of the WB. Assuming approximately that $A_{21} \simeq 10^9$ s^{-1} and $\sigma_{2i} \simeq 10^{-18}$ cm^2, we obtain the following estimate:

$$W_{1i}^a \simeq W_{1i}^b \simeq 0.2 \cdot \exp\left(-\frac{h\nu_\alpha}{kT_{\text{eff}}^\alpha}\right) \exp\left(-\frac{h\nu_\gamma}{kT_{\text{eff}}^\gamma}\right) \text{ [s}^{-1}\text{]} \quad (15.27)$$

Figure 15.5 presents the relationship between the total rate of the RETPI process involving the two pathways, $W_{1i} = W_{1i}^a + W_{1i}^b$, and the effective temperatures T_α^{eff} and T_γ^{eff} in the range $(10\text{--}20) \cdot 10^3$ K. The approximate estimate of the rate of the RETPI of Si II accompanied by the excitation into its ionized continuum close to the triplet state of Si III lies in the range $10^{-9}\text{--}10^{-6}\,\text{s}^{-1}$.

To use this estimate to get an explanation of the intensity observed for the Si III λ 1892 line (Fig. 15.3) seems quite natural. However, such an estimate would be rather approximate, since the volumes of the WBs and the stellar wind regions wherein this

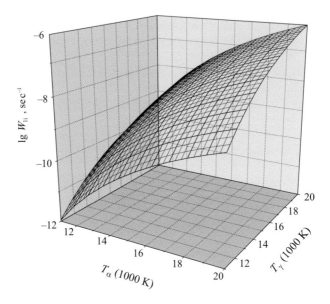

Fig. 15.5 Rate of RETPI (s^{-1}) of Si II as a function of the effective (spectral) temperatures of Lyα and Lyγ.

intercombination line is generated by the RETPI process are unknown. The abundance of Si in the WB is only approximately known, as is the degree of coupling between the triplet state of Si III and the ionization continuum of Si II. Nevertheless, one can note that, with the total volume of the WB and the surrounding stellar wind region being $V \simeq 10^{47}$ cm^3, the Si abundance $N_{\text{Si}} = 5 \cdot 10^{-5} N_{\text{H}}$, with $N_{\text{H}} \simeq 10^8$ cm^{-3}, and the coupling constant of the order of unity, the Si II RETPI rate $W_{1i} \simeq 10^{-7}$–10^{-8} s^{-1} explains fairly well the observed intensity of the 1892 Å Si III line, from the blobs which is $7 \cdot 10^{-12}$ erg·cm^{-2} s^{-1} Å^{-1} (Gull et al., 2001).

Johansson et al. (2006) made the first attempt to use the rich observational data on emission line spectra of the radiation-rich Weigelt blobs and searched for evidence for resonance-enhanced two-photon ionization (RETPI) in an astrophysical object. This object is special for this purpose for several reasons. First, the availability of spectral data from HST/STIS with excellent spatial resolution (no overlap with the radiation from the central star). Second, the effect of a periodical reduction of the ionizing radiation from the central star during the "spectroscopic events" makes it possible to distinguish between the radiative (collision-free) and recombination (collisional) excitation mechanisms (Hartman et al., 2005). It is rather tempting to search for other anomalies in the spectra of the Weigelt blobs to reveal possible contributions from the RETPI process in the ionization of elements.

15.4 Successive RETPI schemes for some light elements

Analysis of the allowed quantum transitions in the carbon atom and subsequent carbon ions, on the basis of the data published in Bashkin and Stoner (1975), points to the

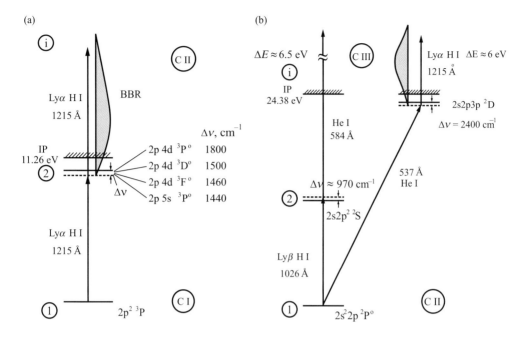

Fig. 15.6 Chain of the successive RETPI schemes for carbon: C I → C II (left), C II → C III (right). (For C III → C IV and C IV → CV, see Fig. 15.7.)

possibility of RETPI with suitable pairs of intense trapped lines of H I, He I, and He II in the EUV region extending from H Lyα (1215 Å) to He III Lyγ (243 Å). Out of a large number of suitable pairs of lines, Johansson and Letokhov (2001d) included only those which have close coincidences with allowed transitions in the carbon atom or the carbon ions. We have not considered transitions characterized by either a large frequency detuning $\Delta\nu$ or a low transition probability A_{21} as the RETPI rate could be low according to (15.10) and (15.13).

The RETPI schemes for carbon (Figs. 15.6, 15.7) provide a path to C V. However, there is no suitable quasi-resonance for C V with any of the EUV spectral lines of H and He, which means that this ion is the end ion in the successive RETPI chain C I → C II → C III → C IV → C V. For each step in the chain, the relative frequency detuning $(\Delta\nu/\nu_{21})$ does not exceed $1.5 \cdot 10^{-2}$. At an effective radiative temperature of the spectral lines in the range $(15-20) \cdot 10^3$ K the total RETPI chain achieves a relatively high rate of 10^{-6}–10^{-4} s^{-1}. In some steps, C I → C II for example, the photoionization of the virtually excited state 2 can be carried out by radiation over a very broad spectral region: $\lambda_2 < 1.2\,\mu$m. This means that the rate $W_{\rm ph}^{2i}$ defined by Eq. (15.8) can be high enough even for photoionization caused by the diluted black-body radiation from the central star. However, this rate depends on the radiation dilution factor, i.e., the distance from the central star.

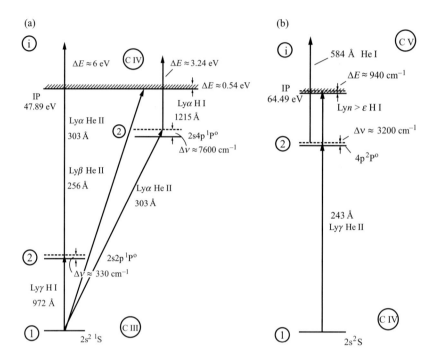

Fig. 15.7 Chain of successive RETPI schemes for carbon: C III → C IV (left) and C IV → C V (right). (For C I and C II, see Fig. 15.6.)

Analogous to the carbon scheme, RETPI of nitrogen can lead to NV. Moreover, for the RETPI cases N I → N II and N II → N III, the detuning from the exact resonance frequency is very small, 300–400 cm^{-1} and ±100 cm^{-1}, respectively. Therefore, one can expect that the relative concentration of the N I and N II ions will be low because of the significant shift of the photoionization balance toward multiply-ionized nitrogen ions. As for carbon, the virtually excited states in the low ionization stages of nitrogen are close enough to the ionization limit to be entirely photoionized by the black-body radiation from the central star.

Similar RETPI schemes for oxygen can terminate with the O V photoion. Specific features of oxygen are the existence of several very close resonances between O I and the lines H Lyβ, γ, and δ, and the close proximity of virtually excited oxygen states to the ionization limit. Obviously the ionization balance O I → O II should be substantially shifted toward O II.

All the excitation and ionization steps for the RETPI process in carbon, nitrogen, and oxygen are summarized in Table 15.1, which shows the chains for the atoms and successive ions of these elements. The atoms and ions are exposed to pairs of spectral lines with progressively increasing photon energies: H I + H I, H I + He I, H I + He II, He I + He I, and finally He II + He II.

Table 15.1 Chains of subsequent schemes of RETPI for C, N, and O based on strong EUV lines of H I, He I and He II

	Pairs of photoionizing spectral lines of H I, He I and He II				
Element	H I + H I	H I + He I	H I + He II	He I + He II	He II + He II
C	C I → C II (11.26 eV) 1215 Å + 1215 Å C II → C III 1026 Å + 972 Å	C II → C III (24.38 eV) 1026 Å + 584 Å 537 Å + 1215 Å	C III → C IV (47.89 eV) 972 Å + 303 Å 303 Å + 1215 Å C IV → C V 243 Å + 919 Å	C IV → C V (64.49 eV) 243 Å + 584 Å	
N	N I → N II (14.53 eV) 972 Å + 1215 Å 949 Å + 1215 Å	N II → N III (29.60 eV) 537 Å + 1215 Å	N III → N IV (47.45 eV) 303 Å + 1215 Å	N II → N III (29.60 eV) 1085 Å + 584 Å	N IV → N V (77.47 eV) 243 Å + 303 Å
O	O I → O II (13.62 eV) 1026 Å + 1215 Å (972 Å) + 1215 Å (949 Å) + 1215 Å (937 Å) + 1215 Å	O II → O III (35.12 eV) 537 Å + 1026 Å		O III → O IV (54.93 eV) 303 Å +584 Å	O IV → O V (77.41 eV) 237 Å +303 Å

It seems reasonable that the proposed elementary process of photoionization by the intrinsic EUV radiation trapped in radiation-rich gas condensations should be taken into consideration when calculating the ionization balance of those elements, whose observed spectral lines are essentially the only information available on the processes occurring inside the blobs. Since the RETPI rate $W_{ph}^{(2)}$ apparently can exceed the collisional electron ionization rate W_e, RETPI will enhance the corresponding recombination lines, which are frequently interpreted in the framework of ionization by electron collisions as being the result of an anomalous abundance effect.

From this standpoint, one can understand that the existing determinations of the chemical abundance from spectral data frequently disagree. For example, an anomalous ratio often observed for the abundances of C IV and C II can be explained by the RETPI of C III, carried out jointly by the H Lyγ line (972 Å) and He II Lyα at 303 Å or directly by He II Lyβ at 256 Å (Fig. 15.7).

Analysis of the permitted photon transitions for neutral Ne and Ar atoms and their ions indicates the possibility of RETPI under the action of appropriate lines or pairs of lines of H I, He I, and He II in transitions to the ground state in the EUV; these lines usually have large optical depths, intensities, and line widths. Using the Grotrian diagrams for neon and argon (Bashkin and Stoner, 1975) and original data for Ne II (Person, 1971), Ne III (Person *et al.*, 1991), Ne IV (Kramida *et al*, 1999), Ar I and Ar II (Norlén, 1973), Ar III (Hansen and Persson, 1987), and Ar IV (Bredice *et al.*, 1995). Johansson and Letokhov (2004b) selected from among the large number of quasi-resonant coincidences those having the closest resonances with some permitted photon transitions for Ne and Ar atoms and their ions.

All these schemes are summarized in Tables 15.1 and 15.2, which present the chains of successive RETPI processes active on the atoms and ions that have been studied so far. In the processes considered the atoms (ions) interact with radiation of one or two spectral lines with progressively increasing photon energies: H I + He I, H I + He II, He I + He II, and He II + He II. The resulting set of successive-RETPI schemes for Ar strongly resembles that for Ne, although it includes other combinations of intense EUV spectral lines. Johansson and Letokhov (2001a, b, d; 2004b) concluded that the elementary RETPI processes considered should be taken into account in ionization-balance calculations for specific spectral lines, especially for radiation-rich regions. One would expect some non-monotonic appearance of certain ionization states owing to the existence of favorable near coincidences between the frequencies of intense emission lines and of absorption lines of the atoms and ions. The resulting non-monotonic appearance of successive ions can lead to deviations from standard models if electron temperatures are estimated without including the effects of RETPI. The same is true for estimates of the abundances of elements based on emission lines, which arise via recombination of ions whose formation rate is enhanced due to RETPI. In contrast to electron-collisional ionization, the probability of the RETPI process displays a non-monotonic dependence even on the effective temperature of the radiation, since $W_{2i}^{(2)}$ is sensitive to the offset $\Delta\nu$, which has various values for various ions. RETPI is only a purely collision-free photon process when it is initiated from the ground state. However, in some of the schemes considered, RETPI is initiated from a low-lying excited level with an energy less than a few eV. In this case, some links in the RETPI

Table 15.2 Chains of successive-RETPI schemes for Ne and Ar based on H I, He I, and He II lines in the EUV

	Pairs of photoionizing H I, He I, and He II lines			
Element	H I + He I	H I + He II	He I + He II	He II + He II
Ne	Ne I → Ne II (21.56 eV) 584 Å + 1215 Å Ar I → Ar II (15.76 eV) 584 Å Ar II → Ar III (27.63 eV) 930 Å + 584 Å 584 Å +1215 Å 522 Å + 1215 Å 515 Å + 1215 Å	Ne II → Ne III (40.96 eV) 303 Å +1215 Å Ne III → Ne IV (63.45 eV) 237 Å + 1025 Å	Ne III → Ne IV (63.45 eV) 256 Å + 584 Å	Ne IV → Ne V (97.11 eV) 234 Å + 303 Å
Ar		Ar I → Ar II (15.76 eV) 303 Å Ar IV → Ar V (59.81 eV) 1025 Å +256 Å	Ar III → Ar IV (40.74 eV) 584 Å + 584 Å 303 Å Ar IV → Ar V (59.81 eV) 522 Å + 303 Å Ar V → Ar VI (75.92 eV) 522 Å + 237 Å	

chain rely on collisions. However, it is unlikely that these links will become a bottle neck for the chain, since the cross-sections for the electronic excitation of the low-lying levels are much greater than the cross-sections for the electronic ionization of the corresponding ions in the collision-dependent chain. Nonetheless, this remains a subject for a specialized analysis of specific models for ionization balance in radiation-rich astrophysical plasmas.

16
Laser and interstellar communications

The astrophysical laser is a natural quantum-electronic "device" of an extreme size. Its location can be seen at the end of the size-active laser particle density diagram shown in Fig. 16.1. In the centre of this diagram, the first laboratory masers and lasers are located. At the upper end of the diagram, the astrophysical masers and lasers are located. The scale of the diagram in Fig. 16.1 covers 20 orders of magnitude (!) both in size and density. The astrophysical lasers discovered are based on optical pumping, and they operate according to a scheme similar to the one suggested for the first laboratory laser 50 years ago by A. Schawlow and C. Townes (1958). We believe that laser amplification and stimulated emission of radiation is a fairly common and widespread phenomenon, at least for gas condensations nearby bright stars.

However, the revelation of the laser effect in space is less evident in the optical region of the spectrum than is the maser effect in the microwave region, where the maser gives rise to strong radiation lines of an exceptionally high brightness temperature (10^{10}–10^{15} K). In the optical region, the laser effect raises the intensity of weak forbidden lines up to the intensity of strong allowed lines providing that there is a selective photoexcitation of the upper levels in the lasing transitions. This is explained by the huge difference ($\approx 10^{15}$–10^{18} times) between the spontaneous emission rates and the inverse population production mechanisms in the optical and microwave wavelength regions.

The main difficulty in detecting the laser effect is the presence of intense spontaneous emission, which masks the effect of stimulated emission. Therefore, the observation of forbidden transitions is a most promising way of detecting the laser effect in space. The intensity of the laser line should become comparable with the intensity of lines that are due to allowed spontaneous emission of radiation and resulting from the optical pumping (direct or indirect) of the upper level. This takes place only in the volume of inverted population with a size $L \gg 1/\alpha$, where α is the amplification coefficient per unit length. In such a case, a spectral line, expected to be weak, must appear as a spectral line of normal intensity, typical for an allowed transition. This is exactly what we have found to be the case with several spectral lines of Fe II pumped indirectly by the intense H Lyα radiation (1215 Å) in a dense gas condensation (blob) in the vicinity of the variable blue star Eta Carinae.

Observation of the narrowing of some spectral lines originating from the lasing zone would be the best proof of the occurrence of astrophysical lasers. It can be done

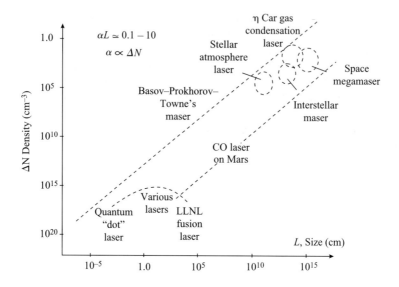

Fig. 16.1 Various laboratory and astrophysical masers and lasers on the diagram: "active media particle density – active medium size".

with a Brown–Twiss–Townes interferometer with a joint laser local heterodyne (diode lasers) having a distance between two telescopes of several meters. Progress of very large optical telescopes with laser heterodyne reception of photons will make this task more realistic in the next decade.

The problem of observation of an astrophysical laser with required ultra-high angular and spectral resolutions is very close to the problem of the optical SETI (Search for Extraterrestrial Intelligence). This problem is a conceptual extension of the SETI program in the microwave range into the optical range. The idea to use a laser for interplanetary and interstellar communications (Schwartz and Townes, 1960) was born immediately after the invention of the laser. It is quite natural because an operating powerful laser emits a beam of exceptionally directed, coherent, intense light, i.e. this beam is the "galactic" lighthouse. Actually, this laser lighthouse can be detected (visible) on interstellar distances. Townes (1983) raised the question: at those wavelengths should we search for signals from extraterrestrial intelligence? He concluded that, apart from the microwave range, experiments at much shorter wavelengths are needed. The detection by photon counting in the optical range has more advantages than the linear detection in the microwave range. There are other advantages too (size of the telescopes, power of extraterrestrial transmitter etc.) (Betz, 1985). The possibility of the detection of signals at distances of about 100 light years can be considered as a first step, and the detection at distances of 100–1000 light years would be a second step in perspective. The first experiments have been performed (Reines et al., 2002) and new ones will be performed in the future. Planned experiments (Dravins et al., 2005; Dravins, 2007) with high temporal resolution and heterodyne direction with spatially separated telescopes are especially important as regards the detection of extraterrestrial laser light.

The observation of a laser effect in the atmospheres of Mars and Venus (Chapter 8) of the CO_2 molecule at the wavelength 10 µm lead to the project of the planetary laser for interstellar communications (Sherwood *et al.*, 1992). We are witnesses of an initial stage of the birth of a new astrolaser technology, and it is very difficult to foresee the future results and consequences of this trend. It is acceptable now to present fantastic ideas, including the possibility of a natural amplification of the travelling laser signal in some sites of the interstellar medium, gravitational lensing of a laser signal with corresponding enhancement of intensity due to focusing of the optical beam, and many others new conceptions.

References

Abbas, M. M., Mumma, M. J., Kostiuk, T., and Buhl, D. 1976. Sensitivity limits of an infrared heterodyne spectrometer for astrophysical applications. *Applied Optics* **15**(2), 427–436.

Allen, C. W. 1973. *Astrophysical Quantities*. Athlone Press, London.

Allen, L. and Peters, G. I. 1972a. Amplified spontaneous emission and OH molecules. *Nature* **235**, 143–144.

Allen, L. and Peters, G. I. 1972b. Spectral distribution of amplified spontaneous emission. *Journal of Physics A: General Physics* **5**, 695–704.

Aller, L. H. 1956. *Gaseous nebulae*. Chapman and Hall, Ltd., London.

Aller, L. H. 1984. *Physics of thermal gaseous nebulae*. D. Riedel Publ. Co., Dordrecht.

Alzetta, G. A., Gozzini, A. L., Moi, L., and Orriois, G. 1976. An experimental method for the observation of RF transitions and laser beat resonances in oriented Na vapor. *Nuovo Cimento* **B36**, 5–20.

Ambartzumian, V. A. 1933. On the radiative equilibrium of a planetary nebula. *Izvestia glavnoi astronomicheskoi observatorii v Pulkovo* (Russian) **13**, 3.

Ambartzumian, R. V., Basov, N. G., Kryukov, P. G., and Letokhov, V.S. 1970. Non-resonant feedback in lasers. Stevens, K. W. N., Sanders, J. H. (eds), in *Progress in Quantum Electronics*, Vol. 1, Pergamon Press, London, 105–193.

Andrillat, Y. and Swings, J. P. 1976. 8200 + 11200 Å spectra of peculiar emission-line objects with infrared excess. *Astrophysical Journal* **204**, L123–L125.

Arecchi, F. T. 1965. Measurement of the statistical distribution of Gaussian and laser sources. *Physical Review Letters* **15**, 912–916.

Arimondo, E. 1996. Coherent population trapping in laser spectroscopy. Wolf, E. (ed.), in Progress in Optics. Vol. XXXV, Elsevier, Amsterdam, 257–354.

Auer, L. H. 1968. Transfer of Lyman alpha in diffusive nebulae. Astrophysical Journal **153**, 783–796.

Avrett, E. H. Ed. 1976. *Frontiers of Astrophysics*. Harvard University Press, Cambridge.

Avrett, E. H. and Hummer, D. G. 1965. Non-coherent scattering: II, line formation with frequency independent source function. *Monthly Notices of Royal Astronomical Society* **130**, 295–331.

Barrett, A. H. and Rogers, A. E. E. 1966. Observations of circularly polarized OH emission and narrow spectra features. *Nature* **210**, 188–190.

Bashkin, S. and Stoner, Jr., J. O. 1975. *Atomic energy levels and Grotrian diagrams*, Vol. 1, North-Holland, Amsterdam.

Basov, N. G. and Prokhorov, A. M. 1954. Applications of molecular beams for radiospectroscopic study of molecular rotational spectra. *Journal of Experimental and Theoretical Physics* (Russian) **27**, 431–438.

Basov, N. G. and Prokhorov, A. M. 1955. On possible methods of preparation of active molecules for molecular generator. *Journal of Experimental and Theoretical Physics* (Russian) **29**, 249–250.

Basov, N. G., Ambartzumian, R. V., Zuyev, V. S., Kryukov, P. G., and Letokhov, V. S. 1966. Nonlinear amplification of light pulse. *Journal of Experimental and Theoretical Physics* **50**, 23–34.

Bennett, W. R. Jr., Faust, W. L., McFarlane, R. A., and Patel, C. K. H. 1962. Dissociative excitation transfer and optical maser oscillation in Ne-O_2 and Ar-O_2 RF discharges. *Physical Review Letters* **8**, 470–473.

Bergeson, S. D., Mullman, K. L., Wickliffe, M. E., Lawler, J. E., Litzén, U., and Johansson, S. 1996. Branching fractions and oscillator strengths for Fe II from the 3d $^6(^5D)$4p subconfiguration. *Astrophysical Journal* **464**, 1044–1049.

Bethe, H. 1964. *Intermediate quantum mechanics*. W. A. Benjamin, Inc. New York, Amsterdam.

Bettwieser, E. 1981. Interstellar masers: the influence of the geometrical shape on the radiation properties. *Journal of Astrophysics and Astronomy* **2**, No. 2, 187–199.

Betz, A. L. 1985. A direct search for extraterrestrial laser signals. *IAF, International Astronautical Congress,* 36th, Stockholm, Sweden, Oct. 7–12, p. 10.

Betz, A. L., McLaren, R. A., Sutton, F. C., and Johnson, M. A., 1977. Infrared heterodyne spectroscopy of CO_2 in the atmosphere of Mars. *Icarus* **30**, 650–662.

Bowen, I. S. 1928. The origin of the nebular lines and the structure of planetary nebulae. *Astrophysical Journal* **67**, 1–15.

Bowen, I. S. 1934. The excitation of the permitted O III nebular lines. *Publications of astronomical society of the Pacific* **46**, 146–150.

Bowen, I. S. 1935. The spectrum and composition of the gaseous nebulae. *Astrophysical Journal* **81**, 1–16.

Bowen, I. S. 1947. Excitation by line coincidence. *Publications of the Astronomical Society of the Pacific* **59**, 196–198.

Boyarchuk, A. A. 1957. Some characteristics of Be star shells. *Soviet Astronomical Journal* **1**, 192–201.

Boyarchuk, A. A. 1969. Symbiotic stars. L. Detre (ed.), in *Non-periodic phenomena in various stars*. Budapest Academia Press, Budapest, 395–410.

Bredice, F., Gallardo, M., Reyna Almandos, J. G., Trigueiros, A. G., and Pagan, C. J. B. 1995. Revised analysis of triply ionized argon (Ar IV). *Physica Scripta* **51**, 446–453.

Brocklehurst, M. 1970. Level populations of hydrogen in gaseous nebulae. *Monthly Notices of the Royal Astronomical Society* **148**, 417–434.

Brown, A., Ferraz, M. C. de M., and Jordan, C. 1984. The chromosphere and corona of T Tauri. *Monthly Notices of the Royal Astronomical Society* **207**, 831–859.

Burbidge, E. N. 1952. The spectrum of CHI Ophiuchi. *Astrophysical Journal* **118**, 418–422.

Cao, H., Xu, J. Y., Ling, Y., Burin, A. L., Seeling, E .W., Lu, X., and Chang, R. H. 2003. Random lasers with coherent feedback. *IEEE Journal of Selected Topics in Quantum Electronics* **9**, No. 1, 111–118.

Carlsson, M. and Judge, P. G., 1993. O I lines in the Sun and stars I – understanding of resonance lines. *Astrophysical Journal* **402**, 344–357.

Carpenter, K. G., Pesce, J., Stencel, R., Johansson, S., and Wing, R. F. 1988. The ultraviolet spectrum of noncoronal late-type stars – The Gamma Crucis (M3, 4III) reference spectrum. *Astrophysical journal. Supplement* **68**, 345–369.

Casperson, L. W. 1977. Threshold characteristics of mirrorless lasers. *Journal of Applied Physics* **48**, 256–262.

Carperson, L. W., and Yariv, A. 1972. Spectral narrowing in high-gain lasers. *IEEE Journal of Quantum Electronics* **8**, 80–85.

Cheung, A. C., Rank, D. M., Townes, C. H., Thornton, O. D., and Welch, W. I. 1969. Detection of water in interstellar regions by its microwave radiation. *Nature* **221**, 626–628.

Clegg, A. W. and Cordes, J. M. 1991. Variability of interstellar hydroxyl masers. *Astrophysical Journal* **374**, 150–168.

Clegg, A. W. and Nedoluha, G. E. (eds). 1993. *Astrophysical masers*. Springer-Verlag, Berlin.

Cohen, M. H., Jancey, D. L., Kellerman, K. T., and Clark, B. G. 1968. Radio interferometry at one-thousandth second of arc. *Science* **162**, 88–94.

Condon, E. U. and Shortley, G. 1963. *The theory of atomic spectra*. Cambridge University Press, Cambridge.

Cook, A. H. 1977. *Celestial masers*. Cambridge University Press, Cambridge.

Cowley, C. R. 1970. in *The theory of stellar spectra*, Topics in astrophysics and space physics. Gordon & Breach, New York, 61.

Damineli, A. 1996. The 5.52 year cycle of Eta Carinae. *Astrophysical Journal* 460, L49-L52.

Damineli, A. 2001. The O I 8446 Å, Fe II 8490 and Fe II 9997 lines in hot luminous star in *Eta Carinae and other mysterious stars. The hidden opportunities of emission line spectroscopy*. ASP Conf. Series, Vol. 242, San Francisco, 203–210.

Das Gupta, S. and Das Gupta, S. R. 1991. Laser radiation in active amplifying media treated as a transport problem: transfer equation derived and exactly solved. *Astrophysics and Space Sciences* **184**, 77–142.

Davidson, K. 2001a. Unique spectroscopic problems related to Eta Carinae. Gull, T., Johansson, S. and Davidson, K. (eds), in *Eta Carinae and other mysterious stars. The hidden opportunities of emission line spectroscopy*. ASP Conf. Series, vol. 242, 3–13.

Davidson, K. 2001b. Ultraviolet and optical spectroscopy. Ferland, G. and Wolf, D. (eds), in *Spectroscopic challenges of photoionized plasmas*. ASP Conf. Series, Vol. 247, 465–478.

Davidson, K. and Humphreys, R. M. 1997. Eta Carinae and its environment. *Annual Reviews of Astronomy* **35**, 1–32.

Davidson, K., Ebbets, D., Weigelt, G., Humphreys, R. M., Hajjan, A. R., Walborn, N. R., and Rosa, M. 1995. HST/FOS spectroscopy of Eta Carinae: the star itself, and ejecta within 0.3 arcsec. *Astronomy and Astrophysics* **109**, 1784–1796.

Davidson, K., Ebbets, D., Johansson, S., Morse, J. A., Hamann, F. W., Balick, B., Humphreys, R. M., Weigelt, G., and Frank A. 1997. HST/GHRS observations of the compact slow ejecta of Eta Carinae. *Astrophysical Journal* **113**, 335–345.

Davies, A. J., Rowson, B., Booth, R. S., Cooper, A. J., Gent, H., Adgie, R. L., and Crowther, J. H. 1967. Measurements of OH emission sources with an interferometer of high resolution. *Nature* **213**, 1109–1110.

Deming, D. and Mumma, M. J. 1983. Modeling of the 10 μm natural laser emission from the mesospheres of Mars and Venus. *Icarus* **55**, 356–368.

Dirac, P. A. M. (1930, 1957). *The principles of quantum mechanics.* Oxford University Press, New York.

Ditchburn, R. W. and Opik, O. 1962. Photoionization processes, in *Atomic and molecular processes*, D. R. Bates (ed.), Academic Press, New York.

Doel, R. C., Gray, M. D., Field, D., and Jones, K. N. 1993. Physical conditions for far-infrared laser emission from dense OH maser regions. *Astronomy and Astrophysics* **280**, 592–598.

Dravins, D. 2001. Quantum-optical signatures of stimulated emission. Gull, T., Johansson, S., Davidson, K. (eds), in *Eta Carinae and other mysterious Stars.* ASP Conf. Series, Vol. 242, 339–345.

Dravins, D. 2007. Photonic astronomy and quantum optics. D. Phelan, O. Ryan, and A. Shearer (eds), in *High time-resolution astrophysics*, (Springer).

Dravins, D., Barbieri, C., Fosbury, R. A. E., Naletto, G., Nillsson, R., Occhipinti, T., Tamburini, F., Uthas, H. and Zampieri, L. 2005. Astronomical quantum optics with extremely large telescopes. *Proceedings of the International Astronomical Union* (Cambridge University Press) **1**, 502–505.

Einstein, A. 1916. Strahlungs – emission und absorption nach der quanten theorie. *Verhandl. Dtsch. Phys. Ges.* **18**, 318–323.

Einstein, A. 1917. Zur quantentheorie der strahlung. *Physikalische zeitschrift* **18**, 121–128 [English translation is available in the book *"The old quantum theory"*, D. Ter Haar. 1967. Pergamon Press, Elmsford, New York, p. 167].

Elitzur, M. 1982. Physical characteristics of astronomical masers. *Review of Modern Physics* **54**, 1225–1260.

Elitzur, M. 1991. Polarization of astronomical maser radiation. *Astrophysical Journal, Part 1*, **370**, 407–418.

Elitzur, M. 1992. *Astronomical maser.* Kluwer, Dordrecht.

Emerson, D. 1996. *Interpreting astronomical spectra.* John Wiley and Sons, Chichester.

Eriksson, M., Johansson, S., Wahlgren, G. M., Veenhuizen, H., Munari, U., and Siviero, A. 2005. Bowen excitation of N II lines in symbiotic stars. *Astronomy and astrophysics* **434**, 397–404.

Eriksson, M., Johansson, S., and Wahlgren, G. 2006. The nature of ultraviolet spectra of AG Pegasus and other symbiotic stars: locations, origins, and excitation mechanisms of emission lines. *Astronomy and Astrophysics* **451**, 157–175.

Evans II, N. J., Hill, R. E., Rydbeck, O. E. H., and Kollberg, E. 1972. Statistics of the radiation from astronomical masers. *Physical Review* **76**, 1643–1647.

Feibelman, W. A., Bruhweiler, F. C., and Johansson, S. 1991. Ultraviolet high-excitation Fe II fluorescence lines excited by O VI, C IV, and H I resonance emission as seen in IUE spectra. *Astrophysical Journal* **373**, 649–656.

Ferland, G. J. 1992. N III line emission in planetary nebulae – continuum fluorescence. *Astrophysical Journal* **389**, L63–L65.

Ferland, G. J. 1993. A masing [Fe XI] line. *Astrophysical Journal. Supplement* **88**, 49–52.

Fermi, E. 1932. Quantum theory of radiation. *Review of Modern Physics* **4**, 87–133.

Ficek, Z., Seke, J., Soldatov, A. V., and Adam, G. 2000. Saturation of a two-level atom in polychromatic fields. *Journal of optics B: quantum semiclassical optics* **2**, 780–785.

Filippenko, A. V. 1985. New evidence for photoionization as the dominant excitation mechanism in Liners. *Astrophysical Journal* **289**, 475–489.

Glauber, R. 1963a. The quantum theory of optical coherence. *Physical Review* **130**, 2529–2539.

Glauber, R. 1963b. Coherent and incoherent states of the radiation field. *Physical Review* **131**, 2766–2788.

Goldberg, L. 1966. Stimulated emission of radio-frequency lines of hydrogen. *Astrophysical Journal* **144**, 1225–1229.

Gonzalez-Alfonso, E., Cernicaro, J., Van Dishockk, E. F., Wright, C. M., and Heras, A. 1998. Radiative transfer models of emission and absorption in the H_2O 6 micron vibrational band toward Orion-BN-KL. *Astrophysical Journal* **502**, L169–L172.

Gordiets, B. F. and Panchenko, V. Y. 1983. Nonequilibrium infrared emission and the natural laser effect in the Venus and Mars atmospheres. *Kosmitcheskie issledovaniia* (Russian) **21**, 929–939.

Gordiets, B. F. and Panchenko, V. Ya. 1986. Gas lasers with solar pumping. *Soviet Physics – Uspekhi* **29**(7), 703–719.

Gordon, J. R., Zeiger, H. J., and Townes, C. H. 1954. Molecular microwave oscillator and new hyperfine structure in the microwave spectrum of NH_3. *Physical Review* **95**, 282–284.

Gordon, J. P., Zeiger, H. J., and Townes, C. H. 1955. The maser – new type of microwave amplifier, frequency standard, and spectrometer. *Physical Review* **99**, 1264–1274.

Gould, G. 1965. Collisional lasers. *Applied optics. Suppl. on chemical lasers*, No. 2, 56–64.

Grandi, S. A. 1975. Starlight excitation of permitted lines in the Orion Nebula. *Astrophysical Journal* **196**, 465–472.

Grandi, S.A. 1980. O I 8446 Å emission of Seyfert 1 galaxies. *Astrophysical Journal* **238**, 10–16.

Gray, M. D. 1999. Astrophysical masers. *Philosophical Transactions of the Royal Society A* **357**, 3277–3298.

Gray, M. D., Doel, R. C., and Field, D. 1991. A model for O H masers in star-forming regions. *Monthly Notices of Royal Astronomical Society* **252**, 30–48.

Gray, M. D., Field, D., and Doel, R. C. 1992. An analysis of intense OH maser emission in star-forming regions. *Astronomy and Astrophysics* **262**, 555–569.

Greenhouse, M. A., Feldman, U., Smith, H. A., Klapisch, M., Bhatia, A. K., and Bar-Shalom, A. 1993. Infrared coronal emission lines and the possibility of their laser emission in Seyfert nuclei. *Astrophysical Journal. Supplement* **88**, 23–48.

Greenhouse, M. A., Smith, H. A., and Feldman, U. 1996. Possibility of infrared coronal line laser emission in Seyfert nuclei. Chiao, R. Y. (ed.), in *Amazing light*, Springer, Berlin, 295–305.

Greenstein, J. L. 1960. (ed.) *Stellar atmospheres*. The University of Chicago Press, Chicago.

Gudzenko, L. I. and Shelepin, L. A. 1963. Negative absorption in a nonequilibrium hydrogen plasma. *Journal of Experimental and Theoretical Physics* **18**, 998–1002.

Gudzenko, L. I. and Shelepin, L. A. 1965. Amplification in a recombining plasma. *Doklady of Academy of Sciences of USSR* **160**, 1296–1299.

Gull, T., Goad, L., Chin, N. Y., Maran, S. P., and Hobbs, R. W. 1973. An emission-line object found in the Orion Nebula. *Publications of the Astronomical Society of Pasific* **85**, 526–527.

Gull, T., Ishibashi, K., Davidson, K., and the Cycle 7 STIS GO Team. 1999. First observations of Eta Carinae with space telescope imaging spectrograph Morse, J. A., Humphreys, R. M., and Damineli, A. (eds), in *Eta Carinae at the millennium*. ASP Conf. Series, Vol. 179, 144–154.

Gull, T., Ishibashi, K., Davidson, K., and Coolins, N. 2001. STIS observations of the Eta Carinae central source and Weigelt blobs BD. Gull, T. R., Johansson, S., and Davidson, K. (eds), in *Eta Carinae and other mysterious stars: The hidden opportunities of emission line spectroscopy*. ASP Conf. Series, Vol. 242, San Francisco, 391–459.

Gwinn, C. R., Moran, J. M., and Reid, M. J. 1992. Distance and kinematics of the W49N H_2O maser outflow. *Astrophysical Journal* **393**, 149–164.

Haisch, B. M., Linsky, J. L., Weinstein, A., and Shine, R. A. 1977. Analysis of the chromospheric spectrum of O I in Acrturus. *Astrophysical Journal*, Part 1, **214**, 785–797.

Haken, H. 1970. Laser theory in *Encyclopedia of physics*, Vol. XXV/2c, Flugge S. (ed.), Springer-Verlag, Berlin.

Hamann, F., De Poy, D. L., Johansson, S., and Elias, J. 1994. High-resolution 6450–24500 Å spectra of Eta Carinae. *Astrophysical Journal* **422**, 626–641.

Hamann, F., Davidson, K., Ishibashi, K., and Gull, T. R. 1999. Preliminary analysis of HST-STIS spectra of compact ejecta from Eta Carinae. Morse J. A., Humphreys, R. M., and Damineli, A. (eds), in *Eta Carinae at the millennium*, ASP Conf. Series, Vol. 179, 116–122.

Hanbury Brown R. 1974. *The intensity interferometer*. Taylor and Francis Ltd., London.

Hanbury Brown, R., and Twiss, R. G. 1957. Interferometry of the intensity fluctuations in light. III. Applications to astronomy. *Proceedings of the Royal Society*, London, **A248**, 199–221.

Hanbury Brown, R. and Twiss, R. G. 1958. Interferometry of the intensity fluctuations in light. IV. A test of an intensity interferometer on Sirius A. *Proceedings of the Royal Society*, London, **A248**, 222–237.

Hansen, J. E. and Persson, W. J. 1987. A revised analysis of the spectrum of Ar III. *Journal of Physics. B. Atomic and Molecular Physics*, **20**, 693–706.

Hardy, A. and Treves, D. 1979. Amplified spontaneous emission in spherical and disk-shaped laser media. *IEEE Journal of Quantum Electronics* **QE-15**, No. 9, 887–895.

Harper, G. M., Wilkinson, E., Brown, A., Jordan, C., and Linsky, J. L. 2001. Identification of Fe II emission lines in fuse stellar spectra. *Astrophysical Journal* **551**, 486–494.

Harris, S. 1997. Electromagnetically induced transparency. *Physics Today* **50**(7), 36–42.

Hartman, H. and Johansson, S. 2000. Ultraviolet fluorescence lines of Fe II observed in satellite spectra of the symbiotic star RR Telescopic. *Astronomy and Astrophysics* **359**, 627–634.

Hartman, H., Damineli, A., Johansson, S., and Letokhov, V. S. 2005. Time variations of the narrow Fe II and H I spectral emission lines from the close vicinity of Eta Carinae during the spectral event of 2003. *Astronomy and Astrophysics* **436**, 945–952.

Heitler, W. 1954. *The quantum theory of radiation*. Oxford University Press, Oxford.

Hillier, J. D., Davidson, K., Ishibashi, K., and Gull, T. 2001. On the nature of the central source in Eta Carinae. *Astronomy and Astrophysics* **553**, 800–837.

Hiltner, V. A. 1947. Infrared spectra of early-type stars. *Astrophysical Journal* **105**, 212–214.

Holstein, T. 1947. Imprisonment of resonance radiation in gases. *Physical Review* **72**, 1212–1233.

Hummer, D. G. and Rybicki, G. 1971. The formation of spectral lines. *Annual Reviews of Astronomy and Astrophysics* **9**, 237–270.

Ivanov, V. V. 1971. *Transfer of line radiation*. Washington D.C., US GPO.

Javan, A., Bennett, W. R. Jr., and Herriott, D. R. 1961. Population inversion and continuous optical maser oscillation in gas discharge containing a He-Ne mixture. *Physical Review Letters* **6**, 106–110.

Johansson, S. 1977. New Fe II identifications in the infrared spectrum of Eta Carinae. *Monthly Notices of the Royal Astronomical Society* **178**, 17–20.

Johansson, S. 1978. The spectrum and term systems of Fe II. *Physica Scripta* **18**, 217–265.

Johansson, S. 1983. Strong Fe II fluorescence lines in RR Tel. and V1016 Cyg excited by C V in Bowen mechanism. *Monthly Notices of the Royal Astronomical Society* **327**, 71P–75P.

Johansson, S. 1988. Indirect IUE observations of O VI from photoexcited fluorescence lines of Fe II, present in the spectrum of RR Telescopic. *Astrophysical Journal* **327**, L85–L88.

Johansson, S. and Hamann, F., 1993. Fluorescence lines in ultraviolet spectra of stars. *Physica Scripta* **T47**, 157–164.

Johansson, S. and Jordan, C. 1984. Selective excitation of Fe II in the laboratory and late-type stellar atmospheres. *Monthly Notices of the Royal Astronomical Society* **210**, 239–256.

Johansson, S. and Letokhov, V. S. 2001a. Resonance-enhanced two-photon ionization of ions by Lyman α radiation in gaseous nebulae. *Science* **291**, 625–627.

Johansson, S. and Letokhov, V. S. 2001b. Multiple two-photon resonance-enhanced ionization of elements in natural conditions. *Letters to the Journal of Experimental and Theoretical Physics* **73**, 135–137.

Johansson, S. and Letokhov, V. S. 2001c. Mysterious UV lines of Fe II from η Car blobs. Gull, T. R., Johansson, S., Davidson, K. (eds), in *Eta Carinae and other mysterious stars*, ASP Conf. Series **242**, 309–324.

Johansson, S and Letokhov, V. S. 2001d. Successive resonance-enhanced two-photon ionization of elements abundant in nebulae. I. Atoms and ions of C, N, and O. *Astronomy and Astrophysics* **375**, 319–327. [Erratum: 2002. *Astronomy and Astrophysics* **395**, 345].

Johansson, S. and Letokhov, V. 2001e. A model for the origin of the anomalous and very bright UV lines of Fe II in gaseous condensations of the star Eta Carinae. *Astronomy and Astrophysics* **378**, 266–278.

Johansson, S. and Letokhov, V. S. 2002. Laser action in a gas condensation in the vicinity of a hot star. *Letters to the Journal of Experimental and Theoretical Physics* **75**, 591–594.

Johansson, S. and Letokhov, V. S. 2003a. Radiative cycle with the stimulated emission from atoms and ions in an astrophysical plasma. *Physical Review Letters* **90**(1), 011101–4.

Johansson, S. and Letokhov, V. S. 2003b. Anomalously bright UV lines of Fe II as a probe of gas condensations in the vicinity of hot stars. *Astronomy and Astrophysics* **412**, 771–776.

Johansson, S. and Letokhov, V. S. 2003c. Astrophysical lasers with radiation pumping by accidental resonance. *Publications of the Astronomical Society of the Pacific* **115**, 1375–1382.

Johansson, S. and Letokhov, V. S. 2004a. Anomalous Fe II spectral effects and high H I Lyα temperature in gas blobs near Eta Carinae. *Letters to the Astronomical Journal* (Russian) **30**, 67–73.

Johansson, S. and Letokhov, V. S. 2004b. The possibility of resonance-enhanced two-photon ionization of Ne and Ar atoms in astrophysical plasma. *Astronomical Journal* (Russian) **81**, 438–446.

Johansson, S. and Letokhov, V. S. 2004c. Astrophysical lasers operating in optical Fe II lines in stellar ejecta of Eta Carinae. *Astronomy and astrophysics* **428**, 497–509.

Johansson, S. and Letokhov, V. S. 2005a. Possibility of measuring the width of narrow Fe II astrophysical laser lines in the vicinity of Eta Carinae by means of Brown–Twiss–Townes heterodyne correlation interferometry. *New Astronomy Journal* **10**, 361–369.

Johansson, S. and Letokhov, V. S. 2005b. Astrophysical lasers in optical Fe II lines in gas condensations near Eta Carinae. Mareassa, L., Helmerson, K., and Bagnato, V. (eds), in *Atomic Physics*-19, AIP, New York, 399–410.

Johansson, S. and Letokhov, V. S. 2005c. Astrophysical laser operating in the OI 8446 Å line in the Weigelt bobs of η Carinae. *Monthly Notices of the Royal Astronomical Journal* **364**, 731–737.

Johansson, S. and Letokhov, V. S. 2006. Astrophysical laser spectroscopy. *Proceedings of 17th international conference on laser spectroscopy*, 19–24 June, 2005, Scotland. Hinds, E. A., Ferguson, A., Riis, E. (eds), World Scientific, Singapore, 80.

Johansson, S. and Letokhov, V. S. 2007. Astrophysical lasers and nonlinear optical effects in space. *New Astronomy Reviews* **51**, 443–523.

Johansson, S. and Zethson, T. 1999. Atomic physics aspects in previously and newly identified iron lines in the HST spectrum of Eta Carinae. Morse, J. A., Humphreys R. M., and Damineli, A. (eds), in *Eta Carinae at the millenium*, ASP Conf. Series, Vol. 179, 171–183.

Johansson, S., Davidson, K., Ebbets, D., Weigelt, G., Balick, B., Frank, A., Hamann, F., Humphreys, R. M., Morse, J., and White, R. L. 1996. Is there a dischromatic UV laser in Eta Carinae. in *Science with Hubble telescope*-II. Benvenuti, P., Macehetto, F. D., Schreier, F. J. (eds), Space Telescope Institute, 361–365.

Johansson, S., Zethson, T., Hartman, H., and Letokhov, V. 2001. Fluorescence lines in Eta Carinae and other objects. Gull, T., Johansson, S., Davidson, K. (eds), in *Eta Carinae and other mysterious stars*, ASP Conf. Series, 242, 297–307.

Johansson, S., Gull, T. R., Hartman, H., and Letokhov, V. S. 2004. Metastable hydrogen absorption in ejecta close to Eta Carinae. *Astronomy and Astrophysics* **435**, 183–189.

Johansson, S., Hartman, H., and Letokhov, V. S. 2006a. Resonance-enhanced two-photon ionization (RETPI) of Si II and an anomalous, variable intensity of the λ1892 Si III] line in the Weigelt blobs of Eta Carinae. *Astronomy and Astrophysics* **457**, 253–256.

Johnson, M. A., Betz, A. L., and Townes, C. H. 1974. 10 micron CO_2 emission lines in the atmospheres of Mars and Venus. *Physical Review Letters* **33**, 1617–1620.

Johnson, M. A., Betz, A. L., McLaren, R. A., Sutron, E. C., and Townes, C. H. 1976. Nonthermal 10 micron CO_2 emission lines in the atmospheres of Mars and Venus. *Astrophysical Journal* **208**, L145–L148.

Jordan, C. 1988a. Ultraviolet spectroscopy of cool stars. *Journal of Optical Society of America* **B5**, 2252–2263.

Jordan, C. 1988b. Ultraviolet Fe II emission cool star chromospheres. Viotti, R., Vittone A., and Friedjung, M. (eds), in *Physics of formation of Fe II lines outside LTE*, D. Reidel Publ. Co., Dordrecht.

Kastner, S. O. and Bhatia, A. K. 1986. PAR: Photoexcitation by accidental resonance. *Comments on Atomic and Molecular Physics* **18**, 39–45.

Kastner, S. O. and Bhatia, A. K. 1991. N III line emission in planetary nebulae – not Bowen fluorescence. *Astrophysical Journal*, Part 2, **381**, L59–L62.

Kastner, S. O. and Bhatia, A. K. 1995. The neutral oxygen spectrum. II. Pumping by hydrogen Lyman-beta under the optically thin condition: a first application to the classical novae. *Astrophysical Journal*, Part 1, **439**, 346–356.

Kenyon, S. J. 1986. *The symbiotic stars*. Cambridge University Press, Cambridge.

Kimble, R. A., Woodgate, B. E., Bowers, C. W., et al. 1998. The on-orbit performance of the space telescope imaging spectrograph. *Astrophysical Journal* **492**, L83–L93.

Klimov, V. V., Johansson, S., and Letokhov, V. S. 2002. The origin of the anomalous intensity ratio between very bright UV Fe II lines and their satellites in gaseous condensations close to the star Eta Carinae. *Astronomy and Astrophysics* **385**, 313–327.

Knowles, S. H., Mayer, C. H., Sullivan, W. T. III, and Cheung A. C. 1969. Galactic water vapor emission: further observations of variability. *Science* **166**, 221–224.

Kramida, A. E., Bastin, T., Biemont, E., Dumont, P. D., and Garnir, H. P. 1999. A critical compilation and extended analysis of the Ne IV spectrum. *European Physics Journal* D, **7**, 525–546.

Krolik, J. H. and McKee, C. F. 1978. Hydrogen emission-line spectra in quasars and active galactic nuclei. *Astrophysical journal. Supplement* **37**, 459–483.

Kurucz, R. L. 2002. http://cfaku5.harvard.edu/atoms.html

Kwon, S. 2000. *The origin and evolution of planetary nebulae*. Cambridge University Press, Cambridge.

Kwong, H. S., Johnson, B. C., Smith, P. L., and Parkinson, W. H. 1983. Transition probability of the Si III 182.9 nm intersystem line. *Physical Review* **A27**, 3040–3043.

Lamb, W. E. Jr. and Scully, M. O. 1967. Quantum theory of an optical maser. I. General theory. *Physical Review* **159**, 208–226.

Lamb, W. E., Schleich, W. P., Scully, M. O., and Townes, C. H. 1999. Laser physics: quantum controversy in action. *Review of Modern Physics* **71**, s263–s273.

Lamers, H. and Cassinelli, J. 1999. *Introduction to stellar wind*. Cambridge University Press, Cambridge, UK.

Landau, L. D. and Lifshitz, E. M. 1958. *Quantum mechanics*. Addison-Wesley, Reading, Mass.

Lavrinovich, N. N. and Letokhov, V. S. 1974. The possibility of the laser effect in stellar atmospheres. *Journal of Experimental and Theoretical Physics* **67**, 1609–1620.

Lavrinovich, N. N. and Letokhov, V. S. 1976. Detection of narrow "laser" lines masked by spatially nonhomogeneous broadening in radiation emitted from stellar atmospheres. *Kvantovaya Elektronika* (Russian) **3**, 1948–1954.

Lawandy, N. M., Balachandran, R. M., Gomes, A. S. L., and Sauvaiian, E. 1994. Laser action in strongly scattering media. *Nature* **368**, 436–438.

Letokhov, V. S. 1965. Spatial effects in optical wave heterodyning. *Radiotechnika i Elektronika* (Russian) **10**, 1143–1145.

Letokhov, V. S. 1966. Stimulated radioemission of the interstellar medium. *Letters to the Journal of Experimental and Theoretical Physics* **4**, 477–481.

Letokhov, V. S. 1967a. Light generation by a scattering medium with a negative resonant absorption. *Journal of Experimental and Theoretical Physics* **53**, 1442–1452 [English translation: *Soviet Physics – JETP* **16**, 835–840 (1968)].

Letokhov, V. S. 1967b. Quantum statistics of multi-mode radiation from an ensemble of atoms. *Journal of experimental and theoretical physics* **53**, 2210–2222 [English translation: *Soviet Physics – JETP* **53**, 1246–1251 (1968)].

Letokhov, V. S. 1972a. Laser action in stellar atmospheres. *IEEE Journal of Quantum Electronics* **QE8**, 615.

Letokhov, V. S. 1972b. Space maser with a feedback. *Astronomical Journal* (Russian) **49**, 737–743.

Letokhov, V. S. 1987. *Laser Photoionization Spectroscopy.* Academic Press Inc., Orlando.
Letokhov, V. S. 1996. Noncoherent feedback in space masers and stellar lasers. In *Amazing light.* A volume dedicated to Charles Hard Townes on his 80th birthday, Chiao, R. Y. (ed.), Springer, Berlin, 409–443.
Letokhov, V. S. 2002. Astrophysical lasers. *Kvantovaya Elektronika* (Russian) **32**(12), 1065–1079.
Letokhov, V. S. 2007. *Laser control of atoms and molecules.* Oxford University Press, Oxford.
Letokhov, V. S. and Chebotayev, V. P. 1977. *Nonlinear laser spectroscopy.* Springer-Verlag, Berlin.
Letokhov, V. S. and Ustinov, N. D. 1983. *Power lasers and their applications.* Harwood Academic Publishers, Chur.
Litvak, M. M. 1969. Infrared pumping of interstellar OH. *Astrophysical Journal* **156**, 471–492.
Litvak, M. M. 1970. Linewidths of a Gaussian broadband signal in a saturated two-level system. *Physical Review A* **2**, 2107–2115.
Litvak, M. M. 1974a. Coherent molecular radiation. *Annual Review of Astronomy and Astrophysics* **12**, 97–112.
Litvak, M. M. 1974b. Radiative transport in interstellar maser. *Astrophysical Journal* **182**, 711–730.
Litvak, M. M., McWhorter, A. L., Meeks, M. L., and Zeiger, H. J. 1966. Maser model for interstellar OH microwave emission. *Physical Review Letters* **17**, 821–826.
Loudon, R. 1983. *The quantum theory of light.* Clarendon Press, Oxford.
Maeda, H. and Yariv, A. 1973. Narrowing and rebroadening of amplified spontaneous emission in high-gain laser media. *Physics Letters* **43A**, 383–385.
Maiman, T. 2000. *The Laser Odyssey.* Laser Press, Blaine, WA.
Makarov, A. A. 1983. Excitation of atoms by light pulses with off-resonance frequencies. *Journal of Experimental and Theoretical Physics* **85**, 1192–1202.
Mandel, L. and Wolf, E. 1965. Coherence properties of optical fields. *Reviews of Modern Physics* **37**, 231–287.
Markushev, V. M., Zolin, V. F., and Briskina, C. M. 1986. Luminescence and stimulated emission of neodymium in sodium lanthanym molybdate powders. *Quantum Electronics* **16**, 281–284.
Martin-Pintado, J., Bachiller, R., Thum, C., and Walmsley, C. M. 1989. A radio recombination line maser in MWC 349. *Astronomy and Astrophysics* **215**, L13–L16.
Massey, H. S. W. and Burhop, E. H. S. 1952. *Electron and ion impact phenomena.* The Clarendon Press, Oxford.
Menzel, D. H. 1969. Laser action in non-LTE atmospheres. Groth, H. G. and Wellmann, R. (eds), in *Spectrum formation in stars with steady-state extended atmospheres.* Proc. of IAU Colloq. 2, April, NBS Special Publ., Munich, 332, 134.
Merrill, P. W. 1951. The spectrum of XX Ophiuchi in 1949 and 1950. *Astrophysical Journal* **114**, 37–46.

Merrill, P. W. 1952. Spectroscopic observations of stars of class S. *Astrophysical Journal* **116**, 21–26.

Merrill, P. W. 1956. *Lines of the chemical elements in astronomical spectra*. DC Carnegie Inst. Wash., Washington.

Meystre, P. and Sargent III, M. 1990. *Elements of quantum optics*. Springer, Berlin.

Michael, E. A., Keoshian, C. J., Wagner, D. R., Anderson, S. K., and Saykally, R. J. 2001. Infrared water recombination lasers. *Chemical Physics Letters* **338**, 277–284.

Mihalas, D. 1978. *Stellar atmospheres*. W. H. Freeman and Co., New York.

Mihalas, D. and Athay, R. 1973. The effects of departures from LTE in stellar spectra. *Annual Reviews of Astronomy and Astrophysics* **11**, 187–218.

Minogin, V. G. and Letokhov, V. S. 1987. *Laser light pressure on atoms*. Gordon and Breach, New York.

Molish, A. F. and Oehry, B. P. 1998. *Radiation trapping in atomic vapours*. Clarendon Press, Oxford.

Moore, C. E. 1945. *A multiplet table of astrophysical interest*. Contr. Princeton Univ. Obs. No. 20 (reprinted 1959, National Bureau of Standards, Technical Note 36).

Moore, C. E. 1949. *Atomic energy levels*, Volume 1. National Bureau of Standards (US) Circular 467.

Moore, C. E. 1952. *Atomic energy levels*, Volume 2 (reprinted as National Standards Reference Data, Seria (US) **35**, 1971).

Moore, C. E. 1958. *Atomic energy levels*, volume 3 (reprinted as National Standards Reference Data, Seria (US) **35**, 1971).

Moore, C. E. 1962. *An ultraviolet multiplet table*. National Bureau of Standards, Circular 488, Section 1-5.

Morse, J. A., Humphreys, R. M., and Damineli, A. (eds). 1999. *Eta Carinae at the millennium*. ASP Conf. Series, Vol. 179.

Mumma M. J. 1993. Natural lasers and masers in the solar system, in *Astrophysical masers*, Clegg, A. W. and Nedoluha, G. E. (eds), Springer, Berlin, 455–467.

Mumma, M. J. Buhl, D., Chin, G., Deming, D., Espenak, F., Kostiuk, T., and Zipoy, D. 1981. Discovery of natural gain amplification in the 10-micrometer carbon dioxide laser bands on Mars: a natural laser. *Science* **212**, 45–49.

Münch, G. and Taylor, K. 1974. On the spectrum of neutral oxygen in the Orion Nebula. *Astrophysical Journal* **192**, L93–L95.

Mustel, E. R. 1941. The distribution of energy in the continuous spectrum of stars of early spectral classes. *Astronomical Journal* (Russian) **18**, 297–310.

Nahar, S. N. and Pradhan, A. K. 1994. Atomic data for opacity calculations: XX. Photoionization cross-sections and oscillator strengths for Fe II. *Journal of Physics B: Atomic and Molecular Physics* **27**, 429–446.

Netzer, H. 1998. The formation of Fe II emission lines. Viotti, R., Vittone, A., Frieddjung, M. (eds), in *Physics of formation of Fe II lines outside LTE*, D. Reidel Publ. Co., Dordrecht, 247–257.

Norlen, G. 1973. Wavelengths and energy levels of Ar I and Ar II based on new interferometric measurements in the region 3400–9800 Å. *Physica Scripta* **8**, 249–268.

Nussbaumer, H. and Schild, H. 1981. A model for V1016 CYG based on the ultraviolet spectrum. *Astronomy and Astrophysics* **101**, 118–131.

Osterbrock, D. E. 1962. The escape of resonance-line radiation from an optically thick medium. *Astrophysical Journal* **135**, 195–211.

Osterbrock, D. E. 1989. *Astrophysics of gaseous nebulae and active galactic nuclei.* Univ. Science Books, Sausalito, California.

Osterbrock, D. E. and Ferland, G. J. 2006. *Astrophysics of gaseous nebulae and active galactic nuclei.* University Science Book, Sausalito, California.

Padmabandu, G. G., Welch, G. R., Shubin, I. N., Fry, E. S., Nikonov, D. E., Lukin, M. D., and Scully, M. O. 1996. Laser oscillation without population inversion in a sodium atomic beam. *Physical Review Letters* **76**, 2053–2056.

Penston, M., Benvenutti, P., and Cassatella, A. *et al.* 1983. IUE and new observations of slow nova RR Tel. *Monthly Notices of Royal Astronomical Society* **202**, 833–857.

Persson, W. 1971. The spectrum of singly ionized neon, Ne II. *Physica Scripta* **3**, 133–135.

Persson, W., Wahlström, C.-G., Jönsson, L., and di Rocco, H. O. 1991. Spectrum of doubly ionized neon. *Physical Review* **A43**, 4791–4823.

Pottasch, S. R. 1984. *Planetary nebulae.* D. Reidel Publ. Comp., Dordrecht.

Prochazka, I., Hamal, K., and Sopko, B. 2004. Recent achievements in single photon detectors and their applications. *Journal of Modern Optics* **51**, 1289–1213.

Prokhorov, A. M. 1958. Molecular amplifier and generator for submillimeter waves. *Journal of Experimental and Theoretical Physics* **34**, 1658–1659.

Prokhorov, A. M. 1963. Amplifying properties of a fiber. *Optica and Spectroscopia* **14**, 73–77.

Raassen, A. J. J. 2002. http://www.science.uva.nl/pub/orth/iron

Reid, M. J. and Moran, J. M. 1981. Masers. *Annual Review on Astronomy and Astrophysics* **19**, 231–276.

Reines, A. E. and Marey, G. W. 2002. Optical search for extraterrestrial intelligence: a spectroscopic search for laser emission from nearby stars. *Publications of the Astronomical Society of the Pacific* **114**, 416–426.

Risken, H. 1984. *The Fokker–Planck equation.* Springer-Verlag, Heidelberg.

Robson, I. 1986. *Active galactic nuclei.* Wiley, Chichester.

Rothermel, H., Kaufl, H. U., and Yu, Y. 1983. A heterodyne spectrometer for astronomical measurements at 10 micrometers. *Astronomy and Astrophysics* **126**, 387–392.

Rowland, P. R. and Cohen, R. J. 1986. Rapid variability of H2O masers in Cepheus A. *Monthly Notices of the Royal Astronomical Society* **220**, 233–251.

Rudy, R., Rosano, G. S., and Puetter, R. C. 1989. The near-infrared oxygen I lines of the planetary nebula IC 4997. *Astrophysical Journal*, Part I, **346**, 799–802.

Rybicki, G. B. and Lightman, A. P. 1979. *Radiative Processes in Astrophysics.* John Wiley and Sons, New York.

Saha, S. 2002. Modern optical astronomy technology and impact on interferometry. *Review of Modern Physics* **74**, 551–600.

Sargent III, M., Scully, M. O., and Lamb Jr., W. E. 1974. *Laser physics.* Addison-Wesley Publ. Co., Reading.

Schawlow, A. L. and Townes, C. H. 1958. Infrared and optical masers. *Physical review* **112**, 1940–1949.

Schrödinger, E. 1926. Der sletigeubergang von micro-zur markomechanik. *Naturwissenschaften* **14**, 664–666.

Schwartz, R. N. and Townes, C. H. 1960. Interstellar and interplanetary communications by optical masers. *Nature* **190**, 205–208.

Schmid, H. 1989. Identification of the emission bands at $\lambda\lambda 6830, 7088$. *Astronomy and astrophysics* **211**, L31–L34.

Scoville, N., Kleinmann, S., Hall, D., and Ridgeway, S. 1983. The cirmustellar and nebular environment of the Becklin-Neugebauer object: $\lambda = 2-5$ micron spectroscopy. *Astrophysical Journal* **275**, 201–224.

Scully, M. and Zubairy, M. S. 1997. *Quantum optics*. Cambridge University Press, Cambridge.

Seaton, M. J. 1959. Radiative recombination of hydrogenic ions. *Monthly Notices of the Royal Astronomical Society* **119**, 81–89.

Seaton, M. J. 1960. Planetary nebulae. *Reports on progress in physics* **23**, 313–354.

Seaton, M. J. 1964. Recombination spectra. III. Populations of highly excited states. *Monthly Notices of Royal Astronomical Society* **127**, 177–184.

Selvelli, P., Danziger, J., and Bonifacio, P. 2007. The He II Fowler lines and the O III and N III Bowen fluorescence lines in the symbiotic nova RR Tel. *Astronomy and Astrophysics* **464**, 715–734.

Sherwood, B., Mumma, M. J., and Donaldson, B. P. 1992. Engineering planetary lasers for interstellar communications. *The second conference on Lunar bases and space activities of the 21st century*. NASA, Johnson Space Center, **2**, 635–637.

Shklovskii, I. S. 1991. *Five billion vodka bottles to the Moon: tales of a Soviet scientist*. W. W. Norton & Company.

Shore, B. W. and Menzel, D. H. 1968. *Principles of atomic spectra*. J. Wiley and Sons, New York.

Siegman, A. 1966. The antenna properties of optical heterodyne receivers. *Applied Optics* **5**, 1588–1594.

Siegman, A. 1986. *Lasers*. University Science Books, California.

Skelton, D. J. and Shine, R. A. 1982. Formation of the O I resonance triplet and intercombination doublet in the solar chromosphere. *Astrophysical Journal* Part 1, **259**, 869–879.

Slettebak, A. 1951. Lines of neutral oxygen in the infrared spectra of Be Star. *Astrophysical Journal* **113**, 436–437.

Smith, H. A. 1969. Population inversions in ions of astrophysical interest. *Astrophysical Journal* **15**, 371–383.

Smith, H. A., Larson, H. P., and Fink, U. 1979. The spectrum of the Becklin–Neugebauer source in Orion from 3.3 to 5.5 microns. *Astrophysical Journal* **233**, 132–139.

Smith, H. A., Strelnitskii, V. S., and Ponomarev, V. O. 1996. Hydrogen masers and lasers in space in *Amazing light*. Chiao, R. (ed.). Volume dedicated to Charles Hard Townes on his 80th Birthday. Springer, Berlin, 603–611.

Smith, H. A., Strelnitskii, V., Miles, J. W., Kelly, D. M., and Lacy, J. H. 1997. Mid-infrared hydrogen recombination line emission from the maser star. *Astrophysical Journal* **114**, 2658–2663.

Smith, N., Davidson, K., Gull, T. R., Ishibashi, K., and Hillier J. D. 2003. Latitude-dependent effects in the stellar wind of Eta Carinae. *Astrophysical Journal* **586**, 432–450.

Snow, T. P. 1995. Comments on two-photon absorption by H_2 molecules as a source of diffuse interstellar bands. *Chemical Physics Letters* **245**, 639–642.

Sobel'man, I. I. 1979. *Atomic spectra and radiative transition.* Springer, Berlin.

Sobel'man, I. I., Vainstein, L. A., and Yukov, E. A. 1981. *Excitation of atoms and broadening of spectral lines.* Springer-Verlag, Berlin.

Sobolev, V. V. 1962. *Radiative energy transfer in stellar and planetary atmospheres.* Van Nonstrand, Princeton, N. J.

Sobolev, V. V. 1963. *A treatise of radiation transfer.* Van Nostrand, Princeton, N. J.

Sorokin, P. P. and Glownia, J. H. 1995. Nonlinear spectroscopy in astronomy: assignment of diffuse interstellar absorption bands to $L(\alpha)$-induced, two-photon absorption by H_2 molecules. *Chemical Physics Letters* **234**, 1–6.

Sorokin, P. P. and Glownia, J. H. 2000. Nonlinear optics in space. *Canadian Journal of Physics* **78**, 461–481.

Sorokin, P. P. and Glownia, J. H. 2002. Lasers without inversion (LWI) in space: a possible explanation for intense, narrow-band, emissions that dominate the visible and/or far UV (FUV) spectra of certain astronomical objects. *Astronomy and Astrophysics* **384**, 350–363.

Stepanova, G. I. and Shved, G. M. 1985. The natural 10 μm CO_2 laser in the atmospheres of Mars and Venus. *Soviet Astron. Letters* **11**, 390–394.

Strelnitskii, V. 2002. The puzzle of natural lasers. Migenas, V. and Reid, M. J. (eds), in IAU Symp. 206, *Cosmic masers: from photostars to black holes*, Cambridge Univ. Press, Cambridge, 479–481.

Sternberg, A., Dalgarno, A., and Roueff, E. 1988. The Bowen mechanism and charge transfer. *Comments on Astrophysics* **13**, 29–37.

Storey, P. J. and Hummer, D. G. 1995. Recombination line intensities for hydrogenic ions – IV. Total recombination coefficients and machine-readable tables for $Z = 1$ to 8. *Monthly Notices of the Royal Astronomical Society* **272**(1), 41–48.

Strelnitskii, V. S., Smith, H. A., Haas, M. R., Colgan, S. W. J., Erickson, E. F., Geis, N., Hallenbach, D. J., and Townes, C. H. 1995. Haas, M. R., Davidson, J. A., and Erickson, E. F. (eds), in *Proceedings of Airborne Astronomy Symposium on the Galactic Ecosystem: From Gas to Stars.* ASP, San Francisco, 271.

Strelnitskii, V., Haas, M. R., Smith, H. A., Eriksson, E F., Colgan, S. W. J., and Hollenbach, D. J. 1996. Far-infrared hydrogen lasers in the peculiar star MWC 349 Å. *Science* **272**, 1459–1461.

Strittmatter, P. A., Woolf, N. J., Thompson, R. I., Wilkerson, S., Angel, J. R. P. Stockman, H. S., and Gilbert, G. 1977. The spectral development of Nova Cygni. 1975. *Astrophysical Journal*, Part 1, **216**, 23–32.

Strömgren, B. 1939. The physical state of interstellar hydrogen. *Astrophysical Journal* **89**, 526–545.

Svelto, O. 1998. *Principles of lasers*, 4rd edition Plenum Press, New York.

Szeifert, T., Humphreys, R. M., Davidson, K., Jones, T. J., Stahl, O., Wolf, B., and Zickgraf, F.-J. 1996. HST and groundbased observations of the Hubble-Sandage variables in M 31 and M 33. *Astronomy and Astrophysics* **314**, 131–145.

Thackeray, A. D. 1935. The emission line $\lambda 4511$ in late-type variables. *Astrophysical Journal* **81**, 467–473.

Thackeray, A. D. 1969. The spectrum of Eta Carinae in the 10000Å region. *Monthly Notices of the Astronomical Society of Southern Africa* **28**, 37–42.

Thum, C., Strelnitskii, V. S, Martin-Pintado, J., Matthews, H. E., and Smith, H. A. 1995. Hydrogen recombination β-lines in MWC 349. *Astronomy and Astrophysics* **300**, 843–850.

Thum, C., Martin-Pintado, J., Quirrenbach, A., and Matthews, H. C. 1998. ISO study of the recombination line maser star MWC 349. Yun, J. L., Liseau, R. (eds), in *Star formation with infrared space observatory*, ASP Conf. Series, Vol. 132, 107–112.

Townes, C. H. 1983. At what wavelengths should we search for signals from extraterrestrial intelligence? *Proceedings of the National Academy of Sciences of the United States of America* **80**(4), 1147–1151.

Townes, C. H. 1993. The early years of research on astronomical lasers. In *Astronomical masers*, A. W. Clegg and G. E. Nedohula (eds) Springer-Verlag, Berlin, 3–11.

Townes, C. H. 1994. *Making waves*. American Institute of Physics, New York.

Townes, C. H. 1997. Space maser and lasers. *Kvantovaya Elektronica* (Russian) **27**, 1063–1066.

Townes, C. H. 1999. *How the laser happened.* Oxford University Press, Oxford.

Townes, C. H., Sutton, E. C., and Storey, J. W. V. 1978. Infrared heterodyne interferometry. Pacini, F., Richter, W., Wilson, R. N. (eds), in ESO Conference, *Optical telescopes of the future*, 409–426.

Traving, G. 1968. Broadening and shift of spectral lines. In *Plasma diagnostics*, by Lochthe-Holtgreven, W. (ed.) North-Holland, Publ. Co., Amsterdam, Chapter 2, 57.

Truitt, P. and Strelntiski, V. 2000. Transition to oscillation regime in flaring water vapor masers. *Bulletin of the American Astronomical Society* **32**, 1484.

Unsold, A. *Physik der Sternatmosphëren.* 1968. Springer-Verlag, Berlin.

van Boekel, R., Kervella, P., Schöller, M., Herbst, T., Brandner, W., de Koter, A., Waters, L. B. F. M., Hillier D. J., Paresce, F., Lenzen, P., and Lagrange, A.-M. 2003. Direct measurements of the size and shape of the present-day stellar wind of Eta Carinae. *Astronomy and Astrophysics* **410**, L37–L40.

Varshni, Y. P. and Lam, C. S. 1976. Laser action in stellar atmospheres. *Astrophysics and Space Science* **45**, 87–97.

Varshni, Y. P. and Nasser, R. M. 1986. Laser action in stellar envelopes II, He I. *Astrophysics and Space Science* **125**, 341–360.

Vavilov, S. I. 1960. *Microstructure of light* (Russian). Publ. House of USSR Acad. Sci., Moscow.

Verner, E. M., Gull, T. R., Bruhweiler, F., Johansson, S., Ishibashi, K., and Davidson, K. 2002. The origin of Fe II and [Fe II] emission lines in the 4000–10000 Å range in the BD Weigelt blobs of Eta Carinae. *Astrophysical Journal* **581**, 1154–1167.

Viotti, R., Rossi, L., Casatella, A., Altamore, A., and Baratta, G. B. 1989. The ultraviolet spectrum of Eta Carinae. *Astrophysical journal. Supplement* **71**, 983–1009.

Vlemmings, W. H. T. 2007. A review of maser polarization and magnetic fields. In *Astrophysical masers and their environments*, J. M. Chapman and W. A. Bann (eds), Proceedings of the IAU Symposium, No. 242, 119–128.

Watson, W. D. and Wyld, H. W. 2003. Apparent sizes and spectral line profiles for spherical and disk masers: solution to the full equations. *Astrophysical Journal* **598**, 357–368.

Weaver, H., Williams, D. R. W., Dieter, N. H., Hannilon, H., and Lum, W. E. 1965. Observations of a strong unidentified microwave line and of emission from the OH molecule. *Nature* **208**, 29–30.

Weigelt, G. and Ebersberger, J. 1986. Eta Carinae resolved by speckle interferometry. *Astronomy and Astrophysics* **163**, L5-L6.

Weinberg, A. M. and Wigner, E. P. 1958. *Physical theory of neutron chain reactors*. University of Chicago, Chicago.

Weinreb, S., Meeks, M. L., Carter, J. C., Barrett, A. H., and Rogers, A. E. E. 1965. Observations of polarized OH emission. *Nature* **208**, 440–441.

Weisskopf, V. 1933a. Die strahlung des lichtes an angeregaten atomen. *Zeitschrift der Physik* **85**, 451–481.

Weisskopf, V. 1933b. The intensity and structure of spectral lines. *The Observatory* **56**, 291–308.

Weisskopf, V. and Wigner, E. 1930. Berechnung der natürlichen linienbreite auf grund der Diracschen lichttheorie. *Zeitschrift fur Physik* **63**, 54–73.

Whittet, D. C. B. 1992. *Dust in the galactic environment*. Institute of Physics Publishing, Bristol.

Wiese, W. L., Fuhr, J. R., and Deters, T. M. 1996. *Atomic transition probabilities of carbon, nitrogen, and oxygen*. J. Phys. Ref. Data. Monograph. NIST, Washington D.C.

Woodbury, G. J. and Ng, W. R. 1962. Stimulated Raman emission in a normal ruby laser. *Proceedings of the IRE* **50**, 1236–1237.

Wooley, R. 1938. Non-coherent formation of absorption lines. *Monthly Notices of the Royal Astronomical Society* **98**, 624–632.

Yariv, A. 1976. *Introduction to optical electronics*, 2nd edition. Holt, Reinhart and Winston, New York.

Yariv, A. and Leite, R. C. C. 1963. Super radiant narrowing in fluorescent radiation of inverted population. *Journal of Applied Physics* **34**, 3410–3411.

Zanstra, H. 1949. On scattering with redistribution and radiation pressure in a stationary nebula. *Bulletin of the Astronomical Institute of the Netherlands* **11**, 1–10.

Zethson, T. 2001. Hubble space telescope spectroscopy of Eta Carinae and Chi Lupi. Ph.D. *Thesis Lund University*, Lund, Sweden.

Zibrov, A. S., Lukin, M. D., Nikonov, D. E., Hollberg, K., Scully, M. O., Velichansky, V. L., and Robinson, H. G. 1995. Experimental demonstration of laser oscillation without population inversion via quantum interference in Rb. *Physical Review Letters* **75**, 1499–1502.

Zoccali, M., Lecureur, A., Barbuy, B., Hill, V., Renzini, A., Minnetti, D., Momany, Y., Gomez, A., and Ortolani, S. 2006. Oxygen abundances in the Galactic bulge: evidence for fast chemical enrichment. *Astronomy and Astrophysics*, **457**, L1–L4.

Zoller, P. 1979. Saturation of two-levels atoms in chaotic fields. *Physical Review* **A20**, 2420–2423.

Zyuzin, A. Ya. 1994. Weak localisation in backscattering from an amplifying medium. *Europhys. Lett. 26*, 517–520.

Appendix A
Useful units

Fundamental and other constants

Speed of light in vacuum	c	$2.997\,924\,58 \cdot 10^8$ cm/s
Planck constant	$h = 2\pi\hbar$	$6.626\,068\,9 \cdot 10^{-27}$ erg/s
	$\hbar$	$1.054\,571\,6 \cdot 10^{-27}$ erg/s
Electron charge	e	$1.602\,176\,487 \cdot 10^{-19}$ C
Electron rest mass	m	$9.109\,382 \cdot 10^{-28}$ g
Proton mass	m_p	$1.672\,621 \cdot 10^{-24}$ g
Atomic mass unit	a.u.m.	$1.660\,538 \cdot 10^{-24}$ g
Rydberg constant	R_∞	$1.097\,373\,157$ cm^{-1}
Stefan–Boltzmann constant	σ	$5.674\,400 \cdot 10^{-5}$ erg/s $\cdot$ cm$^2 \cdot$ K^4
Boltzmann constant	k	$1.380\,650\,4 \cdot 10^{-16}$ erg/K
Bohr radius	a_0	$0.529\,177 \cdot 10^{-8}$ cm

Astronomical Constants

Mass of Sun	$M_\odot$	$1.989 \cdot 10^{33}$ g
Radius of Sun	$R_\odot$	$6.9598 \cdot 10^{10}$ cm
Luminosity of Sun	$L_\odot$	$3.82 \cdot 10^{33}$ erg/s
Effective temperature of Sun	T_eff	5770 K
Astronomical unit	A.U.	$1.495\,979 \cdot 10^{13}$ cm
Parsec	pc	$3.0857 \cdot 10^{18}$ cm $= 3.2616$ l.y.
Light year	l.y.	$9.4607 \cdot 10^{17}$ cm $= 0.3066$ pc
Mass of Earth		$5.98 \cdot 10^{27}$ g

Other Units

Length	Fermi	Fm	10^{-13} cm
	Ångström	Å	10^{-8} cm
	micron	μm	10^{-4} cm
Time	day	d	86400 s
	year	y	$3.156 \cdot 10^7$ s
Angle	degree	1°	$\pi/180 = 1.74 \cdot 10^{-2}$ rad
	minute	1′	$1' = \pi = 2.91 \cdot 10^{-4}$ rad
	second	″	
	rad		$1'' = 4.85 \cdot 15^6$ rad
			57.296°

Spectral and Energy units

$1\,\text{cm}^{-1} = 1.23968 \cdot 10^{-4}$ eV $= 1.9857 \cdot 10^{-16}$ erg
$1\,\text{erg} = 0.50360 \cdot 10^{16}$ cm^{-1} $= 6.2426 \cdot 10^{11}$ eV
$1\,\text{eV} = 8065.541$ cm^{-1} $= 1.60217653 \cdot 10^{-12}$ erg $\to 11604.50$ K.

Index

Absorption cross-section 21, 30, 198
angular momentum and spin 36
autoionization resonances 52

Binding energy 41, 45, 47, 48, 149, 209
Boltzmann distribution 6, 57, 58, 128, 129
Boltzmann relation 103
Born approximation 55, 56
bottlenecks 147
Bowen mechanism 4, 46, 70, 73, 104, 132, 208, 214
broadening 21–26, 29, 30, 34, 66, 72, 80–84 etc.
Brown–Twiss–Townes interferometer 228

CO_2 laser(s) 10, 121, 123, 186
coherent pumping 7
coherent states 95, 96
collisional broadening 23, 24
cosmic abundance 39, 41, 42, 45, 59, 125
cross-section(s) 21, 24, 30, 54–56, 61, 66, 215, 219, 226

Damping factor 26, 34, 66
degeneracy 79, 110, 151
dipole 16–18, 20, 21, 38, 39, 55, 69, 86, 148, 149, 155, 211
Dirac law 20
Doppler 22, 25,
    broadened line 29
    broadening 22, 26, 30, 211
    effect 22, 25, 26, 29, 31
    half-width 26, 34
    redistribution 29, 31, 185, 198
    shift 25, 29
    width 26, 27, 72, 79, 81–83, 112, 124, 125, 141, 142, 152, 161, 175, 178, 182 etc.

Eclipsing binaries 62
Eddington's approximation 204
Edlen, Bengt v
Einstein coefficients 12, 16, 102, 103, 110, 111, 129, 146, 158, 175
electromagnetically induced transparency 8
η Carinae 1, 5, 6, 10, 35, 43, 46, 71, 73–75, 132–135, 137, 139–141, 143, 144, 149, 170–172, 184, 186, 187, 205, 207, 215, 217, 227

Fabry–Perot cavity 85, 92, 190
forbidden Fe^{6+} lines vi
forbidden lines 39, 68–70, 118, 139, 147, 149, 151, 227
forbidden transitions vi, 8, 17, 37, 55, 111, 116

Gaussian profile 22, 26, 30, 31
Gaussian shape 122

H_2O molecule 2, 118
Hamiltonian 16, 18
homogeneous broadening 25, 26, 80–82, 84, 198
homogeneous spectral width 81
homogeneous width 25, 26
Hubble Space Telescope (HST) 1, 6, 43, 67, 73, 74, 127, 132–134, 140, 144, 151, 156, 158, 160, 166, 167, 168, 170–173, 175, 189, 216, 217, 220

Inhomogeneous broadening 26, 81, 100, 198
ionization 5, 11, 38, 40, 41, 44–49, 53–58 etc.
    potential(s) 51–54, 105, 151, 170, 211, 213, 214, 216
    stages vi, 57, 68, 117
isotropic radiation 81–82, 84, 131, 182–185, 210, 212

Lamb shift vi
laser clusters 205
Lorentz contour 22, 24, 26, 30, 32
Lorentz homogeneous width 26
Lorentz wings 28, 34, 141
Lyman continuum 54, 63–66, 74, 118, 135, 139, 140–144, 146, 162–165, 170, 208, 216

Maxwell–Boltzmann distribution 54
Maxwellian distribution 25, 57
metastable states 39, 47, 51, 55, 69, 70, 116, 126 etc.
microparticles 206
Milne relation 54
multiplet numbers 36

Natural linewidth 93

OH maser(s) 4, 79, 118, 182, 183, 199, 201, 203

OH radical 2, 115, 118
optical heterodyne interferometer 186
optical pumping 5, 106, 111, 142, 179, 227
oscillation 18, 81, 85–87, 89–93, 95, 102, 112, 190, 191, 205, 210

Photoexcitation 104, 105, 115, 124, 127, 132, 152, 154, 157, 160, 168, 169, 171, 180, 227
photoionization 51–55, 63–66, 68, 70, 71, 74, 75, 106 etc.
photoselective absorption 59
Planck distribution 13, 57, 58, 61, 130, 141, 180
Planck radiation 15, 45, 71, 103, 125–127, 144, 213, 216
Planck relation 14

Quantization 12, 17
quantum number 49

Radiative (natural) 22
radiative broadening 22–24, 34, 210
radiative equilibrium 102, 111, 118
radiative width 27, 135
Raman lines 9
Raman scattering 7, 9, 207
recoil effect 27, 128
recombination 4, 40, 51–55, 63–65, 67–69, 71, 72, 75, 104, 110, 118 etc.
relative strengths 37
resonance transition 34, 42, 44, 125, 132
resonant cavity 2, 84, 97, 113
RETPI (resonance enhanced two-photon ionization) 5, 6, 43, 63, 207, 209, 210, 213–225
Ring Nebula 67

Saturation parameter 81
scattered radiation 27, 29, 33, 34
scattering area 31, 34, 35
scattering cross-section 27, 29–31, 135, 192, 197, 198
Shklovskii 4
Space Telescopes Imaging Spectrograph (STIS) 1, 132–134, 140, 144, 153, 156, 158, 160, 170–173, 175, 189, 215, 216, 220
spectral line broadening 82
spin 36-38, 42, 44, 50
spontaneous decay 22, 23, 110, 120, 146
spontaneous emission 2, 13, 17, 20–22, 86, 87, 97, 98, 100–104, 107–114, 118, 120, 122, 123, 131, 159, 181, 184, 202, 203, 210, 227
spontaneous transitions 150, 152
stimulated emission 1–3, 5, 8, 10, 12–15, 18–20, 36–40, 84, 99
stimulated transitions 23, 79, 80, 89, 109, 113
Strömgren radius 64, 65, 68, 135, 162
Strömgren zones 68

Thermodynamical equilibrium 5, 12, 55, 57–59, 102, 103
thermonuclear fusion 1
threshold condition 195, 200, 204
threshold of oscillation 87, 90
triplet system(s) 37, 168

Weigelt blobs 73, 74, 76, 132–134, 137, 139, 141, 143–146, 156, 162, 165, 167, 170–172, 174–179, 184, 186, 189, 205 etc.